Examples and Theorems in Ana

Springer
London
Berlin
Heidelberg
New York
Hong Kong
Milan
Paris
Tokyo

Peter Walker

Examples and Theorems in Analysis

With 19 Figures

Springer

Peter Leslie Walker, BA, PhD
College of Arts and Science, American University of Sharjah, PO Box 26666, Sharjah, United Arab Emirates

British Library Cataloguing in Publication Data
Walker, Peter
Examples and theorems in analysis
1. Mathematical analysis 2. Mathematical analysis – Problems, exercises, etc.
I. Title
515
ISBN 1852334932

Library of Congress Cataloging-in-Publication Data
Walker, P. L. (Peter Leslie), 1942-
Examples and theorems in analysis / Peter Walker.
p. cm.
Includes bibliographical references and index.
ISBN 1-85233-493-2 (acid-free paper)
1. Mathematical analysis. 2. Calculus. 3. Fourier analysis. I. Title.
QA303.2.W35 2004
515—dc22 2003058510

ISBN 1-85233-493-2 Springer-Verlag London Berlin Heidelberg
a member of BertelsmannSpringer Science+Business Media GmbH
springeronline.com

Printed in the United States of America

The text was prepared using Scientific Word, © MacKichan Software, Inc., 19307 8th Avenue, Suite C, Poulsbo, WA 98370-7370, USA. www.mackichan.com

Typesetting: Camera-ready by Thomas Unger
12/3830-543210 Printed on acid-free paper SPIN 10835376

Preface

Recent years have seen a number of worthy attempts to increase the accessibility of several branches of mathematics. In the library at AUS we find *A Friendly Introduction to Number Theory*, *An Adventurer's Guide to Number Theory*, and even *Rings and Things* (a volume of abstract algebra). In this vein we can anticipate *Rambling through Fields* (a sequel to the above), *Topology for Toddlers*, or perhaps *Euclidean Geometry on less than 200 Drachmas a Day*.

This said, it is not our intention now to write either *Trouble with Epsilons*, or *The Weekend Gardener's Guide to Uniform Convergence*. Mathematical analysis has its own intrinsic character, associated with inequalities and estimation, and with the balancing of one or more (potentially) small quantities – the epsilons of the above non-existent title. At the same time a number of the results concerning the processes of calculus – the chain rule for differentiation, for instance – are already familiar from previous informal courses and the instructor has to face the objection that 'we know all this; why are you repeating it in a more difficult way?'

To face these pedagogical difficulties is by no means easy, and we make no claim to have discovered the unique and infallible solution. The aim in this book is to try to give the subject concreteness and immediacy by giving the examples equal status with the theorems, as the title implies. The results are introduced and motivated by reference to examples which illustrate their use, as well as further examples to show how far the assumptions may be relaxed before the result fails. Indeed it is many of these examples which first arrest our attention, and perhaps even admiration. For a novice the sum of the geometric series

$$1-\frac{1}{2}+\frac{1}{4}-\frac{1}{8}+\cdots+\left(\frac{-1}{2}\right)^{n}+\cdots=\frac{2}{3}$$

is interesting and a little unexpected, while the numerous classical series for π

such as Gregory's

$$1-\frac{1}{3}+\frac{1}{5}-\frac{1}{7}+\cdots+\frac{(-1)^{n}}{2n+1}+\cdots=\frac{\pi}{4}$$

and Euler's

$$1+\frac{1}{2^2}+\frac{1}{3^2}+\frac{1}{4^2}+\cdots+\frac{1}{n^2}+\cdots=\frac{\pi^2}{6}$$

retain their surprise centuries after their discovery. At a more advanced level we can mention for instance the limit of the iterated sine

$$\lim_{n\to\infty}\sqrt{n}\sin_n(x)=\sqrt{3},\quad 0<x<\pi$$

(where $\sin_n(x)$ denotes the n-fold composition $\sin(\sin(\ldots\sin(x)))$) which is proved in Section 7.4, and the value of the theta function

$$\sum_{n=-\infty}^{\infty} e^{-n^2\pi}=\frac{\Gamma(1/4)}{\sqrt{2}\pi^{3/4}}$$

(which is discussed in Section 5.3 on the Gamma function) as formulae which give analysis its special interest, in addition to its numerous and powerful theoretical results.

The material of Chapters 1 to 6 is largely traditional, though with some novelties of presentation. We have chosen to base the proofs of results which are normally thought of as 'compactness arguments' on either the method of bisection or the existence of convergent subsequences. Indeed compactness – the existence of certain finite open coverings – belongs more in a course on metric spaces or topology rather than in elementary analysis on the real line.

We begin with sequences since there is only one small quantity to deal with and the structure of the definitions is clearer.

The discussion of Newton's method in Section 3.5 is a little more extended than usual, partly due to the author's exasperation with the common misunderstanding that if two terms in an iterative scheme agree to some number of decimal places, then they must equal the value of the limit to that accuracy.

We choose to treat integration using the set of regulated functions (those which have finite left- and right-hand limits at every point) since the technicalities are less than for the general Riemann approach, and at the same time the class of integrable functions is easily identified and sufficient for most uses at this level. For advanced applications, Lebesgue's theory and its generalisations are essential, but that is not our business here.

We consider improper integrals in some detail in Chapter 5 since many of the most interesting examples, for instance

$$\int_0^{\pi/2}\ln(\sin x)\,dx=-\frac{\pi}{2}\ln 2,\qquad \int_0^{\infty}\frac{\sin x}{x}dx=\frac{\pi}{2}$$

are of this type, and improper integrals motivate the discussion of distributions in Chapter 7.

We close with a number of applications in Chapter 7 since without them the question of the purpose of analysis goes largely unanswered. We have chosen Fourier theory, distributions and asymptotics, since they provide ideal settings in which to learn to use the results we have developed. Each of these topics has been the subject of long and detailed investigation and it is not our intention to be at all complete in our treatment; we shall simply attempt to indicate enough for the reader to see both what the subject is about, and what can be done with it.

The book *Introductory Mathematics: Algebra and Analysis* by Geoff Smith [14] is a useful preliminary to this one, and spares us the need for the customary 'Chapter 0'; in particular we shall refer to it for properties of the real and complex number systems.

The end of a proof is signalled as usual by ■; the end of the discussion following an example is signalled by ♦.

The exercises at the ends of chapters are at all levels with no particular indication of difficulty. Those marked with one star develop new ideas in some way, for instance convexity or the use of one-sided derivatives in relation to the Mean Value Theorem. Those with two stars are open questions – if you have solved one of these, please let the author know!

A book is rarely the product of a single person's efforts, and I should begin by thanking Professor John Toland, to whom I owe the original suggestion that I should write something in analysis for Springer UK. Professor Toland also showed me the striking proof of Chebyshev's theorem which is Exercise 13, Chapter 2. Two Dutch friends and colleagues, Dr. Adri Olde Daalhuis in Edinburgh and Dr. Jaap Geluk in Rotterdam read innumerable drafts with patience and good sense. They made valuable suggestions for additional results and examples, as well as helping me to avoid embarrassing errors. And most importantly the editorial staff of Springer UK were invariably helpful and patient with the long delays while I was involved with other projects. The text was prepared using Scientific Word.[1] The figures for which no other attribution is given were generated using Maple™.

To the student reader (who, it is said, never reads prefaces anyway) we shall say only that analysis is a challenge which will be rewarding in proportion to the amount of careful thinking which is devoted to it – Good Luck and Happy Problem Solving!

Peter Walker
Sharjah, Edinburgh, 2000/3

[1] ©MacKichan Software, Inc., 19307 8th Avenue, Suite C, Poulsbo, WA 98370–7370, USA. www.mackichan.com

Contents

1
Sequences

A sequence is simply a set whose elements are labelled by the positive integers (though a more formal definition is given in the next section). We write a sequence in the form $s = (s_1, s_2, s_3, \ldots)$ where the dots indicate that the list of terms continues indefinitely, so that any term, for instance s_{491}, is available for consideration if required. More explicitly we write $s = (s_1, s_2, s_3, \ldots, s_n, \ldots)$ to indicate the nth term, or simply $s = (s_n)_1^\infty$. But s_n alone (no parentheses!) is not the name of a sequence, it is the name of a number which is the nth term of a sequence. For instance $s = (2n-1)_1^\infty$ is the sequence of odd integers, $s = (1, 3, 5, \ldots)$ whose n^{th} term is $s_n = 2n - 1$.

Many more examples of sequences will be given shortly; for now we shall say only that a question of the form

$$\text{find the next terms in the sequence} \quad (1, 2, 4, 8, \ldots)$$

has no meaning, since the natural inclination to continue the sequence in powers of 2 to give $(1, 2, 4, 8, 16, 32, \ldots, 2^{n-1}, \ldots)$ could as well be replaced by $(1, 2, 4, 8, 15, 26, \ldots)$ in which the nth term is $n\left(n^2 - 3n + 8\right)/6$. Or indeed any continuation whatever could equally be presented as an answer to the question.

We shall feel free to start the labelling of the sequence at either 0 or 1 (or occasionally at other integers) as convenient, so that for instance $s = \left(2^{n-1}\right)_1^\infty$, in which $s_n = 2^{n-1}$ for $n \geq 1$, determines the same sequence as $s = (2^n)_0^\infty$, in which $s_n = 2^n$ for $n \geq 0$. We occasionally allow ourselves, particularly in examples, to write (s_n) (omitting the range of values) for a sequence when the range of n is clear (or irrelevant). But s_n alone is always a single term.

The properties of sequences which are of interest to us are usually those which are determined by conditions which apply *to all terms* of the sequence, or perhaps *to all terms from some point on* (say for all $n \geq 100$), or possibly *to infinitely many terms* (for instance all the even numbered ones). A condition which applies for instance to *only the first* 100 terms is unlikely to be useful.

1.1 Examples. Formulae and Recursion

We begin by looking at some sequences in an informal way, to get an idea of the properties we shall be considering.

Example 1.1

(i) $(1, 1/2, 1/3, \ldots, 1/n, \ldots) = (1/n)_1^\infty$,

(ii) $(-1, -2, -4, \ldots, -2^n, \ldots) = (-2^n)_1^\infty$,

(iii) $(1, -1, 1, -1, \ldots, (-1)^{n-1}, \ldots) = \left((-1)^{n-1}\right)_1^\infty = ((-1)^n)_0^\infty$,

(iv) $(1, 2, 6, 24, \ldots, n!, \ldots) = (n!)_1^\infty$,

(v) $(-1, 2, -3, 4, -5, \ldots, (-1)^n n, \ldots) = ((-1)^n n)_1^\infty$,

(vi) $s = (1, 1/4, 1/3, 1/16, 1/5, 1/64, \ldots, s_n, \ldots)$, where $s_n = 1/n$ if n is odd, $s_n = 1/2^n$ if n is even,

(vii) $t = (2, 3, 5, 7, 11, \ldots, t_n, \ldots)$, where t_n denotes the nth prime number,

(viii) $v = \left(0, \sqrt{2}, \sqrt{2+\sqrt{2}}, \sqrt{2+\sqrt{2+\sqrt{2}}}, \ldots, v_n, \ldots\right)$, where $v_0 = 0$ and for $n \geq 0$, $v_{n+1} = \sqrt{2 + v_n}$.

The first of these examples, the sequence $(1, 1/2, 1/3, \ldots, 1/n, \ldots)$, is very familiar, and has several of the properties we shall need. In the first place its terms are getting smaller (in the sense of the ordering of the real numbers), that is for each n, $s_{n+1} < s_n$; we shall say that the sequence is *decreasing*. At the same time the terms are getting smaller in the sense of absolute value; we shall say that they *tend to zero* in a sense to be made precise later (Figure 1.1).

The second example, $(-2^n)_0^\infty$ is also decreasing, but not to zero. Instead it decreases without bound or *tends to* $-\infty$ (Figure 1.2). The third example $((-1)^n)_0^\infty$ neither increases nor decreases, nor does it approach any limit, but at least we can say that its terms are always at most 1 in absolute value; we shall say that it is *bounded*. The fourth example increases without bound, while the

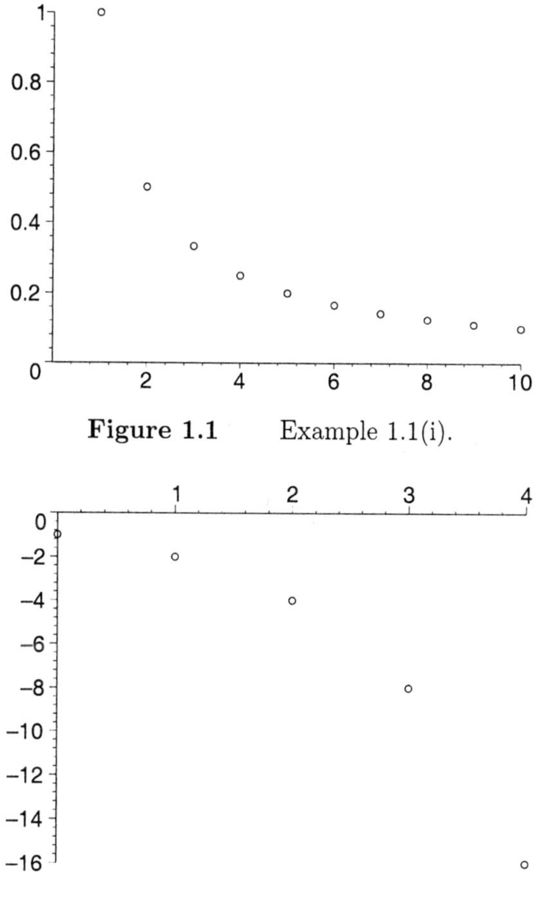

Figure 1.1 Example 1.1(i).

Figure 1.2 Example 1.1(ii).

fifth appears to be trying to go to both $+\infty$ and $-\infty$. The sequence in example (vi) is getting closer to zero, but not in decreasing order. The prescription in (vii) that t_n denotes the nth prime number is allowed, since although there is no easy formula for p_n, there do exist algorithms to determine it. (Notice at this point an important notational matter: in (vi) and (vii) we must not call the first sequence s_1 and the second s_2, etc. since these are already the names of the *terms* of some sequence s.)

The last example (viii) shows a sequence defined recursively, that is by defining each term in terms of one or more of its predecessors. ♦

On a more formal level, as in [14], a sequence is a function s from $\mathbb{Z}^+ = \{1, 2, 3, \ldots\}$ to some set (for instance the real or complex numbers), where we agree to write the values of the function as $(s_1, s_2, \ldots, s_n, \ldots)$ instead of the usual $(s(1), s(2), \ldots, s(n), \ldots)$. This point of view is useful when we want to distinguish between the sequence $s = (s_n)_1^\infty$ and the point set $\{s_n : n \geq 1\}$ which is the range of the function. For instance in example (iii) above in which s_n (or $s(n)$) is equal to $(-1)^n$, the range of the function is the two-point set $\{-1, 1\}$. All our sequences will be real valued unless explicitly stated otherwise.

All of these ideas will be developed and made precise in the remaining sections of this chapter.

1.2 Monotone and Bounded Sequences

Our job is to clarify the ideas which were considered informally in the first section. We begin with the notion of increasing or decreasing sequences.

Definition 1.2

(i) A sequence s of real numbers is called decreasing if $s_{n+1} \leq s_n$ for all $n \geq 1$. (Other phrases used to describe this property are *weakly decreasing* or *non-increasing*.) A sequence s is called strictly decreasing if $s_{n+1} < s_n$ for all $n \geq 1$.

(ii) A sequence s of real numbers is called increasing if $s_{n+1} \geq s_n$ for all $n \geq 1$. (Other phrases used to describe this property are *weakly increasing* or *non-decreasing*.) A sequence s is called strictly increasing if $s_{n+1} > s_n$ for all $n \geq 1$.

(iii) A sequence of real numbers is called monotone (or monotonic) if it is either increasing or decreasing.

If these distinctions seem too fussy, we can only say that we need a careful language to describe the examples we shall be looking at, and that the distinction between for instance decreasing and strictly decreasing is often irrelevant, but occasionally vital.

In Example 1.1, we see that (iv) and (vii) are increasing while (i) and (ii) are decreasing. These assertions are easy to justify – for instance to show that (i) is decreasing we simply observe that $1/n - 1/(n+1) = 1/(n(n+1))$ is positive. Notice also that examples (v) and (vi) show that a sequence need be neither increasing nor decreasing.

Example (viii) is also increasing, though this is less evident. If challenged on the point we could say that clearly each v_n is positive (we take the square root of a positive number) and that $v_1 = \sqrt{2} > v_0 = 0$. This is then the basis for an induction argument: having shown that $v_1 > v_0$, we suppose that $v_n > v_{n-1}$ for some $n \geq 1$ and deduce that $v_{n+1} > v_n$. To do this we write $v_{n+1}^2 = v_n + 2$ and similarly $v_n^2 = v_{n-1} + 2$. Hence for $n \geq 2$ we have

$$\begin{aligned} v_{n+1}^2 - v_n^2 &= v_n - v_{n-1}, \\ v_{n+1} - v_n &= \frac{v_n - v_{n-1}}{v_{n+1} + v_n}, \end{aligned}$$

using the elementary identity $a^2 - b^2 = (a-b)(a+b)$. Since we have assumed that $v_n - v_{n-1} > 0$ and we know that $v_{n+1} + v_n > 0$, it follows that $v_{n+1} - v_n > 0$, and the result follows by induction.

The next examples are useful observations which the reader may verify as exercises.

Example 1.3

(i) For fixed real p, the sequence $(n^p)_1^\infty$ is increasing if $p > 0$, decreasing if $p < 0$ and constant if $p = 0$.

(ii) For fixed positive x, the sequence $(x^n)_0^\infty$ is increasing if $x > 1$, decreasing if $x < 1$ (and constant if $x = 1$). If x is negative, then the sequence is not monotone. ♦

Example 1.4

Show that the sequence $\left(\frac{2n+1}{n+3}\right)_1^\infty$ is increasing, while $\left(\frac{n+2}{4n+1}\right)_1^\infty$ is decreasing.

For the first of these examples, just observe that

$$\begin{aligned} \frac{2n+1}{n+3} - \frac{2(n+1)+1}{(n+1)+3} &= \frac{(2n^2+9n+4) - (2n^2+9n+9)}{(n+3)(n+4)} \\ &= \frac{-5}{(n+3)(n+4)} \end{aligned}$$

is clearly negative since $n \geq 1$. The second example is similar. (See Exercise 2 for the general case in which $s_n = \frac{an+b}{cn+d}$.) ♦

We shall often come across results of the type 'if two sequences (or in the next chapter, two functions) s, t have some property, for instance that of being monotone, then the sum $s + t$ (or possibly the difference $s - t$, or the product $s.t$, or the quotient s/t) has (or does not have!) the corresponding property'. We meet this for the first time in the following proposition. (A proposition is a result which, while not interesting or important enough to be called a theorem, is useful to state for later reference, and useful to prove for the practice in using the definitions. Our policy will be to prove some parts of these propositions, and to leave the rest as exercises.) To be precise about these operations on sequences, we say

Definition 1.5

Let $s = (s_n)$, $t = (t_n)$ be sequences. Then the sum $s + t$, difference $s - t$, and product $s.t$ are defined by

$(s+t)_n = s_n + t_n$, $(s-t)_n = s_n - t_n$, $(s.t)_n = s_n.t_n$, respectively, while the quotient s/t is defined by

$(s/t)_n = s_n/t_n$ whenever $t_n \neq 0$ for all n.

Proposition 1.6

(i) If the sequences s, t are both increasing (or both decreasing) then the sum $s + t$ is also increasing (or decreasing).

(ii) If the sequences s, t are both increasing (or both decreasing) and positive, then the product $s.t$ is also increasing (or decreasing).

Proof

(An essential: analysis, and more generally mathematics as a whole, is the art of deductive reasoning – 'if we can't prove it, we can't use it' must be our motto.)

(i) If both are increasing, then $s_{n+1} \geq s_n$ and $t_{n+1} \geq t_n$. It follows that $s_{n+1} + t_{n+1} \geq s_n + t_n$, so the sum is increasing. (Note that it does *not* follow that $s_{n+1} - t_{n+1} \geq s_n - t_n$; there is no result for the difference of two monotone sequences as the example below shows.)

(ii) If both are increasing and positive, then

$$s_{n+1}t_{n+1} - s_n t_n = (s_{n+1} - s_n)\, t_{n+1} + s_n\,(t_{n+1} - t_n)$$

and the right-hand side is clearly positive. (Again there is no result for the quotient.) ■

Example 1.7

(i) The sequences $(4n)$, and (n^2) are both increasing but their difference $(4n - n^2)$ is not monotone.

(ii) The sequences (n), and $(n-4)$ are both increasing but their product (n^2-4n) is not monotone.

These examples show that there can be no corresponding result for differences, or for products of sequences which change sign. Notice that to show that a sequence (s_n) is not monotone requires examples of indices j, k with $s_j < s_{j+1}$ and $s_k > s_{k+1}$; in the above example, if $s_n = 4n - n^2$ then $s_1 = 3 < s_2 = 4 > s_3 = 3$ is sufficient to show that the sequence is not monotone. ♦

The second important idea which we have to define is the notion of a *bounded sequence*. Informally a sequence is bounded above if there is some number which is greater than all terms of the sequence. Thus for instance the sequence $\left(\frac{n}{n+1}\right)$ is clearly bounded above by 1 and below by 0 since n is positive. The numbers 1 and 0 which occur here are called upper and lower bounds for the sequence. Notice that bounds are not unique; we could say (truthfully but less precisely) that $\frac{n}{n+1} < 2$ for all n, so that 2 is also an upper bound for this sequence. More generally if M is an upper bound and $M' > M$ then M' is also an upper bound (the reader should at once supply a proof of this easy fact). The existence of a *unique least upper bound* (of a bounded sequence) is something we shall comment on further below. To show that a sequence is *not* bounded above requires a proof that *no* number satisfies the definition; i.e. that for *all* real M, there is some n with $s_n > M$. For instance the sequence $(2n+1)$ is not bounded above, for if we suppose that some M is given, then $2n+1 > M$ is satisfied when $n > (M-1)/2$ and so M is not an upper bound. With this preamble we can proceed to a formal definition.

Definition 1.8

A sequence (s_n) is said to be bounded above if for some real number M we have $s_n \leq M$ for all n. The number M is called an upper bound for the sequence. Similarly a sequence is said to be bounded below if for some real number m, $s_n \geq m$ for all n. The number m is called a lower bound. A sequence is said to be bounded if it is bounded both above and below.

Example 1.9

(i) The sequence $(2n+1)$ is bounded below (by zero) but not above. More generally the sequence $(an+b)$ is bounded below but not above for any real $a > 0$, and is bounded above but not below if $a < 0$.

(ii) Any sequence given by a polynomial, $s_n = a_k n^k + a_{k-1} n^{k-1} + \cdots + a_0$

where $a_k > 0$ is bounded below but not above.

(iii) Any sequence of the form $\left(\frac{an+b}{cn+d}\right)_0^\infty$ where $a, b, c, d > 0$ is bounded.

(iv) The sequence of Example 1.1(viii), given by $v_0 = 0$, $v_{n+1} = \sqrt{2 + v_n}$, is bounded above by 2.

(i) The statements about $(an + b)$ are left to the reader, and (ii) is Exercise 4. The sequence $\left(\frac{an+b}{cn+d}\right)$ in (iii) is bounded below by zero since a, b, c, d are all positive. To find a number M with $an+b < M\,(cn + d)$ for all n it is enough to choose M greater than both a/c and b/d. For (iv) we use induction, as before when we looked at this sequence. Clearly $v_0 = 0 < 2$, and if $v_n < 2$ then $4 - v_{n+1}^2 = 2 - v_n$ is positive and so since v_{n+1} is positive, it must be less than 2. ♦

The next result is another easy technicality.

Proposition 1.10

(i) The sequence (s_n) is bounded if and only if $(|s_n|)$ is bounded.

(ii) If (s_n) and (t_n) are bounded sequences then so are $(s_n \pm t_n)$ and $(s_n.t_n)$ (but not necessarily (s_n/t_n)).

Proof

(i) If (s_n) is bounded then there are M, m with $m \leq s_n \leq M$ for all n. Then for $s_n \geq 0$ we have $|s_n| \leq M$ and for $s_n < 0$ we have $|s_n| \leq |m|$. Hence $(|s_n|)$ is bounded by the larger of $M, |m|$. The converse is easy, as is (ii). ■

In the discussion at the beginning of this section, we noticed that a sequence may have many upper bounds but that in the examples there was often a *smallest* upper bound. It turns out that the existence of such a least upper bound is a characteristic property of the real number system; we refer the reader to [14] for further discussion of the following statement.

Axiom 1.11

Every sequence (and every set) of real numbers which is bounded above, has a least upper bound. The least upper bound m is called the supremum of the sequence and written $m = \sup(s_n)$.

To understand this statement we should observe that a least upper bound m has the property that (i) it is an upper bound, i.e. for all n, $s_n \le m$, and (ii) any other upper bound is bigger, i.e. if M is any upper bound then $M \ge m$. This shows in particular that a least upper bound is *unique*, for if m, m' were two least upper bounds we should have both $m \ge m'$ and $m' \ge m$. An equivalent statement to (ii) is that any number less than m is not an upper bound, i.e. if $t < m$ then there is some n with $s_n > t$. We illustrate this with two familiar examples.

Example 1.12

(i) The least upper bound of the sequence $\left(\frac{n}{n+1}\right)$ is 1.

(ii) The least upper bound of the sequence given by $v_0 = 0$, $v_{n+1} = \sqrt{v_n + 2}$ is ≤ 2.

(i) Clearly 1 is an upper bound. To show that it is the least suppose that $0 < t < 1$. Then we can make $\frac{n}{n+1} > t$ by taking $n > (n+1)\,t$, which is equivalent to $n > \frac{t}{1-t}$ which is possible since $0 < t < 1$. Hence t is not an upper bound so 1 must be the least. The statement of (ii) is trivial since we have already shown that $v_n < 2$ for all n, so 2 is an upper bound. The point is that we cannot yet say that the least upper bound is *equal* to 2; this will come after we have studied convergence in the next section. ♦

An attentive reader might ask at this point, 'What about lower bounds – do we need another axiom for them?' Fortunately one axiom is enough; it is a consequence of Axiom 1.11 that a sequence with a lower bound necessarily has a greatest lower bound. To see this, apply the upper bound axiom to the sequence $(-s_n)$ (if l is an upper bound for $(-s_n)$ then $-l$ is a lower bound for (s_n)). The greatest lower bound of a sequence (s_n) is called the infimum of the sequence and written $\inf(s_n)$. Exercise 3 demonstrates another approach.

1.3 Convergence

We now come to the property of sequences which is central to the whole of analysis, namely convergence. In Example 1.1(i) we noticed that the terms got closer to zero, and that the same was true of the apparently more irregular (non-monotone) example (iv). We have to try to make precise this idea of approaching zero (or any other number) as n gets large.

To begin with, notice that to say as we did above that the terms 'get closer to zero' is insufficient on its own, since the same is true for instance of the sequence given by $s_n = 1/n + 1/2$; in fact the terms of any positive decreasing sequence get closer to zero in this sense. What is needed is not simply that the terms get 'closer to zero'; they must get 'arbitrarily close', or 'as close as we like'. When we analyse this requirement we see that we are asking for more than that the terms should get small; we want to know that they can be made less than any preassigned magnitude (call it ε) by taking n large enough (say $n \geq N$). (The use of ε, Greek epsilon, is traditional here, though other writers use h, d, etc.)

For instance, for a given sequence, if we take $\varepsilon = 1/2$ then it might be necessary to take $n \geq 20$, while if $\varepsilon = 1/200$ then $n \geq 1000$ might be required. This idea, that if a small quantity (which we call ε) is given, then a corresponding value (which we call N) can be found, depending on ε, so that s_n differs from zero (or whatever the appropriate limiting value may be) by less than ε when $n \geq N$, is the essence of the following definition of convergence.

Definition 1.13

We say that a sequence (s_n) is convergent with limit l if for each $\varepsilon > 0$, there exists an integer N such that $|s_n - l| < \varepsilon$ whenever $n \geq N$. When this happens we write $l = \lim_{n\to\infty} s_n$, or $s_n \to l$ as n$\to \infty$.

A sequence which is not convergent will be called divergent.

The inequality $|s_n - l| < \varepsilon$ can be stated equivalently as $l - \varepsilon < s_n < l + \varepsilon$. It embodies the informal idea that 's_n is as close to l as we like'. The statement that $s_n \to l$ as $n \to \infty$, is read as 's_n tends to l as n tends to infinity' (Figure 1.3).

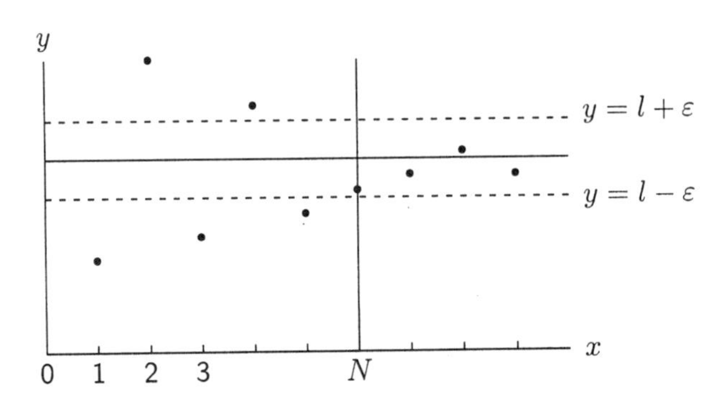

Figure 1.3 Definition 1.13. Adapted from *Pure Mathematics*, G.H. Hardy, 10th edition, 1952. ©Cambridge University Press.

Example 1.14

(i) The sequence $(1/n)$ (or more generally $(1/n^p)$ for $p > 0$) is convergent with limit zero.

(ii) The sequence $(c + 1/n)$ for a positive constant c, does not tend to zero (take $\varepsilon = c/2$) but to c.

(iii) The sequence $\left(\frac{n}{n+1}\right)$ is convergent with limit 1.

(iv) The sequence $(\sqrt{n+1} - \sqrt{n})$ is convergent with limit zero.

(v) The sequence (x^n) with $|x| < 1$, is convergent with limit zero.

To show (i) we suppose that ε is given. Then we make $1/n < \varepsilon$ by making $n \geq N = 1/\varepsilon$. (If an objection is raised that the value of $1/\varepsilon$ might not be an integer, we can reply by taking N as the next larger integer.) If the sequence is $(1/n^p)$ then take $n \geq N = (1/\varepsilon)^{1/p}$.

Part (ii) is a consequence of (i) (the reader can show that $s_n \to c$ is equivalent to $s_n - c \to 0$), and (iii) follows in the same way since $1 - n/(n+1) = 1/(n+1)$.

Example (iv) is an interesting one, since it states that the difference of two sequences which are both getting large may become small. (We can get the same behaviour more obviously by considering $(n + 1/n) - n$.) For $\sqrt{n+1} - \sqrt{n}$ we use the identity

$$\sqrt{a} - \sqrt{b} = \left(\sqrt{a} - \sqrt{b}\right) \frac{\sqrt{a} + \sqrt{b}}{\sqrt{a} + \sqrt{b}} = \frac{a - b}{\sqrt{a} + \sqrt{b}}.$$

Applied to $\sqrt{n+1} - \sqrt{n}$ this gives

$$\sqrt{n+1} - \sqrt{n} = \frac{n + 1 - n}{\sqrt{n+1} + \sqrt{n}} = \frac{1}{\sqrt{n+1} + \sqrt{n}} < \frac{1}{\sqrt{n}}$$

which tends to zero by (i). (The alert reader will have noticed that another small proposition has been assumed here; if $s_n \to 0$ and $0 < t_n < s_n$ then also $t_n \to 0$. This too is an easy exercise, or else a consequence of the 'Sandwich Theorem', Proposition 1.18(iii) below.) Notice finally that sign changes have no effect if a sequence tends to zero; for instance the fact that $(-1)^n / n^p \to 0$ is proved in the same way as (i).

For (v), consider the sequence $(|x|^n)$ which is decreasing and positive and so is bounded below by zero. But then it must have a greatest lower bound, say $t \geq 0$. Then $|x|^{n+1} \geq t$ for all n so $|x|^n \geq t/|x|$, so if $t > 0$, then $t/|x|$ is also a lower bound which is greater than t; this contradiction shows that $t = 0$. Then for any $\varepsilon > 0$, ε is not a lower bound, so there is some N such that $|x|^N < \varepsilon$ and so for $n \geq N$, $|x^n| \leq |x|^N < \varepsilon$ as required to show that $x^n \to 0$. (An alternative method is to use the second solution to Example 1.24(ii) below where it is shown by a different method that if $x > 1$ then $x^n \to \infty$ as $n \to \infty$.) ♦

To show that a sequence does *not* have a limit requires careful thinking. For instance consider $(s_n) = ((-1)^n)$ which was in Example 1.1(iii). Since the values, ± 1 do not appear to be getting nearer to any single value, it is reasonable to assume that the sequence is not convergent. To show this we argue by contradiction, as follows. Suppose that this sequence tends to some limit $l \geq 0$. Notice that for odd n, $s_n = -1$, and so if we take $\varepsilon = 1/2$ then $|s_n - l| \geq 1 > \varepsilon$ for all odd n. Hence the definition of convergence cannot be satisfied for any limit $l \geq 0$. Similarly no negative number satisfies the definition, and so the sequence fails to be convergent.

When we discussed monotone and bounded sequences in the previous section, we found that a sequence could have one property or the other, or both, or neither – that is the two properties were independent of one another. Now when we look at the property of convergence and its relationship to the two others, we find that it is not similarly independent. The most immediate observation is that the convergent sequences are a subset of the bounded ones.

Theorem 1.15

Every convergent sequence is bounded.

Proof

Suppose that (s_n) is convergent with limit l. Then given any $\varepsilon > 0$ (for instance $\varepsilon = 1$) there is N such that for $n \geq N$ we have $l - \varepsilon < s_n < l + \varepsilon$, i.e. the numbers $l \pm \varepsilon$ are upper and lower bounds for that part of the sequence which has $n \geq N$. To find bounds for the whole sequence we simply take the largest (say s_p) and smallest (say s_q) of the finite number of terms with $n < N$, so that $s_p \leq s_n \leq s_q$ for $n < N$. Then the larger of s_q, $l + \varepsilon$ is an upper bound for the whole sequence, and the smaller of s_p, $l - \varepsilon$ is similarly a lower bound. ■

It is tempting to try to short-cut this proof by saying at the outset 'let s_p be the greatest element of the sequence'. But many convergent sequences, for instance the example above with $s_n = n/(n+1)$, do not have a largest element, and the argument cannot get off the ground. For the proof to work it is essential that we first dispose of all terms with $n \geq N$, so that we are left with a finite number of terms from which we can find the largest.

The converse of this theorem is false; for instance the sequence $((-1)^n)_1^\infty$ is bounded but not convergent as we have already shown. Thus the convergent sequences form a *proper* subset of the bounded ones.

The second relation between monotone, bounded and convergent sequences is fundamental for the study of convergence since it allows us to show that a sequence is convergent without having to exhibit a limit.

Theorem 1.16

A sequence which is monotone and bounded must be convergent.

Proof

We can suppose that the sequence is increasing – if not then reverse the role of upper and lower bounds in the rest of the proof. Let M be an upper bound for the sequence. It follows from Axiom 1.11 that there is a *least* upper bound m say. Thus we know that $s_n \leq m$ for all n and that no number less than m has this property. To show convergence, choose $\varepsilon > 0$; since $m - \varepsilon < m$ we see that $m - \varepsilon$ is not an upper bound so there is some N with $s_N > m - \varepsilon$. But the sequence is increasing so for $n \geq N$ we have $m - \varepsilon < s_N \leq s_n \leq m$ $(< m + \varepsilon)$, and the definition of convergence is satisfied. ■

The two results which we have just proved can be illustrated diagrammatically by the Venn diagram in Figure 1.4, where M, B, C are the sets of monotone, bounded and convergent sequences respectively.

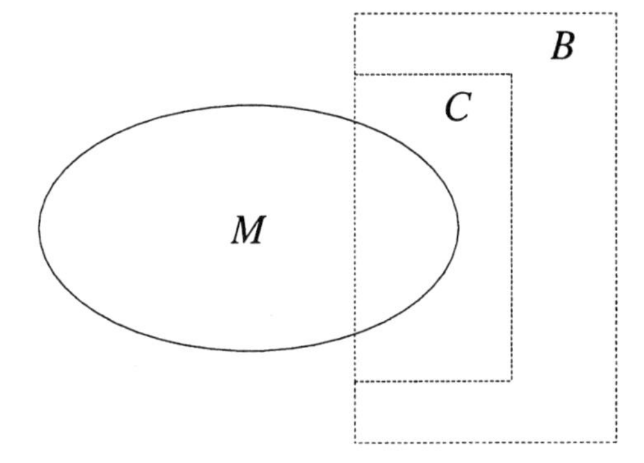

Figure 1.4 Monotone, bounded and convergent sequences.

To illustrate the use of this result, consider the following examples.

Example 1.17

(i) The sequence given by $v_0 = 0$, $v_{n+1} = \sqrt{2 + v_n}$ is convergent with limit equal to 2.

(ii) The sequences $s_n = (1 + 1/n)^n$, and $t_n = (1 + 1/n)^{n+1}$ are respectively increasing and decreasing, and have a common limit.

(i) We have shown already that this sequence is increasing and bounded above by 2. Theorem 1.16 now allows us to deduce that it is convergent and has limit $l \leq 2$. To show that the value is exactly 2 we make use of a result to be proved shortly (a special case of Proposition 1.20 below), that if $s_n \to l$ then $s_n^2 \to l^2$. Then since $v_{n+1}^2 = v_n + 2$ and $v_n \to l$ it follows that $v_{n+1}^2 \to l^2$ and so since limits are unique (Proposition 1.18(ii) below) the limit must satisfy $l^2 = l + 2$. This equation has only 2 and -1 for solutions, and since -1 is impossible (why?) the limit must be 2. The method we have used in considering this example of a sequence defined recursively is very general and we shall use it repeatedly later.

We begin (ii) with a simple observation: we have $s_1 = 2$, $t_1 = 4$ and the ratio $t_n/s_n = 1 + 1/n$ is greater than 1 and tends to 1 as $n \to \infty$. Hence as soon as the monotonicity is proved we shall have $2 \leq s_n < t_n \leq 4$ for all n. It will then follow that the two sequences are bounded and monotone, and so are convergent by the theorem. They have a common limit since $s_n/t_n \to 1$. To prove that (s_n) is increasing we use a little calculus – anyone who finds this out of place before we have proved the necessary facts in Chapter 3 may simply wait until then before reading what follows – but we imagine that all readers will have met calculus at least informally. The fact that the common limit is e, the base of natural logarithms can be deduced too, but we will leave this for now.

To show monotonicity of (s_n) we let $f(x) = (1 + 1/x)^x$, and show that $f'(x) = df(x)/dx > 0$ for all $x > 0$. We have $\ln f(x) = x \ln(1 + 1/x)$ and so

$$\begin{aligned}
\frac{f'(x)}{f(x)} &= \frac{d}{dx} \ln(f(x)) \\
&= \ln(1 + 1/x) - \frac{1}{x+1} \\
&= \ln(1+t) - \frac{t}{1+t} \qquad \text{where } t = 1/x \\
&= g(t) \text{ say.}
\end{aligned}$$

But $g(0) = 0$ and $g'(t) = t/(1+t)^2 > 0$ for $t > 0$ so $g(t) > 0$ for $t > 0$. This shows that $f'(x)/f(x) > 0$ for all $x > 0$, and since $f(x) > 0$ for $x > 0$ it follows that $f'(x) > 0$ for $x > 0$ and thus f is increasing as required. ♦

Having enjoyed ourselves with the examples we must now put a few of the necessary technicalities in place.

Proposition 1.18

(i) Let sequences (s_n), (t_n) be convergent with limits s, t respectively. If for all n (or more generally for all $n \geq$ some n_1), $s_n \leq t_n$ then $s \leq t$.

(ii) The limit of a sequence is unique: if $s_n \to s$ and $s_n \to t$ then $s = t$.

(iii) Let sequences (s_n), (t_n) be convergent with the same limit s and $s_n \leq t_n$ for all n. Let (u_n) be another sequence with $s_n \leq u_n \leq t_n$ for all n. Then (u_n) is convergent with the same limit s.

Proof

(i) Suppose that $s > t$; we will derive a contradiction showing that this is impossible and so we must have $s \leq t$. If $s > t$ then let $\varepsilon = (s - t)/2 > 0$, so that $t < t + \varepsilon = s - \varepsilon < s$. Use this ε in the definition of convergence to get a value of N such that if $n \geq N$ then $t_n < t + \varepsilon = s - \varepsilon < s_n$. This contradicts $s_n \leq t_n$ and the result follows. We leave to the reader the modification of the proof if we assume $s_n \leq t_n$ only for $n \geq n_1$.

(ii) This is a consequence of (i), for clearly $s_n \leq s_n$ and so if $s_n \to s$ and $s_n \to t$ then by (i) we have both $s \leq t$ and $t \leq s$.

(iii) We cannot quite deduce this from (i) since we must first show that (u_n) is convergent. However given $\varepsilon > 0$, there are N_1, N_2 such that

$$\begin{aligned} s - \varepsilon < s_n < s + \varepsilon \qquad & \text{if } n \geq N_1, \text{ and} \\ s - \varepsilon < t_n < s + \varepsilon \qquad & \text{if } n \geq N_2. \end{aligned}$$

It follows that for $n \geq \max(N_1, N_2)$ we have

$$s - \varepsilon < s_n \leq u_n \leq t_n < s + \varepsilon$$

and $u_n \to s$ as required. ■

Note that if we know $s_n < t_n$ in (i), we cannot deduce $s < t$; the limiting process may destroy the strict inequality, as the example $s_n = 0$, $t_n = 1/n$ shows.

Part (iii) of this result is known colloquially as the 'Sandwich Theorem' since the sequence (u_n) is sandwiched between the other two, as in the next example. Notice that it is essential that the limits are the same; we can deduce nothing if $s \neq t$ as the example $s_n = -1$, $u_n = (-1)^n$, $t_n = 1$ shows.

Example 1.19

(i) The sequence $\left(n^{-2}\sin\left(\pi\sqrt{n!+5}\right)\right)$ is convergent with limit zero.

(ii) The sequence $\left((1+1/n)^{n+\alpha}\right)$ is convergent for $0 \leq \alpha \leq 1$.

In (i) the expression inside the sine function is clearly irrelevant – all we need is that $|\sin(x)| \leq 1$ for any x. Then the Sandwich Theorem applies with $s_n,\ t_n = \pm 1/n^2$.

Part (ii) follows from the Sandwich Theorem and Example 1.17 above. ♦

Proposition 1.20

Let sequences (s_n), (t_n) be convergent with limits s, t respectively. Then the sequences $(s_n \pm t_n)$ and $(s_n.t_n)$ are convergent with limits $s \pm t$, $s.t$ respectively.

If in addition t and all t_n are non-zero then the sequence (s_n/t_n) is convergent with limit s/t.

Proof

For a given ε there are N_1, N_2 such that

$$\begin{aligned} s-\varepsilon < s_n < s+\varepsilon \qquad & \text{if } n \geq N_1, \text{ and} \\ t-\varepsilon < t_n < t+\varepsilon \qquad & \text{if } n \geq N_2. \end{aligned}$$

It follows that if n is greater than the larger of N_1, N_2 then both sets of inequalities are satisfied and we have

$$s \pm t - 2\varepsilon < s_n \pm t_n < s \pm t + 2\varepsilon$$

as required.

(This is the place to comment on the fact that the name of the small quantity in the definition of convergence, and how it is arrived at, are irrelevant – all that matters is that it can be made as small as required. In the above argument we ended up with 2ε instead of ε – this does not matter since 2ε can be made small when ε is small. Indeed we could have started the argument with $\varepsilon/2$ in place of ε and ended with ε, but students often complain that this smacks of 'hindsight' and we shall avoid doing so.)

When we consider the product of the sequences we need a further fact, namely that both sequences are bounded, which follows from Theorem 1.15. Suppose that $|s_n| \leq S$ and $|t_n| \leq T$ (and hence also $|t| \leq T$). Then we have

$$\begin{aligned} |s_n.t_n - s.t| &= |s_n(t_n - t) + t(s_n - s)| \\ &\leq S\varepsilon + T\varepsilon \end{aligned}$$

and $(S+T)\,\varepsilon$ may be made small with ε.

For the quotient, we need one additional detail. Since we assume that $t \neq 0$ we can assume that $0 < \varepsilon < |t|/2$. Thus there is some N_0 such that $|t_n - t| < |t|/2$, and so $|t_n| > |t|/2$ for $n \geq N_0$. Then

$$\begin{aligned}\left|\frac{s_n}{t_n} - \frac{s}{t}\right| &= \left|\frac{(s_n - s)\,t + s\,(t - t_n)}{t_n t}\right| \\ &\leq \frac{\varepsilon T + S\varepsilon}{|t|^2/2} = 2\varepsilon\frac{(T+S)}{|t|^2}\end{aligned}$$

which again is small with ε. ∎

Example 1.21

Show

(i) $(5n^2 - 3)/(1 + 2n - n^2) \to -5$ as $n \to \infty$,

(ii) $\left((n+2)^2 - (n+1)^2\right)/n \to 2$ as $n \to \infty$.

To see (i), rewrite the fraction as $(5 - 3/n^2)/(-1 + 2/n + 1/n^2)$. Then $1/n$, $1/n^2$ both $\to 0$ (by Example 1.14(i)), so the numerator and denominator tend to 5, -1 respectively and then the quotient $\to -5$, both from Proposition 1.20. Similarly for (ii) it is enough to simplify the fraction to give

$$\frac{n^2 + 4n + 4 - n^2 - 2n - 1}{n} = \frac{2n+3}{n} = 2 + \frac{3}{n}$$

and the result follows. ♦

This technique of simplifying and making preliminary rearrangements will always be useful, and we shall often use both it and Proposition 1.20 without explicitly mentioning them.

Before we finish this section on convergence, we want to consider one further matter, namely the different ways in which a sequence may fail to be convergent. Consider the following examples, which are all divergent in the sense of Definition 1.13.

Example 1.22

(i) $\left(1, -1, 1, -1, \ldots, (-1)^{n-1}, \ldots\right)$.

(ii) $(1, 4, 9, \ldots, n^2, \ldots)$.

(iii) $(1, 4, 3, 16, 5, 36, \ldots, s_n, \ldots)$ where $s_n = n$ if n is odd, or n^2 if n is even.

(iv) $(1, 0, 3, 0, 5, \ldots, t_n, \ldots)$ where $t_n = n$ if n is odd, or 0 if n is even.

(v) $(1, -2, 3, -4, 5, \ldots, n(-1)^n, \ldots)$.

The first of these stays bounded but does not 'settle down' to any particular limit, while the others are all unbounded. The second example is increasing, while the others are not monotone. Intuitively we think that (ii) and (iii) somehow 'tend to infinity' (wherever that may be) while (iv) and (v) fail to approach anything. This intuitive idea of tending to infinity is made precise in the following definition. Comparing this with Definition 1.13 we notice that in place of the small quantity ε we now have another large number K to deal with. ♦

Definition 1.23

We say that a sequence (s_n) tends to infinity with n (or as n tends to infinity) if for each $K > 0$ there is some N such that $s_n > K$ when $n > N$. In this case we can also write $s_n \to \infty$ as $n \to \infty$. We say that (s_n) tends to minus infinity if the same is true with $s_n < -K$ (or equivalently if $-s_n \to \infty$).

The definition accords with our intuition, since it is satisfied for examples (ii) and (iii) but not for (iv) and (v), as the reader may immediately verify. Note that the symbol ∞ has no meaning on its own, but only as part of a statement such as $s_n \to \infty$.

Example 1.24

(i) The sequence $n^\alpha \to \infty$ as $n \to \infty$ when $\alpha > 0$.

(ii) The sequence $x^n \to \infty$ as $n \to \infty$ when $x > 1$.

(iii) The sequence $n^\alpha x^n \to 0$ when $|x| < 1$, and to ∞ when $x > 1$, irrespective of the value of α.

The proof of (i) is easy from the definition as follows. Suppose that K is given and we want $n^\alpha > K$ when $n > N$. Then we need only take $n > K^{1/\alpha}$ for the definition to be satisfied – compare Example 1.14(i) for sequences which $\to$ 0.

For (ii) we showed in Example 1.14(ii) that $t^n \to 0$ when $|t| < 1$, so with $x = 1/t$ we have $x^n = 1/t^n \to \infty$, using (ii) of Proposition 1.25 below. An alternative method is to put $x = 1 + y$ with $y > 0$, and to prove by induction that $x^n = (1+y)^n \geq 1 + ny$ which clearly tends to infinity with n.

For (iii), which combines the first two we have to work a little harder. Given $x > 1$, choose some y with $1 < y < x$ (for instance $y = (1+x)/2$ would do).

Then writing $s_n = n^\alpha x^n$ we have

$$\frac{s_{n+1}}{s_n} = \left(\frac{n+1}{n}\right)^\alpha x \to x > y \qquad \text{as } n \to \infty,$$

and so there is some N such that $s_{n+1}/s_n > y$ for $n \geq N$. Then for $n > N$ we can write

$$\begin{aligned}\frac{s_n}{s_N} &= \frac{s_n}{s_{n-1}}\frac{s_{n-1}}{s_{n-2}}\cdots\frac{s_{N+1}}{s_N} \\ &> y^{n-N},\end{aligned}$$

or equivalently

$$s_n > \left(\frac{s_N}{y^N}\right) y^n$$

and the result follows since $y^n \to \infty$ as $n \to \infty$. For $|x| < 1$ choose y with $|x| < y < 1$ and let $s_n = n^\alpha |x|^n$. Then $s_{n+1}/s_n \to |x|$ so there is some N with $s_{n+1}/s_n < y$ for $n \geq N$, and the result follows as before since $y^n \to 0$. ♦

Analogues of many (but not all) of the results for convergent sequences which we have established earlier remain true.

Proposition 1.25

(i) If $(s_n), (t_n)$ are sequences with $s_n \to \infty$ while t_n tends to a finite limit or to $+\infty$ then $s_n + t_n \to \infty$. If $(s_n), (t_n)$ are sequences with $s_n \to \infty$ while t_n tends to a finite limit $l > 0$, then $s_n.t_n \to \infty$.

(ii) For a sequence (s_n) of strictly positive terms $s_n \to \infty$ if and only if $1/s_n \to 0$.

(iii) A sequence which is monotone and not bounded must tend to ∞ or $-\infty$.

Before the proof we warn that these are among the most misquoted and misused results in elementary analysis. For instance there is no analogue of (i) for the difference $s_n - t_n$ of sequences which tend to infinity; consider for example the sequences $(n + 1/n) - n$, $(n + 1/n)^2 - n^2$, $(n + 1)^2 - n^2$ which tend to $0, 2, \infty$ respectively. Analogous results for products are in the exercises at the end of the chapter. For (ii) it is certainly *not* true for all sequences that '$s_n \to 0$ if and only if $1/s_n \to \infty$' though generations of students firmly believe it (consider $s_n = (-1)^n/n$); the restriction to strictly positive terms is essential. Having said all this, the proofs are easy.

Proof

(i) If both sequences tend to infinity then given K there are N_1, N_2 such that $s_n > K$, $t_n > K$ when $n \geq N_1$, $n \geq N_2$ respectively, and so $s_n + t_n > 2K$ if $n \geq \max(N_1, N_2)$. If t_n tends to a finite limit l, then replace the second occurrence of K by $l - 1$; the result follows since l is fixed and so $K + l - 1$ is large when K is. The proof of the second statement is similar, and is left as an exercise.

(ii) For positive sequences we can say $s_n > K$ if and only if $1/s_n < 1/K$ so the result follows with $\varepsilon = 1/K$ (or $K = 1/\varepsilon$) in the definition.

For (iii) suppose for definiteness that the sequence is increasing. Then given any $K > 0$, K is not an upper bound, so there is some n with $s_n > K$. But the sequence is increasing, and so $s_m \geq s_n > K$ for $m \geq n$, so $s_n \to \infty$ as stated. ■

1.4 Subsequences

A subsequence has a similar relation to a sequence as a subset does to a set; that is we select some of the elements and discard the others. More precisely, since a sequence is an infinite ordered set, we choose infinitely many elements, keeping them in the same order, and discard the rest. For instance, if (s_n) is the given sequence then

$$(s_1, s_2, s_4, s_8, \ldots, s_{2^{k-1}}, \ldots) \text{ and}$$
$$(s_1, s_5, s_9, \ldots, s_{4k-3}, \ldots)$$

are subsequences.

Thus each subsequence determines, and is determined by, a strictly increasing sequence of integers which tells us which terms of the sequence are selected. If we denote this sequence of integers by $(n_1, n_2, \ldots, n_k, \ldots)$ where

$$1 \leq n_1 < n_2 < \ldots < n_k < n_{k+1} < \ldots$$

then the subsequence is

$$(s_{n_1}, s_{n_2}, \ldots, s_{n_k}, \ldots)$$

where $n_k = 2^{k-1}$, or $n_k = 4k-3$ in the examples. (This double suffix notation is initially off-putting but seems to be unavoidable.) Exercise 21 asks the reader to show that any such strictly increasing sequence (n_k) of integers satisfies $n_k \geq k$, and hence that $n_k \to \infty$ as $k \to \infty$.

Clearly a subsequence will share many properties with the 'parent' sequence from which it is formed. For instance all the properties of monotonicity, boundedness and convergence which we have considered so far are 'inherited' in this way. More precisely we have

Proposition 1.26

(i) If a sequence is monotone then any subsequence is monotone in the same sense (either increasing or decreasing).

(ii) If a sequence is bounded then any subsequence is bounded, with the same (or closer) bounds.

(iii) If a sequence is convergent then any subsequence is convergent with the same limit.

Proof

(i) If (s_n) is increasing then $s_n \leq s_{n+1}$ for all n, and it follows that if $n < m$ then $s_n \leq s_m$. In particular, since $n_k < n_{k+1}$ it follows that $s_{n_k} \leq s_{n_{k+1}}$ as required.

(ii) If (s_n) is bounded above by m, then also $s_{n_k} \leq m$, so m is also an upper bound for the subsequence. (However some smaller number, not necessarily an upper bound for the parent, may also be an upper bound for the subsequence; for instance if $s_n = (-1)^n$ then 1 is the least upper bound, but if we take $n_k = 2k - 1$ then the subsequence is the constant -1 so that the least upper bound is reduced to -1.) The argument for lower bounds is similar, and the proof of (iii) is a nice exercise (Exercise 22 at the end of the chapter). ■

The main interest in studying subsequences, however, does not lie in properties which they may inherit, but rather in additional desirable properties which they may have. For instance the following theorem and its corollary are both spectacular and unexpected, and have the further bonus of an elegant proof.

Theorem 1.27

Every sequence has a monotone subsequence.

At first sight a proof of this (even if we believe it) seems very difficult even to begin; we are given no information about the sequence, and no indication of whether to aim for an increasing or decreasing subsequence. The proof depends on the idea of a 'peak point' and seems to be part of mathematical folklore.

Proof

We shall say that a point s_k of the sequence is a 'peak point' if it is at least equal to all subsequent ones; more formally the requirement for s_k to be a peak point is that $s_k \geq s_l$ for all $l \geq k$. (In terms of the graph of the sequence

(Figure 1.5) we can imagine standing at the point (k, s_k) and looking to the right (the direction of increasing k). Then we are at a peak point if we can 'see to infinity' from our standpoint – all subsequent points are no higher than where we are.)

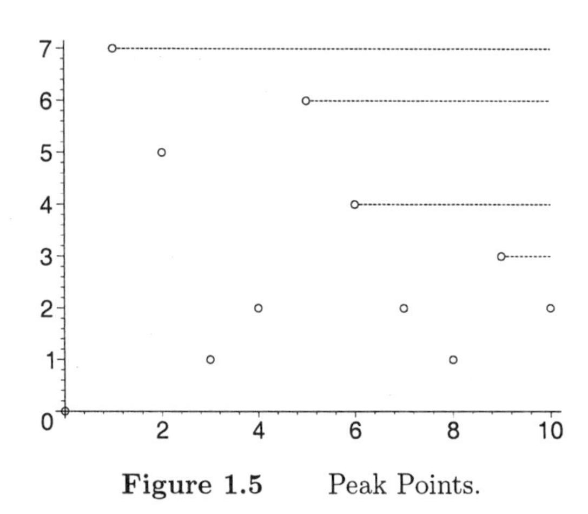

Figure 1.5 Peak Points.

There are evidently two possibilities, firstly that there is an infinite supply of peak points, or secondly that their number is finite.

In the first case let the peak points have indices $(n_1, n_2, \ldots, n_k, \ldots)$. Since there are infinitely many such points this determines a genuine subsequence (s_{n_k}). The definition of a peak point shows that since $n_k < n_{k+1}$ then $s_{n_k} \geq s_{n_{k+1}}$ and so the peak points themselves determine a decreasing subsequence.

In the second case in which the number of such points is finite, there will be some n_1 after which there are no more peak points. (If there are no peak points at all, then we take $n_1 = 1$. If there are a finite number of peak points, say at $n = m_1, m_2, \ldots, m_k$ then take $n_1 = m_k + 1$.) Then for every $n \geq n_1$, s_n is not a peak point so some subsequent term must be greater. For n_2 choose the first n with $s_{n_2} > s_{n_1}$. Similarly choose for n_3 the first n with $s_{n_3} > s_{n_2}$. Continue in this way to choose for each n_k, the first value of n for which $s_{n_k} > s_{n_{k-1}}$. This determines a (strictly) increasing subsequence. ■

This argument constructs a special subsequence which has an additional property: among all convergent subsequences, it has the greatest limit. This is explored in Exercise 23 at the end of the chapter. The following examples show how the proof of Theorem 1.27 works in practice.

Example 1.28

(i) Let $s = ((-1)^n/n)_1^\infty$. The peak points are $(s_2, s_4, \ldots, s_{2n}, \ldots)$ which form a decreasing subsequence.

(ii) Let $s = (n)_1^\infty$. There are no peak points, and the method constructs the whole sequence s as the increasing subsequence.

(iii) Let $s_1 = 1$, $s_n = -1/n$ for $n \geq 2$. Only s_1 is a peak point, so the method constructs $n_1 = 2, \ldots, n_k = k + 1$ which determine an increasing subsequence. ♦

Corollary 1.29

Every bounded sequence has a convergent subsequence.

Proof

By the previous result, there is a monotone subsequence, which must be bounded since the parent sequence is bounded. The result now follows from Theorem 1.16. ∎

Example 1.30

(i) The sequence $(n(-1)^n)_1^\infty$ has the monotone subsequences given by the odd and even terms. (This is an example of a sequence which has no convergent subsequences.)

(ii) The sequence $((-1)^n)_1^\infty$ is bounded and has the convergent (constant) subsequences given by the terms with odd and even indices.

(iii) The sequence $((-1)^n/n)_1^\infty$ is bounded and has the convergent (monotone) subsequences given by the terms with odd and even indices. ♦

1.5 Cauchy Sequences

The definition of convergence in Section 1.3 is open to the objection that to apply it we must first produce a limit, while for a given sequence we may have no idea of what the limit might be. The situation is better for monotone sequences since we know that a monotone sequence is convergent if and only if it is bounded, though the question of identifying the limit still remains, as Example 1.17(ii) shows.

The following method of showing that a sequence is convergent without first finding the limit is due to A.L. Cauchy (1789–1857) and is based on the simple observation that if the terms of a sequence are getting close to a limit then they must at the same time be getting *close to each other.* This idea is embodied in the following definition.

Definition 1.31

A sequence is called a Cauchy sequence if for each $\varepsilon > 0$ there is some N depending on ε such that

$$|s_m - s_n| < \varepsilon \text{ when } m, n \text{ are both } \geq N.$$

The remark above, that if a sequence is convergent then the terms get close to each other now translates into the following proposition.

Proposition 1.32

Every convergent sequence is a Cauchy sequence.

Proof

Let (s_n) be a convergent sequence with limit l, and for $\varepsilon > 0$, choose N such that $|s_n - l| < \varepsilon$ when $n \geq N$. Then for $m, n \geq N$ we have $|s_n - s_m| \leq |s_n - l| + |l - s_m| < \varepsilon + \varepsilon = 2\varepsilon$, as required. ■

This result is not much use on its own – we are really interested in the converse.

Theorem 1.33

Every Cauchy sequence is convergent.

Proof

The proof begins by showing that a Cauchy sequence is bounded. It follows from this by Corollary 1.29 that the sequence has a convergent subsequence, and the proof is completed by showing that the whole sequence is convergent with the same limit.

Suppose then that (s_n) is a Cauchy sequence. For any $\varepsilon > 0$ there is N such that $|s_n - s_m| < \varepsilon$ when $m, n \geq N$. In particular $|s_n - s_N| < \varepsilon$ when $n \geq N$, so $|s_N|+\varepsilon$ is a bound for that part of the sequence with $n \geq N$. It follows now, as in the proof of Theorem 1.15, that the whole sequence is bounded.

By Corollary 1.29 there is a number l and a subsequence (s_{n_k}) with $s_{n_k} \to l$ as $k \to \infty$. Since $n_k \to \infty$ as $k \to \infty$, there is some K such that $n_k > N$ if $k \geq K$, and so if $n \geq N$, $k \geq K$, then $|s_n - s_{n_k}| < \varepsilon$. Now let $k \to \infty$; it follows since $s_{n_k} \to l$ that $|s_n - s_{n_k}| \to |s_n - l|$ (Exercise 6). We deduce that $|s_n - l| \leq \varepsilon$ when $n \geq N$, i.e. the sequence is convergent with limit l. ■

The statement that a sequence is convergent if and only if it is a Cauchy sequence is sometimes called 'the general principle of convergence'. Exercise 16 shows that in fact half of Definition 1.31 is sufficient.

Example 1.34

(i) Let the sequence (x_n) be defined by $x_1 = 1$, $x_2 = 2$, and for $n \geq 3$, $x_{n+1} = (x_n + x_{n-1})/2$. Then (x_n) is a Cauchy sequence.

(ii) The sequence $(x_n)_1^\infty$ given by $x_n = 1+1/2+1/3+\cdots+1/n$ is not a Cauchy sequence.

In (i) we can write the recurrence as

$$\begin{aligned} x_{n+1} - x_n &= (-1/2)(x_n - x_{n-1}) = (-1/2)^2 (x_{n-1} - x_{n-2}) \\ &= \cdots = (-1/2)^{n-1}(x_2 - x_1). \end{aligned}$$

We now use the formula for the sum of a geometric series to write for $m > n$

$$\begin{aligned} x_m - x_n &= (x_m - x_{m-1}) + (x_{m-1} - x_{m-2}) + \cdots + (x_{n+1} - x_n) \\ &= \left[(-1/2)^{m-2} + \cdots + (-1/2)^{n-1}\right](x_2 - x_1) \\ &= \frac{(-1/2)^{n-1}}{1+1/2}\left(1 - (-1/2)^{m-n}\right)(x_2 - x_1), \\ |x_m - x_n| &\leq \frac{4}{3}|x_2 - x_1|\, 2^{-n} \end{aligned}$$

which clearly tends to zero when $m, n \to \infty$. (In fact it isn't difficult here to find an explicit expression for x_n and so to prove convergence and find the limit directly; this is Exercise 17. The second example does not have this disadvantage.)

In (ii) we see for $m > n$ that $x_m - x_n = 1/(n+1) + \cdots + 1/m$ and that each term in this sum is at least $1/m$. Hence $x_m - x_n \geq (m-n)/m$. In particular on putting $m = 2n$, we see that $x_{2n} - x_n \geq 1/2$ which does not tend to zero, so the sequence is not a Cauchy sequence, and so is not convergent. This example will play an important part in Chapter 6 on series. ♦

The definition of a Cauchy sequence involves a special case of what is called a double limit. Here a function of two integer variables m, n is given, and we ask what happens when both of them become large. Looking at Definition 1.31 suggests that we adopt the following definition.

Definition 1.35

The (double) sequence $(t_{m,n})_{m,n=1}^{\infty}$ is said to be convergent, and we write $t_{m,n} \to t$ as $m, n \to \infty$, if for each $\varepsilon > 0$ there exists an integer N such that

$$|t_{m,n} - t| < \varepsilon \qquad \text{when both } m, n \geq N.$$

Example 1.36

(i) $1/(m+n) \to 0$ as $m, n \to \infty$.

(ii) $mn/(m^2 + n^2)$ has no limit as $m, n \to \infty$.

The first of these is immediate, for given ε take $N > 1/\varepsilon$ so that if $m, n \geq N > 1/\varepsilon$ then $m + n > 2/\varepsilon$ and so $1/(m+n) < \varepsilon/2$. For the second, notice that if $m = n$ then $t_{m,n}$ has the constant value $1/2$, while if $m = 2n$ then it is $2/5$, so that no number can satisfy the required condition for a limit. ♦

The definition of a double limit given above, in which both m, n tend to infinity together, should not be confused with the quite different results which occur when the two variables go to infinity one after the other, resulting in so-called repeated or iterated limits.

Definition 1.37

The sequence $(t_{m,n})$ is said to have an iterated (or repeated) limit as $m, n \to \infty$ if either

(i) the limit $u_m = \lim_{n\to\infty} t_{m,n}$ exists for each m, and the sequence (u_m) has a limit u which is written as

$$u = \lim_{m\to\infty} \left(\lim_{n\to\infty} t_{m,n} \right),$$

or

(ii) the limit $v_n = \lim_{m\to\infty} t_{m,n}$ exists for each n, and the sequence (v_n) has a limit v which is written as

$$v = \lim_{n\to\infty} \left(\lim_{m\to\infty} t_{m,n} \right).$$

These limits may or may not have the same value, if indeed they exist at all. In the above example in which $t_{m,n} = mn/\left(m^2+n^2\right)$, both iterated limits exist and are equal to zero, though the double limit does not exist. The following examples illustrate further possibilities.

Example 1.38

(i) $(m-n)/(m+n)$ has distinct iterated limits (± 1) and no double limit.

(ii) $(-1)^m/n$ has a double limit and one iterated limit, but not the other.

(iii) $(-1)^{m+n}/\min(m,n)$ has a double limit, but neither iterated limit.

(iv) $\left(m+n^2\right)/(m+n)$ has one iterated limit only.

Example (i) here is particularly important since it is the first time we come across the situation in which we have two limiting processes, and the result depends on the order in which they are applied. To be precise,

$$\lim_{m\to\infty}\lim_{n\to\infty} t_{m,n} = -1 \neq 1 = \lim_{n\to\infty}\lim_{m\to\infty} t_{m,n}.$$

This dependence of the result on the order in which limiting operations are carried out will be a recurring source of interest. Examples (ii) and (iii) illustrate the following result which shows that relations between these limits are not completely chaotic. ♦

Theorem 1.39

(i) If a sequence has a double limit, and one (or both) iterated limits, then these limits must be equal.

(ii) If the sequence is increasing for fixed m and for fixed n, then either all limits are infinite, or all are finite and equal.

Note that in part (ii) we can replace increasing by decreasing both times it occurs, but that if one is increasing while the other is decreasing then the two iterated limits may be distinct as Example 1.38(i) shows.

Proof

(i) Since the double limit exists, given any $\varepsilon > 0$, there is N such that for $m, n \geq N$ we have $|t_{m,n} - t| < \varepsilon$. Then if say $\lim_{m\to\infty}\lim_{n\to\infty} t_{m,n}$ exists, and $\lim_{n\to\infty} t_{m,n} = u_m$, we can let $n \to \infty$ in $|t_{m,n} - t| < \varepsilon$ to get $|u_m - t| \leq \varepsilon$ if $n \geq N$, and so $u_m \to t$ as required. Part (ii) is Exercise 25. ■

EXERCISES

1. Show that the triangle inequality $|x+y| \leq |x|+|y|$ can be stated more sharply for real numbers as

$$|x+y| \begin{cases} = |x|+|y| & \text{if } \; xy \geq 0, \\ < \max(|x|,|y|) & \text{if } \; xy < 0. \end{cases}$$

2. Let

$$s_n = \frac{an+b}{cn+d}, \text{ for } a,b,c,d > 0.$$

What conditions on a,b,c,d are needed to make the sequence $(s_n)_1^\infty$ increasing, or decreasing?

3. Let (s_n) be a sequence (or a set) which is bounded below, and let S be the set of all lower bounds. Show that S is bounded above, and that the least upper bound of S is the greatest lower bound of (s_n).

4. Show that a sequence s given by a polynomial, $s_n = a_0 + a_1 n + \cdots + a_k n^k$ with $a_k > 0$, $n \geq 0$, is bounded below but not above.

5. Find the limits of the sequences given by

$$\text{(i) } \frac{3n+2}{4n-1}, \qquad \text{(ii) } \sqrt{n^2+n}-n,$$

$$\text{(iii) } n\left(\sqrt{n^2+n+2}-\sqrt{n^2+n}\right), \qquad \text{(iv) } \frac{2^n+n^3}{3^n+n^2}.$$

6. Show that if $s_n \to l$ as $n \to \infty$, then $|s_n| \to |l|$ as $n \to \infty$, but that the converse is false for $l \neq 0$.

7. Show that if (s_n) is bounded and $t_n \to 0$, then $s_n t_n \to 0$.

8. Give an alternative proof of the convergence of the sequence given by $v_0 = 0$, $v_{n+1} = \sqrt{v_n + 2}$ using the inequality (to be established) $2 - v_{n+1} < (2 - v_n)/3$.

9. Show that if (r_n) is a sequence of positive terms which satisfies $r_{m+n} \leq r_m r_n$ for all m, n then $(r_n^{1/n})$ is convergent to its greatest lower bound.

10. Show that the sequence given by $v_0 = 0$, $v_{n+1} = \sqrt{av_n + b}$ is convergent for all $a, b > 0$.

11. (i) Given x, y with $0 < x < y$, show that $x < \sqrt{xy} < (x+y)/2 < y$.

(ii) Let a_0, b_0 be arbitrary real numbers with $0 < b_0 < a_0$ and define a_n, b_n for $n \geq 0$ by $a_{n+1} = (a_n + b_n)/2$, $b_{n+1} = \sqrt{a_n b_n}$.

Show that for $n \geq 1$, $b_0 < b_n < b_{n+1} < a_{n+1} < a_n < a_0$ and that the two sequences have a common limit called the *arithmetic-geometric mean* of a_0, b_0. (See also Exercise 17, Chapter 7.)

12. Show that if $s_n \to l$ then also $(s_1 + s_2 + \cdots + s_n)/n \to l$. Give an example in which the converse fails.

13. Show that $n^{1/n} \to 1$ as $n \to \infty$.

14. If for some c, $0 < c < 1$, and integer $N > 0$, the sequence (s_n) satisfies $|s_{n+1}/s_n| < c$ for $n \geq N$, then $s_n \to 0$. Deduce that for any real x, $x^n/n! \to 0$.

15. For which real values of x do the sequences $(\cos(nx))_0^\infty$, $(\sin(nx))_0^\infty$ converge?

16. Let (s_n) be a sequence which is bounded above and which has the property that for each $\varepsilon > 0$, there is some N such that $s_n > s_m - \varepsilon$ for $n > m > N$. Show that the sequence is convergent.

17. Show that in Example 1.34(i), the sequence is given by $x_n = (5 - (-1/2)^n)/3$, and hence $x_n \to 5/3$ as $n \to \infty$.

18. Show that if $s_n \to \infty$, and $t_n \to l > 0$ or $t_n \to \infty$, then $s_n t_n \to \infty$. Give examples to show that no result is possible if $l = 0$.

19. Show that sequence given by $x_0 = 2$, $x_{n+1} = 2 - 1/x_n$ for $n \geq 0$, is decreasing and find its limit.

    Consider more generally $x_0 = c$, $x_{n+1} = c - 1/x_n$ for other values of $c > 2$.

20. (i) Let $x_n = 1 + 1/2 + \cdots + 1/n$. Deduce from $x_{2n} - x_n \geq 1/2$ (see Example 1.34(ii)) that

$$x_{2^k} \geq 1 + \frac{k}{2}$$

and hence that $x_n \to \infty$ as n$\to \infty$.

(ii) Let

$$y_n = \frac{1}{n+1} + \cdots + \frac{1}{2n}.$$

Show that the sequence (y_n) is decreasing and has a limit which is $\geq 1/2$.

21. Show that if (n_k) is a strictly increasing sequence of positive integers then $n_k \geq k$ for all k.

22. Show that if a sequence is convergent with limit l, then any subsequence is convergent with the same limit.

23.* (i) Given any sequence s, let L be the limit of the convergent subsequence constructed in the proof of Theorem 1.27. Let l be the limit of any other convergent subsequence. Show that $l \leq L$. The limit L is called the *limit superior* (lim sup for short) of the sequence. Similarly $L' = -\limsup(-s)$ is called the *limit inferior* (lim inf for short). See the next exercise for an example of the use of this concept.

(ii) Show that for any $\varepsilon > 0$, we have $s_n > L - \varepsilon$ for infinitely many n, and $s_n < L + \varepsilon$ for all n from some point on.

(iii) Show that if (s_n) is bounded then also $L = \lim_{n\to\infty}\left(\sup_{m\geq n} s_m\right)$.

24. Let (a_n) be any sequence with $a_n > 0$ for all n. Show that

$$\limsup\left(\frac{a_1 + a_{n+1}}{a_n}\right)^n \geq e = \lim_{n\to\infty}\left(1 + \frac{1}{n}\right)^n.$$

25. Let $t_{m,n} = \left((m^2 - n^2) / (m^2 + n^2 + 1)\right)^2$ for all $m, n \geq 0$. Find the double limit and both iterated limits (if they exist) as $m, n \to \infty$.

26. Consider the sequence defined by the following procedure:

The first 3 terms are $-1, 0, 1$ (we go from -1 to 1 in steps of 1).

The next 7 terms are $-2, -3/2, -1, -1/2, 0, 1/2, 1, 3/2, 2$ (we go from -2 to 2 in steps of $1/2$).

The next k_n terms are $-n, -n+1/n, \ldots, -1/n, 0, 1/n, \ldots, n-1/n, n$ (we go from $-n$ to n in steps of $1/n$).

Find the number of terms k_n in the nth block. Show that the sequence takes every rational value infinitely many times. (For an explicit example of a sequence which takes each rational value exactly once see [15].)

27. The Collatz Problem. Define a function ϕ on the integers $n \geq 1$ by

$$\phi(n) = \begin{cases} n/2 & \text{if } n \text{ is even,} \\ 3n+1 & \text{if } n \text{ is odd.} \end{cases}$$

Show that for initial values ≤ 30 say, repeated application of ϕ comes back eventually to 1. ** Does this happen for all initial values?

2
Functions and Continuity

Informally, a function f is described by a formula or a rule which, for a given input (usually a real number x), determines uniquely an output (again typically a real number y). The input is supposed to be an element of a set A called the domain of the function, and the output belongs to a set B called the codomain. When this happens we write $y = f(x)$ and $f : A \to B$. Examples of functions of this sort with $A = B = \mathbb{R}$ (the set of all real numbers) are given for instance by (i) $y = f(x) = x^2 + 1$, (ii) $y = g(x) = 1$ if $x \geq 0$, $= 0$ if $x < 0$, (iii) $y = h(x) =$ the smallest prime number $\geq x$. Many more examples will be given in the first section of this chapter.

More formally, given any non-empty sets A, B, their Cartesian product $A \times B$ is defined to be the set of ordered pairs (a, b) with $a \in A$, $b \in B$. Then a function f from A to B is given by (or is equal to, if one prefers that point of view) a subset $F \subset A \times B$ such that for all $a \in A$ there is a unique $b \in B$ with $(a, b) \in F$. Then $b = f(a)$ is equivalent to $(a, b) \in F$. The subset $\{y \in B : y = f(x) \text{ for some } x \in A\}$ of the codomain is called the range of the function. For instance in example (ii) above the range of the function g is the two-point set $\{0, 1\}$.

For a more extended development of these ideas, the entertaining discussion in Chapter 1 of [14] gives exactly the required background. The distinction between f, the name of a function, and $f(x)$, the value of f at x, corresponds to the distinction between s, the name of a sequence, and s_n, its nth term.

We use the conventional notation $(a, b) = \{x : a < x < b\}$ for an (open) subinterval of the real line in which the end points $a, b \in \mathbb{R}$ are *excluded*, and $[a, b] = \{x : a \leq x \leq b\}$ for the corresponding (closed) interval in which the end points $a, b \in \mathbb{R}$ are *included* (and $(a, b]$ or $[a, b)$ to indicate that only one is included). The set $\{x : a < x\}$ should logically be written $(a, \ )$ but convention requires (a, ∞), adding further confusion to the use of the symbol ∞. Similarly $\{x : x < a\}$ is written $(-\infty, a)$.

As in Chapter 1 we begin with examples and the simpler ideas of boundedness and monotonicity, before going on to define limits and continuity. Some of the examples naturally introduce the concept of countability of sets and we consider this briefly in Section 2.8.

2.1 Examples

To begin, let's consider some examples to illustrate the wide variety of behaviour which a function of a real variable may possess.

Example 2.1

(i) Let $p(x) = x^2+1$, defined for all real x. More generally, consider any polynomial function of degree n given by $p(x) = a_0+a_1x+\cdots+a_nx^n$, $a_n \neq 0$, defined for all real x.

(ii) Let $f(x) = x^2/(x-1)$, defined for all real $x \neq 1$. More generally, consider any rational function $f(x) = p(x)/q(x)$, where p, q are polynomials with q not identically zero. The domain of definition is now the set of all real x with $q(x) \neq 0$.

(iii) Let $f(x) = x^\alpha$, for any real exponent $\alpha \in \mathbb{R}$, defined for all $x > 0$. Alternatively consider $f(x) = |x|^\alpha$, for $\alpha \in \mathbb{R}$, defined for all $x \neq 0$.

(iv) Let $f(x) = x/\sqrt{1-x^4}$, defined for all $x \in (-1, 1)$.

(v) Let $f(x) = \lfloor x \rfloor$, the 'floor' function, where $\lfloor x \rfloor$ is defined to be the unique integer n such that $n \leq x < n+1$. This function is defined for all $x \in \mathbb{R}$.

(vi) Let

$$f(x) = \begin{cases} 0 & \text{if } x \leq 0, \\ 1/2^n & \text{if } 1/2^n \leq x < 1/2^{n-1} \quad \text{for} \quad n = 1, 2, 3, \ldots, \\ 1 & \text{if } x \geq 1. \end{cases}$$

(vii) $f(x) = \cos x$ is defined for all real x.

(viii) $f(x) = \tan x$ is defined for all real $x \neq (k+1/2)\pi$, $k \in \mathbb{Z}$.

(ix) Let $f(x) = 1/q$ when $x = p/q$ is a rational number in lowest terms (i.e. p, q are integers with no common factors and $q > 0$), otherwise $f(x) = 0$.

(x) Let $f(x) = 1$ when x is rational, otherwise $f(x) = 0$.

The first of these examples, $f(x) = x^2 + 1$ has the familiar parabolic shape, decreasing for $x < 0$ down to the minimum point at $x = 0, y = 1$ then increasing 'to infinity' as x becomes large (Figure 2.1).

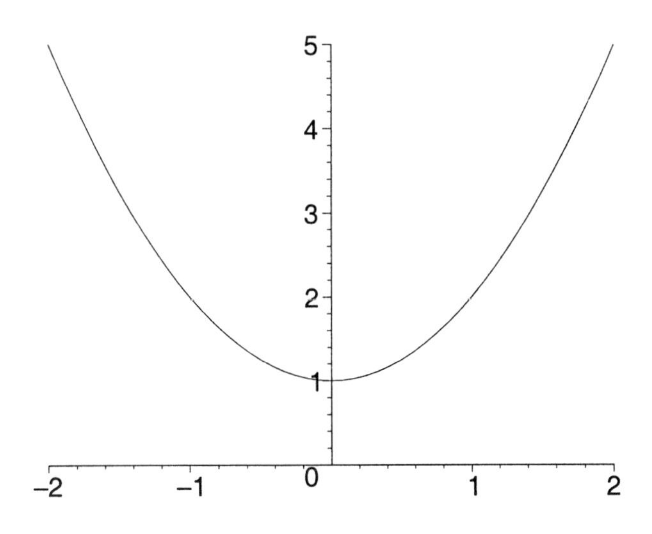

Figure 2.1 Example 2.1(i).

The second example given by $x^2/(x-1)$ is defined at points where its denominator is non-zero; moreover if at such a point, here $x = 1$, we have $q = 0$ but $p \neq 0$, then for nearby values of x, $q(x)$ will be small and the quotient will tend to infinity, the sign being determined by the signs of p, q. In this example $p(x) = x^2$ is positive near $x = 1$, so the quotient tends to $+\infty$ when x is near 1 on the right, and to $-\infty$ when x is near 1 on the left (Figure 2.2).

In example (iii) the non-integer power x^α is defined for $x > 0$, by $x^\alpha = \exp(\alpha \ln x)$. (Readers who do not want to see logarithms until they have been (re-)defined in Chapter 4 may skip to the next paragraph.) For $\alpha \neq 0$, the function has both domain and range equal to $(0, \infty)$ and is increasing or decreasing depending on the sign of α. Example (iv) is algebraic (a combination of polynomials and roots); its domain is $(-1, 1)$ and its range is $\mathbb{R}$.

Example (v) is the well-known 'staircase' function which is constant on each interval $(n, n+1)$ but jumps up by 1 unit at each integer. Example (vi)

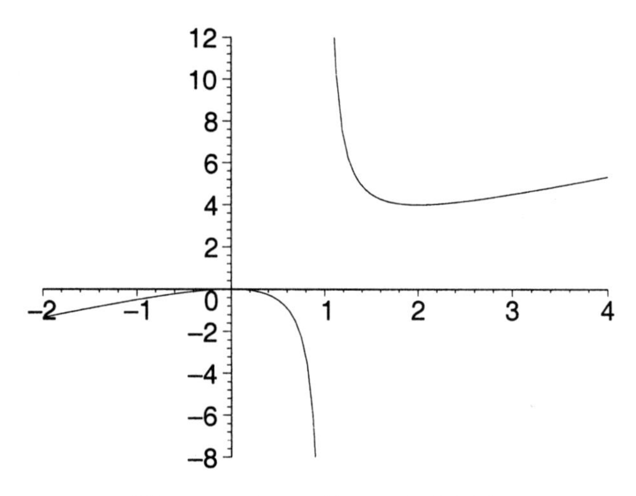

Figure 2.2 Example 2.1(ii).

is similar, except that the jumps are scaled to fit into the square $[0, 1] \times [0, 1]$ (Figure 2.3).

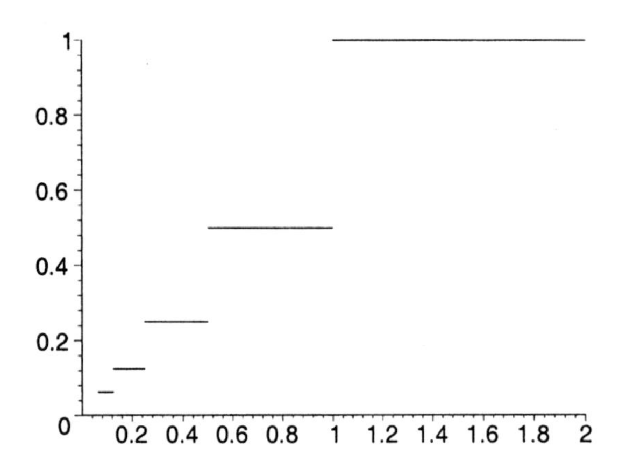

Figure 2.3 Example 2.1(vi).

The trigonometric functions in (vii) and (viii) are periodic, that is to say there is a positive number a such that $f(x + a) = f(x)$ for all real x. The least such value (if it exists) is called the period of the function; the functions sine and cosine have period 2π while tangent has period π. They will be fundamental to our consideration of trigonometric series in Chapter 7.

Example (ix) is a 'pathological' example (so called by people who don't enjoy oddities). It will help us to illustrate some important distinctions when we discuss continuity in the rest of this chapter, since its points of continuity (which are the irrational numbers) and discontinuity turn out to be intimately mixed together (Figure 2.4). It is also an unlikely integrable function in Chapter 4.

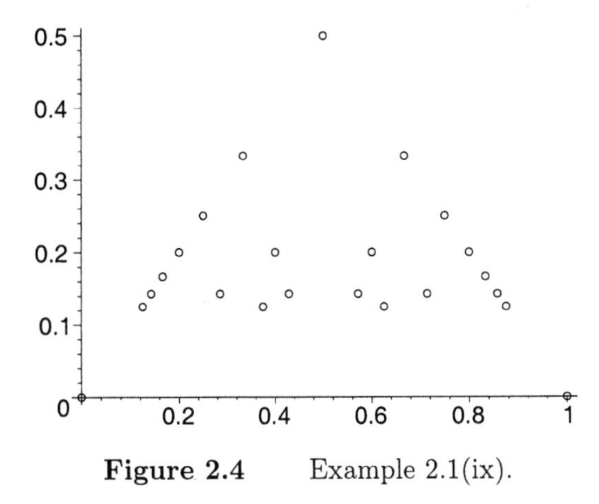

Figure 2.4 Example 2.1(ix).

Example (x) is a more violent version of the same phenomenon; this time the function is discontinuous everywhere. ♦

2.2 Monotone and Bounded Functions

The definition of monotone functions is parallel to that of monotone sequences in Section 1.2.

Definition 2.2

Let f be a function whose domain contains an interval I, which may be open or closed.

(i) We say that f is increasing on I if for all $x, y \in I$ with $x < y$ we have $f(x) \le f(y)$. We say that f is strictly increasing on I if $f(x) < f(y)$ when $x < y$.

(ii) We say that f is decreasing on I if for all $x, y \in I$ with $x < y$ we have $f(x) \ge f(y)$. We say that f is strictly decreasing on I if $f(x) > f(y)$ when $x < y$.

(iii) We say that f is monotone on I if it is either increasing or decreasing on I.

Notice that we do not define simply 'f is increasing/decreasing', but more specifically 'f is increasing/decreasing on I'. The property of monotonicity, and similarly the properties of boundedness and continuity which we shall soon consider, make sense only when related to some interval, or more general set, on which they apply. Notice also that we do not require that I is the whole of the domain of f but only that it is a subset of the domain.

Example 2.3

(i) The function $f(x) = x^3 - 3x$ is strictly increasing on $(-\infty, -1]$ and on $[1, \infty)$, and strictly decreasing on $[-1, 1]$.

(ii) The function $f(x) = (ax + b)/(cx + d)$, where $ad - bc > 0$, $c \neq 0$, is strictly increasing on $(-\infty, -d/c)$ and $(-d/c, \infty)$.

These examples would conventionally be thought of as exercises in calculus, but it is interesting to see how they can be deduced from elementary algebra.

For instance in (i) we have $f(y) - f(x) = (y - x)(x^2 + xy + y^2 - 3)$, and so if $y > x \geq 1$ then $x^2 + xy + y^2 > 3$ and $f(x) - f(y) > 0$, and a similar argument works on the other intervals. In (ii) we see that

$$f(y) - f(x) = \frac{(ad - bc)(y - x)}{(cx + d)(cy + d)}$$

and so for instance if both $x, y \in (-d/c, \infty)$ and $x < y$ then the bracketed terms in the denominator have the same sign while those in the numerator are positive. ♦

Other examples such as the two 'staircase' functions given in Examples 2.1(v) and (vi) show that a monotone function can have 'jump discontinuities', that is points where the right- and left-hand limits (to be defined in Section 2.7) exist but are different.

Notice that in Example 2.3(i), the restriction of the domain of a function to a subinterval may have desirable properties, in this case monotonicity, which are not present in the original function. This parallels Section 1.4 in which we found that a subsequence may have additional properties which the parent sequence does not possess.

Some of the results which we proved for monotone sequences, Proposition 1.6 for example, remain true for monotone functions with proofs which require only verbal alterations and we shall not spend time re-proving them. Another result which holds similarly for both sequences and functions can be proved here.

Proposition 2.4

Let

$$\max(f,g)(x) = \max(f(x), g(x)), \qquad a < x < b,$$
$$\min(f,g)(x) = \min(f(x), g(x)), \qquad a < x < b.$$

Then if f, g are both increasing (or both decreasing) on (a, b), then both $\max(f,g)$ and $\min(f,g)$ are increasing (respectively decreasing) on (a, b).

Proof

Let $x, y \in (a, b)$ with $x < y$ and let $h = \max(f,g)$. If both $h(x) = f(x), h(y) = f(y)$ then $h(x) = f(x) \leq f(y) = h(y)$ as required, with a similar argument if both $h(x) = g(x), h(y) = g(y)$. On the other hand if say $h(x) = f(x)$, $h(y) = g(y)$, then $h(x) = f(x) \leq f(y) \leq h(y)$ which is what we want. ■

The concept of boundedness is defined in the same way as for sequences, except that as noted above, we must supply a subset of the domain on which the property holds.

Definition 2.5

Let A be a subset of the domain of a function f. The function is said to be bounded on A if there is some M such that

$$|f(x)| \leq M \qquad \text{for all } x \in A.$$

We could obviously have defined the separate concepts of boundedness above and below, and show as we did for sequences that a function is bounded on an interval if and only if it is bounded both above and below, but we believe that the reader can manage such things unaided by now.

Note that in Example 2.1(i) the function given by $x^2 + 1$ is bounded (above and below) on say $[-2, 2]$, and is bounded below, but not above, on $[0, \infty)$. The function of Example 2.1(iii) given by x^α is unbounded on $(0, \infty)$ when $\alpha > 0$ when it is increasing, and when $\alpha < 0$ when it is decreasing. As was the case for sequences, it follows from Axiom 1.11 that a function which is bounded above on an interval will have a least upper bound there, and similarly for functions which are bounded below. Results such as the analogue of Proposition 1.10, which says that the sum of two bounded functions is bounded, remain true with only verbal alterations in the proofs.

We shall look at the relation between boundedness (a property of the function on an interval) and continuity (a property of the function at a point) in Section 2.4.

2.3 Limits and Continuity

The motivating idea in this section is that of a continuous function. Intuitively we consider that a function is continuous on an interval if its graph can be drawn 'continuously', that is without lifting the pen from the paper. When we analyse this concept more carefully, we see that what is required is that at each point where the function is defined, it should have a limit which is equal to the value of the function at the point. Examples such as those in Section 2.1 show that a function may have some intervals where it behaves smoothly and other points, or intervals where it does not; we have to frame definitions which will distinguish between these various kinds of behaviour.

We begin with the definition of a limit at a point. To parallel the discussion of convergence of sequences in Chapter 1, we shall say that a function has a limit l at a point c, if the values of f at points near c can be made as close as we wish to l by making x close to c. Thus in the definition which follows we have two small quantities to deal with, namely ε (Greek epsilon) which measures the difference $|f(x) - l|$ between the function and its limit, and δ (Greek delta) which measures how near x is to c. This leads to the following definition.

Definition 2.6

Let f be defined on an interval $(a,b) \subset \mathbb{R}$, and let $c \in (a,b)$. We say that f has a limit equal to l at c if for each $\varepsilon > 0$, there is some $\delta > 0$ such that $|f(x) - l| < \varepsilon$ when $0 < |x - c| < \delta$. In this case we write

$$f(x) \to l \text{ as } x \to c, \text{ or } \lim_{x \to c} f(x) = l.$$

If we compare this with Definition 1.13 for the convergence of sequences, we see that the condition $|s_n - l| < \varepsilon$ has been replaced by $|f(x) - l| < \varepsilon$, while the condition which is required to make this happen is now $0 < |x - c| < \delta$ (x is close but not equal to c) instead of $n > N$ (n is large). So the same kind of consideration, in which we imagine that ε is given (by an imaginary other person, for example) and we have to find the value of δ depending on ε, is helpful here. Notice that the function f may or may not be defined at c, and that if it is, the value $f(c)$ has no effect on the limit. The next examples illustrate this.

Example 2.7

(i) Let $f(x) = 1$ for $x \geq 0$, $f(x) = 0$ for $x < 0$. Then the limit $\lim_{x \to c} f(x)$ is equal to 1 for $c > 0$, to 0 for $c < 0$, and is undefined at $c = 0$.

(ii) Let $f(x) = 1$ for $x = 0$, $f(x) = 0$ for $x \neq 0$. Then $\lim_{x \to c} f(x) = 0$ for all c.

These results come directly from the definition. For instance in (i), if $c > 0$, then choose $\delta \leq c$, when $f(x) = f(c) = 1$, and hence $|f(x) - f(c)| = |1 - 1| = 0 < \varepsilon$ for $|x - c| \leq \delta$, independently of the value of $\varepsilon > 0$. The same argument works if $c < 0$. If $c = 0$ then any interval $(-\delta, \delta)$ contains both positive and negative values of x, and thus contains both points where $f(x) = 0$ and others where $f(x) = 1$, so no limit can exist. For (ii) the argument is similar; in particular when $c = 0$ then all x with $0 < |x - c| < \delta$ have $f(x) = 0$ and so $\lim_{x \to 0} f(x) = 0$. ♦

As another example of a function which does not have a limit, consider the following which is in some way similar to the sequence $((-1)^n)$ which in Section 1.3 we found to have no limit as $n \to \infty$.

Example 2.8

Let f be defined for $x \neq 0$ by $f(x) = \sin(\pi/x)$. Then f has no limit at 0 (Figure 2.5).

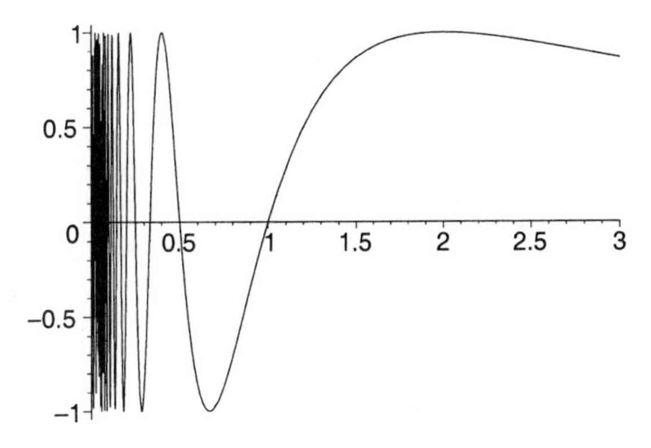

Figure 2.5 Example 2.8.

Notice that the function has the values ± 1 infinitely often in every interval $(-\delta, \delta)$ (to be precise $f(x) = (-1)^n$ when $x = x_n = 1/(n + 1/2)$). Hence if $l \geq 0$ then $|f(x_n) - l| \geq 1$ when n is odd, and so f cannot have any positive limit as $x \to 0$, and similarly it cannot have any negative limit either. ♦

To complete the picture we need a description of what in some sense is the normal (or at least the most desirable) behaviour of a function in which the limit exists *and is equal to the value of the function* at c. Such a function is said to be continuous at c. The following definition makes this precise.

Definition 2.9

Let f be defined on an interval $(a, b) \subset \mathbb{R}$, and let $c \in (a, b)$. We say that f is continuous at c if for each $\varepsilon > 0$, there is some $\delta > 0$ such that $|f(x) - f(c)| < \varepsilon$ when $|x - c| < \delta$.

If we compare this with Definition 2.6 above the differences are first that we have $f(c)$ in place of l, and second that $x = c$ is now allowed where before it was excluded.

Since we should become good at recognising continuity we give some easy examples which illustrate the most basic methods for verifying that a given function is continuous.

Example 2.10

(i) $f(x) = 3x + 1$ is continuous at $x = 2$.

(ii) $f(x) = x^2$ is continuous at $x = 1$.

(iii) $f(x) = 1/x$ is continuous at $x = 3$.

(iv) $f(x) = \sqrt{x}$ is continuous at $x = 4$.

For example (i) we suppose that $\varepsilon > 0$ is given (it seems that most elementary proofs in analysis begin with an ε). We want to find a value of δ (depending on ε) such that $|f(x) - f(2)| < \varepsilon$ when $|x - 2| < \delta$. But

$$|f(x) - f(2)| = |3x + 1 - 7| = 3|x - 2| < 3\delta \qquad \text{when } |x - 2| < \delta.$$

Hence if we take $3\delta = \varepsilon$, or $\delta = \varepsilon/3$ then $|x - 2| < \delta$ is enough to ensure that $|f(x) - f(2)| < \varepsilon$, as required. Not all such arguments are quite as easy as this one, but it still serves as a useful model, since it clearly shows how δ must be found in terms of ε. The other parts illustrate some of the tricks which may be needed.

For (ii) we want to find, for a given ε, a value of δ such that $\left|x^2 - 1\right| < \varepsilon$ when $|x-1| < \delta$. Here we factorise $x^2 - 1 = (x-1)(x+1)$ and since $|x-1|$ is already $< \delta$, we must find a way of dealing with $|x+1|$. But if x is near 1, say $|x-1| < 1$, then $0 < x < 2$ and so $|x+1| < 3$. Hence

$$|f(x) - f(1)| = |(x-1)(x+1)| < 3|x-1| < \varepsilon \qquad \text{when } |x-1| < \varepsilon/3,$$

so to make $|f(x) - f(1)| < \varepsilon$ it is enough to have both $|x-1| < 1$ and $|x-1| < \varepsilon/3$; equivalently it is enough to take $\delta = \min(1, \varepsilon/3)$.

For (iii), given $\varepsilon > 0$, we want $\delta > 0$ such that

$$|f(x) - f(3)| = |1/x - 1/3| = |x-3|/|3x| < \varepsilon \qquad \text{when } |x-3| < \delta.$$

For $|x-3| < 1$ we have $2 < x < 4$, so $|x-3|/|3x| < |x-3|/6$ which is $< \varepsilon$ when $|x-3| < 6\varepsilon$. Hence $\delta = \min(1, 6\varepsilon)$ will do.

For (iv) we use the identity

$$\left|\sqrt{x} - 2\right| = \frac{|x-4|}{\sqrt{x}+2} < \frac{|x-4|}{2}$$

which we found useful in Example 1.14(v). Hence to make $|\sqrt{x} - 2| < \varepsilon$ and $x > 0$, $|x-4| < \delta = \min(2\varepsilon, 4)$ will do. ♦

Example 2.11

The function of Example 2.1(ix), in which $f(p/q) = 1/q$ for p, q integers with no common factor, otherwise zero, is continuous except at rational x.

This follows since for any $\varepsilon > 0$ there are only a finite number of points where $f(x) > \varepsilon$, so for a given irrational c where $f(c) = 0$ we can take δ as the distance to the nearest point where $f(x) > \varepsilon$. Then $f(x) \leq \varepsilon$ on $(x - \delta, x + \delta)$ so f is continuous at c. Similarly at a rational point d, $f(d) > 0$ but there is an interval around d on which $f(x) \leq f(d)/2$ (take $\varepsilon = f(d)/2$ in the preceding argument) so f is discontinuous at d. ♦

It is interesting to note that the reverse situation in which f is continuous only on the rationals is impossible; a proof of this is in [2]. In Example 2.1(x), the function is equal to 1 on the rationals and zero elsewhere. Since these sets are both dense, every interval $(c - \delta, c + \delta)$ contains both points where $f = 1$ and where $f = 0$. Hence f is discontinuous at every point.

For functions we can define the sum, difference, product and quotient, exactly as was done for sequences in Chapter 1, and the content of part (i) of the theorem below will come as no surprise. However there is an additional operation of composition defined as follows.

Definition 2.12

Let $g : A \to B$, and $f : B \to C$ be functions where the range of g is contained in B, the domain of f. Then the composition $f \circ g : A \to C$ is defined by $(f \circ g)(x) = f(g(x))$ for all $x \in A$.

For instance if $f(x) = x^2 - 1$ and $g(x) = x + 1$ (both from $\mathbb{R}$ to $\mathbb{R}$) then both $f \circ g(x) = (x+1)^2 - 1 = x^2 + 2x$ and $g \circ f(x) = x^2$ exist, but they define different functions. Part (ii) of the next theorem shows that the composition of continuous functions gives another continuous function; there are three small quantities to deal with in the proof.

Theorem 2.13

(i) Let f, g be functions which are continuous at c. Then $f + g$, $f - g$ and $f.g$ are continuous at c, while f/g is continuous if in addition $g(c) \neq 0$.

(ii) Let g be continuous at c, and f be continuous at $g(c)$. Then $f \circ g$ is continuous at c.

Proof

(i) The proofs for $f \pm g$ are the same as those for sequences in Chapter 1. For the product, notice that taking $\varepsilon = 1$ there is some $\delta_1 > 0$ such that for $|x - c| < \delta_1$ we have $|f(x) - f(c)| < 1$ and so $|f(x)| < |f(c)| + 1$. Then given $\varepsilon > 0$ we have both $|g(x) - g(c)|$ and $|f(x) - f(c)| < \varepsilon$ for $|x - c| < \delta_2$ say. Hence

$$\begin{aligned}
|f(x)g(x) - f(c)g(c)| &= |f(x)(g(x) - g(c)) + g(c)(f(x) - f(c))| \\
&\leq |f(x)||g(x) - g(c)| + |g(c)||f(x) - f(c)| \\
&< (|f(c)| + 1)\varepsilon + |g(c)|\varepsilon \\
&= (|f(c)| + |g(c)| + 1)\varepsilon
\end{aligned}$$

when both $|x - c| < \delta_1$ and $|x - c| < \delta_2$. But $(|f(c)| + |g(c)| + 1)\varepsilon$ can be made as small as required, and so the proof that $f.g$ is continuous is complete. (Some would prefer to re-cast this argument by taking say $\varepsilon_1 = \varepsilon / (|f(c)| + |g(c)| + 1)$ and following the above reasoning with ε_1 in place of ε, thereby arriving triumphantly with '$< \varepsilon$' at the end. This is open to the objection to 'proof from hindsight' already referred to – but it is simply a matter of taste which one is preferred.)

For the quotient we need another technicality (compare the corresponding result, Proposition 1.20, for sequences). Since $g(c) \neq 0$ we can take

$\varepsilon = |g(c)|/2$ in the definition of continuity, to show that there is some δ such that when $|x-c| < \delta$ then $|g(x) - g(c)| < |g(c)|/2$, and so $|g(c)|/2 < |g(x)| < 3|g(c)|/2$. (Compare Exercise 5 at the end of the chapter.) Then we can write

$$\left|\frac{1}{g(x)} - \frac{1}{g(c)}\right| = \left|\frac{g(x) - g(c)}{g(x)\,g(c)}\right| < \frac{2}{|g(c)|^2}\,|g(x) - g(c)|$$

which can be made small since g is continuous at c. Hence $1/g(x)$ is continuous at c, and the result for f/g now follows since $f/g = f.(1/g)$.

(ii) In this argument we have three small quantities to deal with; we call them ε, δ and η. The continuity of f at $g(c)$ ensures that given $\varepsilon > 0$ there is a value of $\delta > 0$ such that

$$|f(t) - f(g(c))| < \varepsilon \qquad \text{when } |t - g(c)| < \delta. \tag{2.1}$$

But since g is continuous at c, there is a value of $\eta > 0$ such that $|g(x) - g(c)| < \delta$ when $|x - c| < \eta$ (note that we have used δ in place of ε as the 'target error' for g here). Hence if $|x - c| < \eta$ then $g(x)$ can be substituted for t in (2.1) and so $|f(g(x)) - f(g(c))| < \varepsilon$ when $|x - c| < \eta$, as required. ■

We can now go back to the intuitive idea of continuity with which we began the section. The idea was that a function was continuous if its graph could be drawn without lifting the pen from the paper. Thus we should consider not only whether the function is continuous at certain points as in Definition 2.9 (continuous 'locally' in more sophisticated language), but also whether it is continuous on an interval (or 'globally'). This leads us to the next definition.

Definition 2.14

(i) Let f be defined on an open interval $I = (a, b) \subset \mathbb{R}$. We say that f is *continuous on* I if it is continuous at every point of I in the sense of Definition 2.9.

(ii) If f is defined on a closed interval $I = [a, b] \subset \mathbb{R}$, then we say that f is *continuous on* I if firstly it is continuous at every point of the open interval (a, b) as in (i), and in addition for each $\varepsilon > 0$ there is some $\delta > 0$ such that $|f(x) - f(a)| < \varepsilon$ when $a \le x < a + \delta$, and similarly there is some (other) $\delta > 0$ such that $|f(x) - f(b)| < \varepsilon$ when $b - \delta < x \le b$.

Part (ii) of this definition anticipates the definition of continuity on the left and on the right which will be explored in more detail in Section 5.

Example 2.10 can immediately be re-used as follows.

Example 2.15

(i) $f(x) = 3x + 1$ (and more generally $f(x) = ax + b$, for constant a, b) is continuous on $\mathbb{R}$.

(ii) $f(x) = x^2$ is continuous on $\mathbb{R}$.

(iii) $f(x) = 1/x$ is continuous on $\mathbb{R} - \{0\}$.

(iv) $f(x) = \sqrt{x}$ is continuous on $\{x : x \geq 0\}$. ♦

We can also extend Theorem 2.13 to cover the new context.

Theorem 2.16

(i) Let f, g be functions which are continuous on an interval I (which may be open or closed). Then $f + g$, $f - g$ and $f.g$ are continuous on I, and f/g is continuous at all points of I where $g \neq 0$.

(ii) Let g be continuous on I, and f be continuous on J, where J is an interval which contains the range of g. Then $f \circ g$ is continuous on I.

(iii) Any polynomial function is continuous on $\mathbb{R}$. Any rational function $f = p/q$, where p, q are polynomials, is continuous at all points where $q \neq 0$. If g is continuous on I, then $1/g$ is continuous at all points of I where $g \neq 0$, and $\sqrt{g}$ is continuous at all points of I where $g \geq 0$.

Proof

(i) and (ii) are just the corresponding parts of Theorem 2.13 in the new context. For (iii) we notice that repeated multiplication by x (and repeated use of part (i) of the theorem) shows that $a_n x^n$ is continuous on $\mathbb{R}$ for any integer $n \geq 1$ and any real constant a_n. Then repeated addition shows that any polynomial is continuous on $\mathbb{R}$. Also $f(x) = 1/x$ is continuous for $x \neq 0$, so (ii) shows that $1/q(x)$ is continuous when $q \neq 0$; then writing $p/q = p(1/q)$ shows that p/q is continuous when $q \neq 0$. Part (iii) follows in the same way with $f(x) = 1/x$, $\sqrt{x}$ respectively. ■

This theorem allows us to make statements about limits 'by substitution', as in the next example.

Example 2.17

Show that

$$\text{(i)} \qquad \lim_{x\to 2} \frac{x^2 - 2/x}{x^4 - 1} = \frac{1}{5},$$

$$\text{(ii)} \qquad \lim_{x\to -1} \sqrt{x^3 + 5x^2 - 3} = 1.$$

To see (i) we deduce successively from the theorem and previous examples that as $x \to 2$ we have $x^2 \to 4$, $2/x \to 1$, $x^2 - 2/x \to 3$, $x^4 \to 16$, and $(x^2 - 2/x) / (x^4 - 1) \to 3/15 = 1/5$. For (ii) similarly $x^3 + 5x^2 - 3 \to -1 + 5 - 3 = 1$ as $x \to -1$ and part (iii) above gives the final step. ♦

These would evidently be quite unpleasant to try to prove directly. Ways of evaluating less obvious limits will be considered in Section 3.4.

To finish this section, consider the relation between the convergence of a sequence and continuity.

Proposition 2.18

Let f be a function which is continuous at c, and let (x_n) be a sequence with $x_n \to c$ as $n \to \infty$. Then $f(x_n) \to f(c)$ as $n \to \infty$.

Proof

Since f is continuous at c, given $\varepsilon > 0$ there is some $\delta > 0$ such that $|f(x) - f(c)| < \varepsilon$ when $|x - c| < \delta$. Then since $x_n \to c$ there is some N such that $|x_n - c| < \delta$ when $n \geq N$. Combining these statements gives $|f(x_n) - f(c)| < \varepsilon$ when $n \geq N$ as required. ■

The converse of this result is also true; it is Exercise 4.

The next sections look at some deeper consequences of continuity.

2.4 Bounds and Intermediate Values

It is often useful to know that a function is bounded on an interval, even when we have no clear way of finding a bound explicitly. For instance the rational function given by $f(x) = (x^5 - 5x^2 + 1) / (x^6 + 5x^4 + 3)$ is clearly continuous on $\mathbb{R}$ since the denominator is ≥ 3. However to show that it is bounded, for instance on $[-1, 1]$, requires an estimate of the form $|f(x)| \leq (1 + 5 + 1)/3$

which gives a bound which is very far from the best, and for a more complicated example, there might be no obvious way of estimating its size at all. Hence it is important to have indirect ways of showing that a function is bounded, and since continuity is by now easy to recognise, it is good to know that a function which is continuous on a closed bounded interval must be bounded there, as the next theorem shows.

The method of proof by bisection, which we use here for the first time, allows us to construct a sequence of successively smaller intervals on each of which some required property holds. It is widely used (according to rumour) for such diverse purposes as finding lions in the Sahara desert or finding partners on a crowded dance floor.

Theorem 2.19

A function which is continuous on a closed interval $[a, b]$ must be bounded on $[a, b]$.

(The notation $[a, b]$ here implies that a, b are real numbers, so that the interval is also bounded – neither end point is at infinity.)

Proof

Suppose that f is not bounded on $[a, b]$. Denote the given interval $[a, b]$ by $[a_0, b_0]$ and divide it in half at the midpoint $c_0 = (a_0 + b_0)/2$. Since f is not bounded on $[a_0, b_0]$, it must be unbounded on one of the halves $[a_0, c_0]$, $[c_0, b_0]$; call this half $[a_1, b_1]$. (If f is unbounded on both halves, choose say the left half for definiteness.) Continue in the same way to choose $[a_n, b_n]$ as one of the halves of $[a_{n-1}, b_{n-1}]$ on which f is not bounded. This determines two sequences (a_n), (b_n) with $a_{n-1} \leq a_n < b_n \leq b_{n-1}$, so that (a_n) is increasing and bounded above by b, and (b_n) is decreasing and bounded below by a. Thus both sequences are convergent and, since $b_n - a_n = (a_0 - b_0)/2^n \to 0$, they have a common limit, say c. Since $a_0 \leq a_n \leq c \leq b_n \leq b_0$, it follows that c is a point of $[a, b]$, and hence that f is continuous at c. Then for any $\varepsilon > 0$ (fix $\varepsilon = 1$ for definiteness) there is a value of $\delta > 0$ such that $|f(x) - f(c)| < 1$ when $x \in [a, b]$ and $|x - c| < \delta$ (notice that this takes care of both cases, when $x \in (a, b)$, and when $x = a$ or b). Hence $|f(x)|$ is bounded by $|f(c)| + 1$ on the interval $(c - \delta, c + \delta) \cap [a, b]$. But since both a_n, $b_n \to c$, we can choose n large enough to make $[a_n, b_n]$ a subset of $(c - \delta, c + \delta)$, and this contradicts the construction of $[a_n, b_n]$ as an interval on which f was not bounded. ■

This result is the first of several which deduce a conclusion which is global in nature, here that the function is bounded on an interval, from a hypothesis which is local, in this case that the function is continuous at each point of the interval.

It is necessary that the interval should be closed, since otherwise the conclusion may fail. For instance the functions $1/x$ on $(0, 1]$, and x^2 on $[0, \infty)$ are continuous but not bounded (their domains of definition are either not closed or not bounded) and we cannot repair the situation by defining the function by $1/x$ on $(0, 1]$, and $f(0) = 0$, since the resulting function is not continuous at 0.

A consequence of this result is even more useful; a function which is continuous on a closed bounded interval must attain its (closest) bounds. For instance we shall be able to deduce that for the function given by $(x^5 - 5x^2 + 1) / (x^6 + 5x^4 + 3)$ which we looked at above, if m_1, m_2 are its greatest lower and least upper bounds on $[-1, 1]$ respectively, then there are points x_1, $x_2 \in [-1, 1]$ with $f(x_1) = m_1$, and $f(x_2) = m_2$ even though we have no way of finding them explicitly. Notice that the *existence* of such bounds is assured, since the range of f is a bounded set, and so Axiom 1.11 shows that it will have a unique greatest lower and a unique least upper bound. A continuous function on an unbounded interval need not attain its bounds, as is shown by the example $f(x) = \tan^{-1}(x)$ on $\mathbb{R}$.

The proof that closest bounds are attained is a neat example of a proof by contradiction. However it is purely an existence theorem, and gives no indication of how such points might be found. Methods for finding best bounds explicitly will be developed in the next chapter on differentiation.

Corollary 2.20

Let f be continuous on $[a, b]$ and let m_1, m_2 be its greatest lower bound and least upper bound respectively. Then there are points x_1, $x_2 \in [a, b]$ with $f(x_1) = m_1$, and $f(x_2) = m_2$.

Proof

Suppose that m is the least upper bound of f on $[a, b]$ but that there is no x with $f(x) = m$. Then $m - f(x)$ is strictly positive on $[a, b]$ and so $g(x) = 1/(m - f(x))$ is defined, continuous and positive on $[a, b]$. Hence by applying Theorem 2.19 to g, there must be some M which is an upper bound for g on $[a, b]$, that is

$$0 < g(x) = \frac{1}{m - f(x)} \leq M \qquad \text{for } a \leq x \leq b.$$

It follows that

$$\frac{1}{M} \leq m - f(x) \text{ and so } f(x) \leq m - \frac{1}{M}, \qquad \text{for } a \leq x \leq b,$$

and we have found a new upper bound $m - 1/M$ for f on $[a,b]$ which is strictly less than the supposed least upper bound m. Thus our supposing that there is no x with $f(x) = m$ has led to a contradiction and the result follows. ■

The most immediately useful of the results in this group of consequences of continuity is the Intermediate Value Theorem. This says that if a function is continuous on an interval then it must be 'between-valued' – that is for every two values which the function assumes, it also assumes every value in between. An equivalent way of saying this is that a continuous function maps intervals into intervals.

Theorem 2.21 (Intermediate Value Theorem)

Let f be continuous on $[a,b]$, and let y be a real number between $f(a)$ and $f(b)$. Then there is some point $c \in [a,b]$ with $f(c) = y$.

Proof

We can simplify the proof by supposing that $y = 0$ for if not, then we can consider $f - y$ in place of f. Then in addition we can suppose that $f(a) < 0 < f(b)$, since if $f(a) > 0 > f(b)$, we consider $-f$ in place of f. (Of course if $f(a)$ or $f(b) = y$ then we just take $c = a$ or b.) Thus it is enough to show that if f is continuous on $[a,b]$ and $f(a) < 0 < f(b)$ then there is some $c \in (a,b)$ with $f(c) = 0$. The proof is again by bisection, and gives a simple constructive process (an algorithm, in technical language) by which the required value of c can be approximated as closely as required.

Let $a_0 = a$, $b_0 = b$, and $c_0 = (a_0 + b_0)/2$. If $f(c_0) < 0$, let $a_1 = c_0$, $b_1 = b_0$; if $f(c_0) > 0$, let $a_1 = a_0$, $b_1 = c_0$; in either case we have found a subinterval $[a_1, b_1]$ of half the length, with $f(a_1) < 0 < f(b_1)$. (Of course if $f(c_0) = 0$, we can stop the process at once since the desired value of c has been found.)

In general, when $[a_{n-1}, b_{n-1}]$ has been found, let $c_{n-1} = (a_{n-1} + b_{n-1})/2$ and if $f(c_{n-1}) < 0$ let $a_n = c_{n-1}$, $b_n = b_{n-1}$, while if $f(c_{n-1}) > 0$ then put $a_n = a_{n-1}$ and $b_n = c_{n-1}$. (Again, if $f(c_n) = 0$, we can stop the process since the desired value of c has been found.) This constructs a sequence of subintervals $[a_n, b_n]$, each a subset of the one before, and of half the length, on each of which $f(a_n) < 0 < f(b_n)$. As in the proof of Theorem 2.19, both sequences (a_n), (b_n) have a common limit c, and it follows from Proposition 2.18 that since f is continuous at c, both $(f(a_n))$ and $(f(b_n))$ converge to $f(c)$.

Hence since $f(a_n) < 0 < f(b_n)$ we must have $f(c) \leq 0 \leq f(c)$, so $f(c) = 0$ as required. ∎

The usefulness of this result is that if we are trying to solve an equation $f(x) = 0$, and we have found a, b with say $f(a)$ negative and $f(b)$ positive, then we can be sure that there is a solution of the equation in between, and we can use the method used in the proof (the method of bisection) to locate the root as precisely as we wish. The following example shows how this works.

Example 2.22

Given a circle with unit radius, it is required to find a value of r such that if we draw a circle of radius r, centred at a point on the circumference of the unit circle, then the area of the region in which the circles overlap is one half the area of the unit circle. (Usually the problem concerns a farmer who has a circular field, and wants to tether a goat to a point on the edge so that it can graze half the field.)

Consider any circle with radius r and centre C, and a sector CAB of angular opening 2θ. The area of the sector is $r^2\theta$ and the area of the triangle CAB is $r^2 \sin\theta\cos\theta$, so the area of the segment remaining is $r^2(\theta - \sin\theta\cos\theta)$.

Hence in the given situation we have a unit circle centred at C, and another circle centred at a point P on the circumference of the unit circle with the circles intersecting at A, B, as in Figure 2.6.

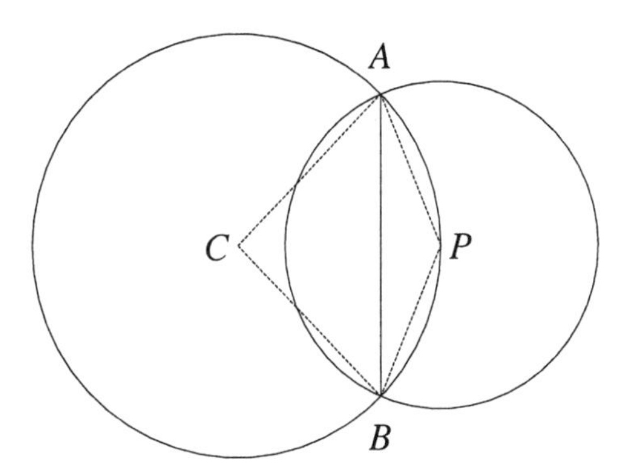

Figure 2.6 Example 2.22.

If AB subtends an angle of 2θ at C then it subtends an angle of $\pi - \theta$ at P, and $AP = r = 2\sin(\theta/2)$. Thus the combined area in which the circles overlap is given by the sum of the areas of two segments, namely

$$\begin{aligned} f(\theta) &= (\theta - \sin\theta\cos\theta) + (2\sin(\theta/2))^2(\pi/2 - \theta/2 - \sin(\theta/2)\cos(\theta/2)) \\ &= (\theta - \sin\theta\cos\theta) + (1 - \cos\theta)(\pi - \theta - \sin\theta) \\ &= \pi - \sin\theta - (\pi - \theta)\cos\theta. \end{aligned}$$

For the overlap to have half the area of the unit circle, we want $f(\theta) = \pi/2$. But $f(0) = 0$ and $f(\pi) = \pi$, and we know (or choose to believe until Chapter 6) that the trigonometric functions sin, cos are continuous, so there must be some $c \in [0, \pi]$ with $f(c) = \pi/2$. (A little calculus will show that f is increasing so that the solution is unique, but we do not need this now.)

To solve the equation, put $g(\theta) = f(\theta) - \pi/2 = \pi/2 - \sin\theta - (\pi - \theta)\cos\theta$, so that we want a value of θ with $g(\theta) = 0$. We have $g(0) = -\pi/2$, $g(\pi) = \pi/2$, so we take $c_0 = 1.5 \simeq \pi/2$, $g(c_0) = 0.46$. Then listing only the successive values of c_n, $g(c_n)$ we find the following values. (Notice that it is not necessary to take for c_n the exact midpoint – any sequence of intervals which contains the required solution and whose length goes to zero will do.)

$$\begin{aligned} c_0 = 1.5,&\quad g(c_0) = 0.46 \\ c_1 = 1.0,&\quad g(c_1) = -0.427 \\ c_2 = 1.2,&\quad g(c_2) = -0.06 \\ c_3 = 1.3,&\quad g(c_3) = 0.11 \\ c_4 = 1.25,&\quad g(c_4) = 0.025 \\ c_5 = 1.23,&\quad g(c_5) = -0.0106 \\ c_6 = 1.24,&\quad g(c_6) = 0.0073 \\ c_7 = 1.235,&\quad g(c_7) = -0.0016 \end{aligned}$$

Hence the required solution is between 1.235 and 1.24 and so is equal to 1.24, correct to 2 decimal places. ♦

More rapidly convergent methods to approximate the root will be discussed in the next chapter.

2.5 Inverse Functions

When we consider functions defined on abstract sets, we pay particular attention to the properties of being one-to-one (an injection), and of being onto (a surjection). A *one-to-one* function has the property that distinct elements of the domain give distinct values of the function: if $x \neq y$ then $f(x) \neq f(y)$. A function is *onto* if its codomain is equal to its range: for all y in the codomain of f, there is some x in the domain with $y = f(x)$. As might be expected, these have interesting relations to the properties of monotonicity and continuity when the domain and codomain are sets of real numbers, and we look at these relations in this section.

Firstly, a function which is strictly monotone on an interval I must be one-to-one there, since if $x, y \in I$, $x \neq y$, then we can assume $x < y$ so $f(x) < f(y)$ (if increasing) or $f(x) > f(y)$ (if decreasing); in either case f is one-to-one. The converse is of course not true; Exercise 14 gives an example of a bijection (both one-to-one and onto) on an interval which is not monotone. However a continuous injection must be monotone as we now show.

Theorem 2.23

Let f be continuous and one-to-one on an interval $I = [a, b]$. Then it is strictly monotone on I.

Proof

Since f is one-to-one we must have $f(a) \neq f(b)$; we shall show that if $f(a) < f(b)$ then f is strictly increasing – it follows similarly that if $f(a) > f(b)$ then f is strictly decreasing.

We first notice that if $f(a) < f(b)$ then we must have $f(a) < f(x) < f(b)$ for all $a < x < b$ since if for instance $f(x) \leq f(a)$ for some $x > a$, then by the Intermediate Value Theorem there would be some $y \in [x, b]$ with $f(y) = f(a)$ and so f would not be one-to-one. Similarly $f(x) \geq f(b)$ is impossible. Then if for any $x < y \in I$ we have $f(x) \geq f(y)$, there would be some $z \in [a, x]$ with $f(z) = f(y)$ and again f would not be one-to-one. Hence for $x < y \in I$ we have $f(x) < f(y)$ as required. ■

We showed in the previous section that a function which is continuous on an interval $[a, b]$ assumes its least upper and greatest lower bounds, say M, m respectively, and the Intermediate Value Theorem shows that it also takes all values in between. Hence the range of a function which is continuous on an interval I, is always an interval $[m, M]$. If we combine this observation with the previous theorem we get the following result.

Corollary 2.24

A continuous function with domain $I = [a, b]$ and range $[m, M]$ is a bijection if and only if it is strictly monotone on I.

The essential property of a bijection f is that it has an inverse f^{-1} for which $f^{-1}(f(x)) = x$ for all x in the domain of f, and $f\left(f^{-1}(y)\right) = y$ for all y in the range of f, which is also the domain of f^{-1}. The above corollary gives us most of the next theorem.

Theorem 2.25

A function f which is continuous and strictly monotone on an interval (which may be open or closed, bounded or unbounded) has an inverse which is also continuous and strictly monotone in the same sense (i.e. increasing/decreasing if f is increasing/decreasing respectively).

Proof

We prove the result first in the case when the domain of f is a closed bounded interval $I = [a, b]$, in which case its range is also closed and bounded, let's say $[m, M]$. The proof in this case has all been done above except for the continuity of f^{-1}. This is not difficult, but requires some small mental gymnastics. To be precise, what we need is that for any point $y \in [m, M]$ and any $\varepsilon > 0$, there must be some $\delta > 0$ such that $\left|f^{-1}(z) - f^{-1}(y)\right| < \varepsilon$ when $|z - y| < \delta$. Suppose that f is increasing, let $x = f^{-1}(y)$ and consider z_1, $z_2 = f(x - \varepsilon)$, $f(x + \varepsilon)$ respectively. Then $z_1 < f(x) = y < z_2$ so if we let $\delta = \min(y - z_1, z_2 - y)$ then $\left|f^{-1}(z) - f^{-1}(y)\right| < \varepsilon$ when $|z - y| < \delta$ as required.

We leave to the reader the modifications when the domain is open or unbounded; it requires only an application of the first part to all subintervals of the form $[a, b]$. ∎

Notice that we could have defined the inverse explicitly for an increasing function by

$$f^{-1}(y) = \sup\{x \in [a, b] : f(x) \leq y\},$$

and that this definition continues to make sense for functions which may not be either continuous or strictly increasing, though the resulting function is not an inverse in the strict sense we have been considering.

The value of Theorem 2.25 is that it allows us to deduce the existence of inverse functions, even when we have no explicit notation for them, as several of the following examples show.

Example 2.26

(i) Let $f(x) = x + x^2$ for $x \geq -1/2$. Then f is strictly increasing and continuous on $[-1/2, \infty)$, with range $[-1/4, \infty)$. Its inverse is given by

$f^{-1}(y) = \left(-1 + \sqrt{1+4y}\right)/2$ for $y \geq -1/4$.

(ii) Let $f(x) = x + x^5$ for all real x. Then f is strictly increasing and continuous on $\mathbb{R}$, with range $\mathbb{R}$. Its inverse is also strictly increasing and continuous on $\mathbb{R}$ (but there is no formula for it).

(iii) Let $f(x) = x/\sqrt{1+x^2}$ for $x \in \mathbb{R}$. Then f is strictly increasing and continuous on $\mathbb{R}$, with range $(-1,1)$. Its inverse is strictly increasing and continuous on $(-1,1)$, given by $f^{-1}(y) = y/\sqrt{1-y^2}$.

(iv) Let $f(x) = xe^x$ for $x \geq 0$. (As usual the reader who does not want to see special functions until their proper place may skip this and the following examples.) Then f is strictly increasing and continuous on $[0,\infty)$, with range $[0,\infty)$. Its inverse is strictly increasing and continuous on $[0,\infty)$ (but there is no formula for it).

(v) Let $f(x) = \tan x$ for $x \in (-\pi/2, \pi/2)$. Then f is strictly increasing and continuous on $(-\pi/2, \pi/2)$, with range $\mathbb{R}$. Its inverse (the elementary inverse tangent $\tan^{-1}$) is strictly increasing and continuous on $\mathbb{R}$ with range $(-\pi/2, \pi/2)$.

(vi) Let $f(x) = \tan^{-1}(\sinh x)$ for $x \in \mathbb{R}$. Then f is strictly increasing and continuous on $\mathbb{R}$, with range $(-\pi/2, \pi/2)$. Its inverse is strictly increasing and continuous on $(-\pi/2, \pi/2)$, given by $f^{-1}(y) = \sinh^{-1}(\tan y)$. This is Gudermann's function, for which we have the alternative expression $f(x) = \sin^{-1}(\tanh x)$ (which the reader is invited to establish), and hence the striking fact that the inverse is given by rotation through 90° in the complex plane:

$$f(ix) = if^{-1}(x).$$

All we need by way of justification in these examples is to show that the functions are monotone, when the rest follows from Theorem 2.25. For instance in (iii) a simple algebraic calculation shows that $a/\sqrt{1+a^2} < b/\sqrt{1+b^2}$ when $a < b$. The other parts are easy calculus exercises which can be done now, or left until Section 6.5. ♦

2.6 Recursive Limits and Iteration

In Chapter 1 we considered sequences (x_n) which were defined by relations of the form $x_{n+1} = f(x_n)$, where each term was determined in terms of one (or possibly more) of the preceding ones. To be precise, we assume that f is a function on an interval (a,b), that $f((a,b)) \subset (a,b)$ and that $x_0 \in (a,b)$ so that each $x_{n+1} = f(x_n)$ is well defined. Now that we can refer to properties of functions we can investigate further, though a more complete study will not come until the next chapter.

We see that if the sequence defined by $x_{n+1} = f(x_n)$ (we shall say that it is defined by *iteration* of f) is convergent with limit c, and if f is continuous at c then the limit must satisfy $c = f(c)$ (this follows from Proposition 2.18). Thus the set of possible limits must be a subset of the solutions of the equation $x = f(x)$.

We look at a few examples, before trying to make any further deductions.

Example 2.27

(i) $x_0 = 0$, $x_{n+1} = \sqrt{x_n + 2}$.

(ii) $x_0 = 2$, $x_{n+1} = x_n/2 + 3$.

(iii) $x_0 = 2$, $x_{n+1} = 2x_n - 1$.

The first of these is Example 1.1(viii), which we showed to be convergent with limit 2, the only positive solution of $x = \sqrt{x+2}$. For (ii) we will try some numerical values; $x_0 = 2$ gives $(2, 4, 5, 5.5, 5.75, \ldots)$ which seems to be heading in the right direction for the known value $x = 6$, the solution of $x = x/2 + 3$. Starting at $x_0 = 10$ gives $(10, 8, 7, 6.5, 6.25, \ldots)$ similarly. For (iii) we find $x_0 = 2$ gives $(2, 3, 5, 9, 17, \ldots)$ which is going away from $x = 1$ which satisfies $x = 2x - 1$, and even if we begin with $x_0 = 0.9$ we get $(0.9, 0.8, 0.6, 0.2, -0.6, \ldots)$ which is again heading in the wrong direction. ♦

The difference between examples (ii) and (iii) is that on writing $f(x)$ for the expressions $x/2+1$, $2x-1$, we see that for any real x, y we have $|f(x) - f(y)| = |x-y|/2$ in the one and $|f(x) - f(y)| = 2|x-y|$ in the other. Thus the action of the function in (ii) is to move points nearer together, while in (iii) they are moved further apart. This motivates the following definition.

Definition 2.28

A function defined on an interval I is said to be a contraction if for some $c \in (0,1)$ and all $x, y \in I$ we have

$$|f(x) - f(y)| \leq c|x-y|. \tag{2.2}$$

Notice that it is essential for a contraction that c is strictly < 1. The condition (2.2) forces $f(y) \to f(x)$ as $y \to x$ and so a contraction is a special kind of continuous function.

The next result shows that iteration behaves as nicely as we could hope for when we use a contraction.

Theorem 2.29

Let f be a contraction which maps an interval $I = [a, b]$ into itself. Then the equation $x = f(x)$ has a unique solution $d \in I$, and for any $x_0 \in I$ the sequence given recursively by $x_{n+1} = f(x_n)$, $n \geq 0$, is convergent with limit d and

$$|x_n - d| \leq \frac{c^n}{1-c} |x_0 - x_1| .$$

Proof

That the equation $x = f(x)$ has a solution follows from the Intermediate Value Theorem. Put $g(x) = f(x) - x$ and note that since f maps $[a, b]$ into itself we have $f(a) \geq a$ and $f(b) \leq b$, i.e. $g(a) \geq 0 \geq g(b)$. Thus $g(x) = 0$, or equivalently $f(x) = x$, must have a solution between a and b. That the solution is unique follows from (2.2) since if d, e are both solutions we must have

$$|d - e| = |f(d) - f(e)| \leq c|d - e|$$

and so $|d - e| = 0$ since $c < 1$.

When we consider the iteration $x_{n+1} = f(x_n)$ we find

$$\begin{aligned}
|x_{n+1} - x_n| &= |f(x_n) - f(x_{n-1})| \leq c\,|x_n - x_{n-1}| \\
&= c\,|f(x_{n-1}) - f(x_{n-2})| \leq c^2\,|x_{n-1} - x_{n-2}| \\
&\dots \\
&\leq c^n\,|x_1 - x_0|
\end{aligned}$$

and hence for $m > n$,

$$\begin{aligned}
|x_m - x_n| &\leq |x_m - x_{m-1}| + |x_{m-1} - x_{m-2}| + \cdots + |x_{n+1} - x_n| \\
&\leq \left(c^{m-1} + c^{m-2} + \cdots + c^n\right) |x_1 - x_0| \\
&= \frac{c^n - c^m}{1-c} |x_1 - x_0| \qquad (2.3)
\end{aligned}$$

which tends to zero as $m, n \to \infty$. Hence (x_n) is a Cauchy sequence, which therefore has a limit d say, and $d = f(d)$ follows from the continuity of f. Finally letting $m \to \infty$ in (2.3) gives

$$|x_n - d| \leq \frac{c^n}{1-c} |x_0 - x_1|$$

as required. ∎

Having shown that iterative sequences which are defined by contractions must converge, we can change our viewpoint and use iteration as a method of solving equations. For instance suppose that we want to solve the quadratic equation $x^2 + x = 3$. Clearly $x^2 + x$ is strictly increasing for $x > 0$, and equal to 2, 6 at $x = 1, 2$ respectively, so the Intermediate Value Theorem shows that there is a unique solution in the interval $[1, 2]$. (Of course for such a simple equation we can solve explicitly to give $x = \left(\sqrt{13} - 1\right)/2 = 1.30277\ldots$, but the point is that the method will work just as well with $x^5 + x = 3$ for which no formula exists.) It remains to rearrange the equation suitably in the form $x = f(x)$ and to see what happens.

Suppose that we make the most obvious arrangement as $x = 3 - x^2$ and begin with $x_0 = 1$. Then we get $(1, 2, -1, 2, -1, 2, \ldots)$ as the successive iterates, and there is no convergence. Even if we begin very close to the known root – at $x_0 = 1.3$ say, we get $(1.3, 1.31, 1.28, 1.35, \ldots)$ which gets progressively worse. We try again, this time rearranging to get $x = \sqrt{3 - x}$. Now starting with $x = 1$ gives $(1, 1.4, 1.25, 1.31, 1.29, 1.305, \ldots)$ which seems to be approaching the correct value. To prove that the sequence is indeed converging we show that $f(x) = \sqrt{3 - x}$ is a contraction, mapping $[1, 2]$ into itself. Clearly f is strictly decreasing and $f(1) = \sqrt{2} = 1.4\ldots$, and $f(2) = 1$ so f maps $[1, 2]$ into itself. Also for $x, y \in [1, 2]$ we have

$$|f(x) - f(y)| = \left|\sqrt{3 - x} - \sqrt{3 - y}\right| = \frac{|(3 - x) - (3 - y)|}{\sqrt{3 - x} + \sqrt{3 - y}} \leq \frac{|x - y|}{2}$$

so f is a contraction with $c = 1/2$. We shall go more deeply into the question of using iterative methods to solve equations in the next chapter; for now we point out only that the two rearrangements considered in the above discussion are mutually inverse – if $y = 3 - x^2$ then $x = \sqrt{3 - y}$, which gives a clue to how we might proceed.

2.7 One-Sided and Infinite Limits. Regulated Functions

It happens, as we have seen in many examples, that not all functions are continuous at all points. For instance the staircase function of Example 2.1(v) is discontinuous at all integer points and continuous elsewhere. Examples 2.1(ix) and (x) show in the first case points of continuity and discontinuity mixed together (as we shall shortly show), while in the second there are no points of continuity at all.

To distinguish between these various possibilities, we shall consider separate limits on the left and right. We also introduce the idea of a regulated function as one which has at every point c, both left-hand and right-hand limits which need not be equal either to each other or to $f(c)$.

Definition 2.30

(i) Let $c \in \mathbb{R}$ and let f be defined on an interval $(c, b) \subset \mathbb{R}$. We say that f has a right-hand limit equal to l at c if for each $\varepsilon > 0$, there is some $\delta > 0$ such that $|f(x) - l| < \varepsilon$ when $c < x < c + \delta$. In this case we write

$$f(x) \to l \text{ as } x \to c+, \text{ or } \lim_{x \to c+} f(x) = l, \text{ or } f(c+) = l.$$

(ii) Let $c \in \mathbb{R}$ and let f be defined on an interval $(a, c) \subset \mathbb{R}$. We say that f has a left-hand limit equal to l at c if for each $\varepsilon > 0$, there is some $\delta > 0$ such that $|f(x) - l| < \varepsilon$ when $c - \delta < x < c$. In this case we write

$$f(x) \to l \text{ as } x \to c-, \text{ or } \lim_{x \to c-} f(x) = l, \text{ or } f(c-) = l.$$

(iii) We say that a function is regulated on an interval I if at all points of $c \in I$, it has both right-hand and left-hand limits (which need not be equal either to each other or to $f(c)$). If c is an end point of the interval, say the left, then we require only the existence of $f(c+)$; similarly at the right end point we require only the existence of $f(c-)$.

The jump $j(f, c)$ of a regulated function f at c is defined to be

$$j(f, c) = \max(|f(c+) - f(c)|, |f(c-) - f(c)|, |f(c+) - f(c-)|)$$

at an interior point, or $|f(c+) - f(c)|$, $|f(c-) - f(c)|$ at left, right end points, respectively.

Evidently f has a limit at c if and only if it has both left and right limits at c, *and they are equal*; similarly f is continuous at c if and only if it has left-hand and right-hand limits at c which are equal *to each other and to* $f(c)$. Hence a regulated function is continuous at c if and only if $j(f, c) = 0$.

The one-sided limits $f(c+)$, $f(c-)$ are sometimes written as $f(c+0)$, $f(c-0)$, but this seems illogical, since $c+0$ already has a meaning, namely $c+0 = c$. Care should be taken when c is negative, however, since for instance $f(-2-)$ looks odd, and is better written $f((-2)-)$. Some writers prefer to use subscripts, so $f(c+)$ becomes $f(c_+)$.

Right- and left-hand limits are most clearly seen in the staircase function which is Example 2.1(v). If for this function we take c equal to an integer n

then the function has a right-hand limit at c equal to n since $f(x) = n$ for $n < x < n+1$, and a left-hand limit equal to $n-1$ since $f(x) = n-1$ for $n-1 < x < n$. At non-integer points $c \in (n, n+1)$ the function has a two-sided limit equal to n, and since this is also the value of the function at c, the function is continuous at c.

We could now prove results for one-sided limits along the lines of 'if f, g have right-hand limits at c, then $f+g$ has a right-hand limit at c, and $\lim_{x \to c+} (f(x) + g(x)) = \lim_{x \to c+} f(x) + \lim_{x \to c+} g(x)$'. However these involve no new ideas or methods and we leave them for the reader to supply when required.

The definition of one-sided continuity now follows naturally.

Definition 2.31

(i) Let f be defined on $[a, b) \subset \mathbb{R}$. We say that f is continuous on the right at a if for each $\varepsilon > 0$, there is some $\delta > 0$ such that $|f(x) - f(a)| < \varepsilon$ when $a \leq x < a + \delta$; equivalently f is continuous on the right at a if $f(a+)$ exists and is equal to $f(a)$.

(ii) Let f be defined on $(a, b] \subset \mathbb{R}$. We say that f is continuous on the left at b if for each $\varepsilon > 0$, there is some $\delta > 0$ such that $|f(x) - f(b)| < \varepsilon$ when $b - \delta < x \leq b$; equivalently f is continuous on the left at b if $f(b-)$ exists and is equal to $f(b)$.

Notice that we already effectively used this definition in the statement and proof of Theorem 2.19 where the function was required to be continuous on a closed interval $[a, b]$ and so continuous on the right at a and on the left at b.

Example 2.32

(i) Let f be defined by $f(x) = \sin(1/x)$ if $x > 0$, $f(x) = 0$ if $x \leq 0$. Then f has no limit on the right at 0, but is continuous on the left at 0.

(ii) The function of Example 2.1(ix), in which $f(p/q) = 1/q$ for p, q integers with no common factor, otherwise is zero, has right- and left-hand limits equal to zero at all points, and is thus regulated.

(i) The argument which shows that f has no limit on the right is the same as in Example 2.8, while for $x \leq 0$, $f(x) = 0$ and so $f(0-) = 0 = f(0)$.

(ii) This continues Example 2.11 in which we found that for any x and any $\varepsilon > 0$ there is a $\delta > 0$ such that $0 \leq f(x) < \varepsilon$ on $(x - \delta, x) \cup (x, x + \delta)$; thus both left and right limits are zero at every point and the function is regulated. ♦

One-sided limits occur most naturally in the study of monotone functions. Suppose for instance that f is increasing on (a,b) so that for $a < x < y < b$ we have $f(x) \leq f(y)$. Then $f(x)$ is a lower bound for $f(y)$ as y decreases to x, so that f must have a right limit at x, with $f(x) \leq f(x+)$ (this is the analogous result to Theorem 1.16 for sequences). Similarly $f(x-)$ exists for all x and $f(x-) \leq f(x)$. In particular, a monotone function can have only simple discontinuities, that is where right- and left-hand limits exist but may not be equal, and so a monotone function is necessarily regulated. Exercise 10 shows that as functions of x, $f(x-)$, $f(x+)$ are also monotone and continuous on the left, right respectively.

Before finishing this section, we consider infinite limits, as we did for sequences in Section 1.3. Here there are two different types of limit to consider, the first of which corresponds to Definitions 1.13 and 1.23 for sequences. We say that $f(x) \to l$ as $x \to \infty$, (or $f(x) \to \infty$ as $x \to \infty$) when for any $\varepsilon > 0$ (respectively for any $K > 0$), there is some N such that $|f(x) - l| < \varepsilon$ (respectively $f(x) > K$) when $x \geq N$. These 'limits at infinity' have properties which are parallel to those for sequences in Chapter 1.

The second kind of limit in which the function becomes large as x approaches a finite point ($f(x) \to \infty$ as $x \to c$) has no analogue for sequences. For instance the function $f(x) = x^2/(x-1)$, which is Example 2.1(ii), becomes large when x is near 1; more precisely it becomes large and positive when $x - 1$ is small and positive, and it becomes large and negative when $x - 1$ is small and negative. We formalise this as follows, with the obvious modifications for right-hand and left-hand limits.

Definition 2.33

(i) We say that $f(x) \to \infty$ (or $-\infty$) as $x \to c$ if for each $K > 0$ there is some $\delta > 0$ such that $f(x) > K$ (or $f(x) < -K$) when $0 < |x - c| < \delta$.

(ii) We say that $f(x) \to \infty$ (or $-\infty$) as $x \to c+$ if for each $K > 0$ there is some $\delta > 0$ such that $f(x) > K$ (or $f(x) < -K$) when $c < x < c + \delta$.

(iii) We say that $f(x) \to \infty$ (or $-\infty$) as $x \to c-$ if for each $K > 0$ there is some $\delta > 0$ such that $f(x) > K$ (or $f(x) < -K$) when $c - \delta < x < c$.

Example 2.34

(i) $(\sin x)/x \to 0$ as $x \to \infty$.

(ii) $2^x/x^5 \to \infty$ as $x \to \infty$. (More generally $a^x/x^\alpha \to \infty$ as $x \to \infty$, for any $a > 1$ and any real α.)

(iii) $\sqrt{x}/\left[(x+1)^2-1\right] \to \infty$ as $x \to 0+$.

Here part (i) follows since $|\sin(x)/x| \leq 1/x \to 0$ as $x \to \infty$ (the analogue for functions of a real variable x, of the Sandwich Theorem 1.18(iii)). Part (ii) follows similarly from the corresponding result, Example 1.24(iii), for sequences. For part (iii) we rearrange to get

$$\frac{\sqrt{x}}{(x+1)^2-1} = \frac{\sqrt{x}}{2x+x^2} = \frac{1}{\sqrt{x}\,(2+x)}$$

which clearly tends to ∞ as $x \to 0+$. ♦

2.8 Countability

When we think of the size of a set, we think informally of the number of elements in the set. The simplest distinction which can be made is between the finite sets, which contain say 5 or 5^{50} elements, and the infinite sets. Formally (but not completely logically) we define a set S to be finite, with n elements, if there is a bijection (a function which is one-to-one and onto) from S to $\{1, 2, \ldots, n\}$. The empty set $\emptyset = \{\}$ is also said to be finite, with 0 elements. Then a set is infinite if it is not finite; equivalently there is no function from $\{1, 2, \ldots, n\}$ onto S for any n.

(The reason for the illogicality is that the construction of the integers requires an Axiom of Infinity and should properly precede the definition of a finite set. However we have been unapologetically using the integers and real numbers so far, and do not intend to turn aside now for details of this construction. For details, see [3].)

The sets of integers (positive and negative) will be denoted by $\mathbb{Z}$, the strictly positive integers by $\mathbb{Z}^+$, the rational numbers (fractions of the form m/n, for integers $m \in \mathbb{Z}$, $n \in \mathbb{Z}^+$) by $\mathbb{Q}$, and the real numbers by $\mathbb{R}$ (for which we have the alternative notation as an unbounded interval $(-\infty, \infty)$). These are infinite sets. (Sample proof: Let f map $\{1, 2, \ldots, n\}$ to $\mathbb{Z}^+$, and let N be the largest of the values $f(1), \ldots, f(n)$. Then $N+1$ is not in the range of f so f is not onto.)

However it will be important for us to distinguish between different sizes of infinite sets, and the concept of countability helps us to do this.

Definition 2.35

A set S is said to be countable if there is a function from $\mathbb{Z}^+$ onto S. A set which is not countable is said to be uncountable.

This definition deserves some comment, and many examples. Informally, an infinite set is countable if its elements can be arranged as the terms of a sequence, $S = (s_1, s_2, \ldots, s_n, \ldots)$ but the definition allows finite sets to be countable too. (To insist on the point, we can refer to a countably infinite set if we want to exclude the possibility of a set being finite.)

Example 2.36

(i) If A is countable, and there is a function from A onto B then B is countable. In particular, any subset of a countable set is also countable.

(ii) Any finite set is countable.

(iii) $\mathbb{Z}$ (and all subsets of $\mathbb{Z}$) are countable.

(iv) $\mathbb{Q}$ (and all subsets of $\mathbb{Q}$) are countable.

For (i), we notice that if f maps $\mathbb{Z}^+$ onto A, and g maps A onto B, then $g \circ f$ maps $\mathbb{Z}^+$ onto B. In particular, if $B \subset A$, then we can map A onto B by mapping all elements of B to themselves, and all elements of $A \setminus B$ onto a fixed arbitrarily chosen element of B. Similarly, to show that a finite set, with say n elements, $S = \{s_1, s_2, \ldots, s_n\}$ is countable, we map $\{1, 2, \ldots, n\}$ onto $\{s_1, s_2, \ldots, s_n\}$ by $j \to s_j$, and the rest of $\mathbb{Z}^+$ onto say s_1.

For (iii) and (iv) it is sufficient to give maps from $\mathbb{Z}^+$ onto $\mathbb{Z}$, $\mathbb{Q}$ respectively. The first is easy – for instance map the even integers onto the positive integers, and the odd ones to the negatives; say $2 \to 1$, $4 \to 2, \ldots, 2n \to n$, and $1 \to 0$, $3 \to -1$, $5 \to -2, \ldots$, $2n+1 \to -n$. An example of a mapping from $\mathbb{Z}^+$ onto $\mathbb{Q}$ was given in Exercise 26 in Chapter 1. ♦

Now that we have lots of examples of countable sets we can tackle the obvious question – aren't all sets of real numbers countable? In fact $\mathbb{R}$, as well as any subinterval (not subset!) is uncountable, as the following result shows. We must get used to a situation in which the set of rational numbers, and indeed any countable set of real numbers, is rather small in relation to the totality of all real numbers.

Theorem 2.37

Any subinterval of $\mathbb{R}$ which contains more than one point is uncountable.

Proof

Suppose first that the interval is $(0, 1)$ and consider any f from $\mathbb{Z}^+$ into $(0, 1)$.

Let

$$\begin{aligned} f(1) &= 0.a_{11}a_{12}\ldots a_{1n}\ldots, \\ f(2) &= 0.a_{21}a_{22}\ldots a_{2n}\ldots, \\ &\vdots \\ f(k) &= 0.a_{k1}a_{k2}\ldots a_{kn}\ldots \end{aligned}$$

be the decimal expansions of the real numbers $f(1)$, etc. (In case of ambiguity, for instance $0.5 = 0.4999\ldots$, we use the non-terminating form.)

Consider the digits $a_{11}, a_{22}, \ldots, a_{nn}, \ldots$ on the diagonal, and form a new number b whose digits are given by $b_n = 3$ if $a_{nn} \neq 3$, $b_n = 7$ if $a_{nn} = 3$. (This is evidently highly arbitrary – we could take any distinct digits in place of 3 and 7, provided we avoid 0 and 9.)

Then the new $b = 0.b_1b_2\ldots b_n\ldots$ is non-terminating and different from $f(n)$ in the nth place and so is different from all $f(n)$. Hence f is not onto $(0,1)$ and so the interval is uncountable.

In the case of a general interval (a,b) with $a < b$ we can map it onto $(0,1)$ by $x \to (x-a)/(b-a)$, so if (a,b) were countable then by Theorem 2.36(i), $(0,1)$ would be countable too. Other intervals, for instance $(a,b]$, $[a,b]$, (a,∞), all contain (a,b) (notice that $[a,a]$ is ruled out since it contains only one point) and so they must be uncountable too. ■

We might imagine that we could somehow form a more numerous set by taking infinitely many infinite sets, and forming their union. But this fails, at least for countable sets and unions, as we now show.

Theorem 2.38

Let $\{A_1, A_2, \ldots, A_n, \ldots\}$ be a countable set each of whose elements is countable. Then $\bigcup_1^\infty A_n$ is also countable.

Proof

Since A_1 is countable we can map $\{1,3,5,\ldots\}$ onto A_1, and since A_2 is countable we can map $\{2,6,10,\ldots\}$ onto A_2. Similarly for each n, since A_n is countable we can map $\{1.2^n, 3.2^n, 5.2^n, \ldots\}$ onto A_n.

But every positive integer can be written uniquely in the form $k.2^n$ with k odd, so it follows that we can combine these mappings to give a function from $\mathbb{Z}^+$ onto the whole of $A_1 \cup A_2 \cup \ldots \cup A_n \cup \ldots$. ■

It is important and useful to know that each interval (a,b) with $a < b$

contains (infinitely many) rationals as the next result shows; a set with this property is said to be *dense* in $\mathbb{R}$. Thus the rationals are dense among the real numbers, despite forming only a rather small subset of them.

Theorem 2.39

Every subinterval of $\mathbb{R}$ which contains more than one point contains a rational number.

Proof

Suppose the interval is (a, b) and choose an integer $n > 0$ with $b - a > 1/n$. Let i be the least integer (in $\mathbb{Z}$ – it may be negative) with $i/n > a$. Then $(i - 1)/n$ must be $\leq a$, since i was the least, and so $i/n \leq a + 1/n < b$, and so $i/n \in (a, b)$ as required. ■

The reader can show similarly that every such subinterval contains an irrational number. Exercise 1 shows that a monotone function can have only a countable number of discontinuities.

When we come to integration, we shall need to know that the functions which we want to integrate do not have too many discontinuities. In particular we shall want the following.

Theorem 2.40

The set of discontinuities of a regulated function is countable.

The idea of the proof is as follows. Suppose that f is regulated on $[a, b]$ and for each n, let A_n be the set of points where the jump of f is greater than $1/n$. We shall show that each A_n is finite, so that their union, which is the set of all discontinuities, must be countable by Theorem 2.38. For the proof that A_n is finite we need the following result, the Bolzano–Weierstrass theorem, which is important in its own right.

Theorem 2.41

Let E be an infinite subset of a closed interval $[a, b] \subset \mathbb{R}$. Then there is some point $c \in [a, b]$ such that for every $\varepsilon > 0$, the interval $(c - \varepsilon, c + \varepsilon)$ contains infinitely many points of E.

Proof

The idea of the proof by bisection should be by now familiar – we repeatedly divide the region in half and choose the half which contains what we are looking for – in this case infinitely many points of E.

More precisely, let $[a_0, b_0] = [a, b]$, which we are given contains infinitely many points of E. Then supposing $[a_0, b_0], \ldots, [a_n, b_n]$ to have been found, let $c_n = (a_n + b_n)/2$. It follows that either $[a_n, c_n]$ or $[c_n, b_n]$ must contain infinitely many points of E, since $[a_n, b_n]$ does. Choose for $[a_{n+1}, b_{n+1}]$ the one of $[a_n, c_n]$ or $[c_n, b_n]$ which contains infinitely many points of E (if both do then choose the left half for definiteness). This constructs two sequences (a_n), (b_n) which are monotone increasing, decreasing respectively with a common limit c since $b_n - a_n \to 0$. Hence for given $\varepsilon > 0$, there must be some N with $c - \varepsilon < a_N \leq c \leq b_N < c + \varepsilon$, and the result follows since $[a_N, b_N]$ contains infinitely many points of E. ■

Notice that the conclusion may be false for unbounded intervals; for instance if $I = [0, \infty)$ and $E = \mathbb{Z}^+$ then no interval $(c - \varepsilon, c + \varepsilon)$ can contain infinitely many points of E.

We can now finish the proof of Theorem 2.40.

Proof

We have to show that for each n, the set A_n where the jumps of f are $> 1/n$ must be finite. Suppose then that some A_n is infinite. By the Bolzano–Weierstrass theorem there is some point $c \in [a, b]$ such that every interval $(c - \varepsilon, c + \varepsilon)$ contains infinitely many points of A_n. But f is regulated at c, so it has both right and left limits there. Choose $\delta > 0$ so that both $|f(x) - f(c+)|$ and $|f(y) - f(c-)|$ are $< 1/(4n)$ for $c - \delta < y < c < x < c + \delta$. Then for any $x, x' \in (c, c + \delta)$ we have $|f(x) - f(x')| < 1/(2n)$ and so the jump of f at any point of $(c, c + \delta)$ must be $\leq 1/(2n)$. It follows that $(c, c + \delta)$, and similarly $(c - \delta, c)$ contain no points of A_n, which gives a contradiction on taking $0 < \varepsilon < \delta$. ■

EXERCISES

1. Show that the jumps (intervals of discontinuity) of a monotone function form a countable set of disjoint open intervals.
2. Let $f(x) = (x^5 - 6x + 5)/(x - 1)$ if $x \neq 1$, $f(1) = k$. Find the value of k to make f continuous at $x = 1$. Consider similarly $f(x) =$

$\left(\sqrt{x^2+7}-4\right)/(x-3)$ if $x \neq 3$, $f(3)=k$.

3. Show that a monotone function is continuous if and only if its range is an interval (in other words the converse of the Intermediate Value Theorem is valid for monotone functions).

4. Show that $f(x_n) \to f(c)$ for all sequences (x_n) with $x_n \to c$ if and only if f is continuous at c.

5. Show that if f is continuous at c, and $f(c) \neq 0$, then there is some interval around c on which $f(x)$ has the same sign as $f(c)$, and satisfies $f(x)/f(c) > 1/2$.

6. Modify the proof of Theorem 2.19 to show that if a function is regulated on a closed interval $[a,b]$ then it is bounded there.

7. Show that if $([a_n, b_n])$ is a sequence of intervals with $[a_{n+1}, b_{n+1}] \subset [a_n, b_n]$ and $|b_{n+1} - a_{n+1}| \leq c^n |b_n - a_n|$ for all n and some $c < 1$, then $\cap_{n=1}^{\infty} [a_n, b_n]$ consists of a single point.

8. Show that $\lim_{x \to 0} x^\alpha \sin(1/x) = 0$ for $\alpha > 0$ and that $\lim_{x \to \infty} x^\alpha \sin(1/x) = 0$ for $\alpha < 1$.

9. Lipschitz Continuity. A function f defined on $[a,b]$ which satisfies $|f(x) - f(y)| \leq M|x-y|$ for some $M > 0$ and all $x, y \in [a,b]$ is said to be Lipschitz continuous on $[a,b]$. (Compare Definition 7.12 for Lipschitz continuity at a point.) Show that $f(x) = x^2$ (or any polynomial) is Lipschitz continuous, but that $\sqrt{|x|}$ is not Lipschitz continuous on any interval containing 0.

10. Show that if $f(x)$ is monotone on (a,b) then $g(x) = f(x+)$ is also monotone and continuous on the right on (a,b).

11. Horizontal Chords. Show that if f is continuous on $[0,1]$ with $f(0) = f(1)$ then there is some $x \in [0,1]$ such that $f(x) = f(x+1/2)$. Show that the same is true if $1/2$ is replaced by $1/3$, but not if it is replaced by $2/5$. Can you state (and prove?) a general result?

12. Rising Sun Lemma. Let f be continuous on $[a,b]$, and define a set G by $x \in G$ if and only if there is some $y > x$ with $f(y) > f(x)$. Show that if $(c,d) \subset G$, but $c, d \notin G$ then $f(c) = f(d)$. (What has this to do with the rising sun?)

13. Chebyshev's Theorem.

    (i) Define a sequence of polynomials (the Chebyshev polynomials) $(T_n(x))_0^\infty$ by $\cos(n\theta) = T_n(\cos\theta)$ (i.e. show by induction that the definition makes sense). Show that T_n has degree n with leading coefficient 2^{n-1}, and that $|T_n(x)| \leq 1$ for $-1 \leq x \leq 1$.

(ii) Let p be any polynomial of degree n with leading coefficient 2^{n-1} such that $|p(x)| \leq 1$ for $-1 \leq x \leq 1$ Show by considering the number of points where the graphs of p and T_n intersect that $p = T_n$.

(iii) Deduce that if q is any polynomial of degree n, with leading coefficient 1, then $|q(x)| \geq 2^{1-n}$ for some $x \in [-1, 1]$. (More precisely, either $q = 2^{1-n}T_n$, or $|q(x)| > 2^{1-n}$ for some $x \in [-1, 1]$.)

14. The converse of the Intermediate Value Theorem 2.21 is not true in general (but see Exercise 3 above for monotone functions). For instance, if $f(x) = x$ for rational x, $f(x) = 1 - x$ for irrational x, then f is a bijection of $[0, 1]$ to itself which has the intermediate value property, but only one point of continuity. Modify this example to find a function which is a bijection of $[0, 1]$ to itself, with no points of continuity.

15. Use the method of iteration (Theorem 2.29) to find solutions of the equations

    (i) $x^3 + x = 5$ on $[1, 2]$, (ii) $x = \tan x$ on $[\pi, 3\pi/2]$.

16. Show that if f, d are as in Theorem 2.29 then $|x_n - d| \leq c^n |x_0 - d|$.

17. Show that $f(x) \to l$ as $x \to \infty$ if and only if $|f(x) - f(y)| \to 0$ as $x, y \to \infty$.

3
Differentiation

The continuous functions which we considered in the previous chapter are characterised geometrically by the requirement that their graphs can be drawn without a break – without lifting the (idealised) pen from the paper. Most of the examples which were considered there have a much stronger property, expressed geometrically by the fact that the graph possesses a direction, or a tangent line at any point; such functions will be called differentiable. It was originally thought that all continuous functions must have a derivative, at least most of the time, so perhaps the example which we shall give in Section 6.6 of a continuous function with no derivative at any point will come as something of a surprise. Such nowhere differentiable functions are however now generally accepted as part of the normal array of mathematical objects, and are used in both physical and behavioural sciences to model processes which are subject to random perturbations.

3.1 Differentiable Functions

The definition of a differentiable function arises in the following way. Suppose we want to describe analytically the property that the graph of a function f has a tangent at a point $(c, f(c))$. Since the tangent is a line through $(c, f(c))$, its equation must be of the form $y = t(x) = f(c) + m(x - c)$ where the slope m is to be found. So far we have that $f(x) - t(x) \to 0$ as $x \to c$, but to reflect the fact that the curve and the tangent have the same direction we need more, namely that the difference $f(x) - t(x)$ should tend to zero *faster than* $x - c$. This requirement is expressed by putting $(f(x) - t(x)) / (x - c) = h$ and requiring that $h \to 0$ as $x \to c$. Writing t in terms of f, this gives $f(x) = f(c) + (m + h)(x - c)$, or

$$\frac{f(x) - f(c)}{x - c} = m + h \to m \qquad \text{as } x \to c.$$

Hence the slope of the tangent must be the limit (if it exists) of the quotient $(f(x) - f(c)) / (x - c)$, and we arrive at the following definition.

Definition 3.1

(i) The function f is differentiable at the point $(c, f(c))$ if the limit

$$\lim_{x \to c} \frac{f(x) - f(c)}{x - c} = \lim_{h \to 0} \frac{f(c + h) - f(c)}{h} \tag{3.1}$$

exists (as a finite real value, not $\pm\infty$). The value of the limit is called the derivative of the function at the point and is denoted by $f'(c)$; it gives the slope of the tangent to the graph of the function at the point.

(ii) If the limits in (i) exist only on the right or left, we write

$$f'_+(c) = \lim_{h \to 0+} \frac{f(c + h) - f(c)}{h}, \quad f'_-(c) = \lim_{h \to 0^-} \frac{f(c + h) - f(c)}{h}$$

respectively

(iii) The function f is differentiable on an interval I if it is differentiable at every point of I. In this case $f'(x)$ exists at all points of I so that f' is a function (the derived function, or simply the derivative) of f on I.

A number of comments are called for here. If the interval in (iii) is closed then we require only one-sided limits at the end points. Notice that $f'(c+) = \lim_{h \to 0^+} f'(c + h)$ means something different from $f'_+(c)$ (though they are often equal, as Exercise 4 shows). The existence of the limit in (3.1) is equivalent to the usual geometrical definition of the derivative as the limit of the slope of a chord (Figure 3.1).

Most of us are first introduced to derivatives in Newton's notation in which $f(x) - f(c)$ is denoted δf (or perhaps δy), $x - c$ is denoted δx and the derivative is df/dx where

$$\frac{df}{dx} = \lim_{\delta x \to 0} \frac{\delta f}{\delta x}$$

which is fine provided that we remember that df/dx is a composite symbol whose constituent parts, df and dx have no independent meaning. Newton's notation has some advantage as an aid to calculation, as illustrated below.

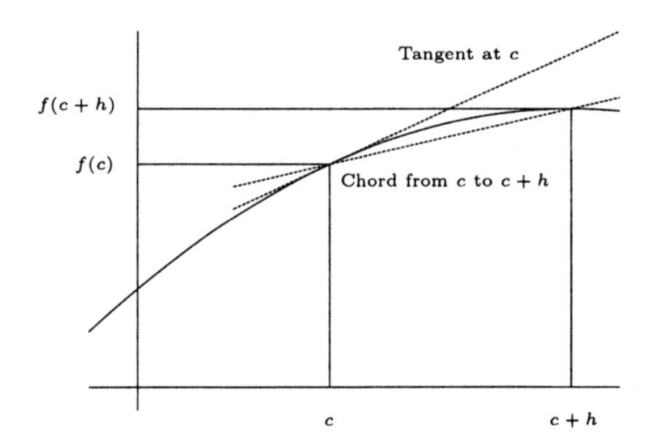

Figure 3.1 The derivative $f'(c)$ as the limit of the slope of a chord.

Example 3.2

(i) Let $f(x) = ax^2 + bx + d$ for constants a, b, d. Then f is differentiable everywhere.

(ii) Let $f(x) = 1/(x-a)$ for constant a. Then f is differentiable everywhere except at $x = a$, with $f'(x) = -1/(x-a)^2$.

(iii) Let $f(x) = \sqrt{x}$ for $x \geq 0$. Then f is differentiable for $x > 0$, with $f'(x) = 1/(2\sqrt{x})$.

(iv) Let $f(x) = |x|$. Then f is differentiable everywhere except at $x = 0$.

For (i) we consider

$$\begin{aligned} \frac{f(x) - f(c)}{x - c} &= a\frac{x^2 - c^2}{x - c} + b\frac{x - c}{x - c} \\ &= a(x + c) + b \to a(2c) + b \end{aligned}$$

as $x \to c$. Thus f is differentiable at c and its derivative is $f'(c) = 2ac + b$. In accordance with part (ii) of the definition, f is differentiable everywhere, with derivative given by $f'(x) = 2ax + b$.

For (ii)

$$\begin{aligned} \frac{f(x) - f(c)}{x - c} &= \frac{1/(x-a) - 1/(c-a)}{x - c} \\ &= \frac{(c-a) - (x-a)}{(x-a)(c-a)(x-c)} = \frac{-1}{(x-a)(c-a)} \to \frac{-1}{(c-a)^2} \end{aligned}$$

as $x \to c$. As in (i) f is differentiable at all points except a (where f is undefined), with $f'(x) = -1/(x-a)^2$.

For (iii), with $x, c > 0$,

$$\frac{f(x) - f(c)}{x - c} = \frac{\sqrt{x} - \sqrt{c}}{x - c} = \frac{1}{\sqrt{x} + \sqrt{c}} \to \frac{1}{2\sqrt{c}}$$

as $x \to c$. Thus f is differentiable for all $x > 0$, with $f'(x) = 1/(2\sqrt{x})$. Notice that f is continuous (on the right) at $x = 0$, but not differentiable there, since with $c = 0$, the quotient $(f(x) - f(c)) / (x - c)$ reduces to $\sqrt{x}/x = 1/\sqrt{x} \to \infty$ as $x \to 0+$.

In part (iv), the function is given by $f(x) = x$ for $x \geq 0$, $= -x$ if $x < 0$. Thus as in (i), the function is differentiable for $x > 0$ with $f'(x) = 1$, and for $x < 0$ with $f'(x) = -1$. At $x = 0$

$$\frac{f(x) - f(0)}{x - 0} = \frac{|x|}{x} = 1 \text{ for } x > 0, \quad = -1 \text{ for } x < 0,$$

which has no limit as $x \to 0$, and it follows that f has no derivative at $x = 0$. ♦

Notice that parts (iii) and (iv) show that differentiability is a stronger requirement that continuity – functions may be continuous but not differentiable. Conversely however, differentiability always implies continuity.

Proposition 3.3

If a function f is differentiable at c then it is continuous there.

Proof

The definition of differentiability requires that $(f(x) - f(c)) / (x - c)$ has a finite limit as $x \to c$. Then $f(x) - f(c)$, which is the product of this with $(x - c)$, must tend to zero as $x \to c$ and so f is continuous at c. ■

The existence of a derivative on an interval does not however imply that the derivative is continuous, as the next example shows.

Example 3.4

Let $f(x) = x^2 \sin(1/x)$ for $x \neq 0$, $f(0) = 0$. Then f is differentiable everywhere, but its derivative is not continuous at $x = 0$.

This function is continuous for all x; in particular the continuity at $x = 0$ is Exercise 12, Chapter 2 (or follows at once from the Sandwich Theorem). Its derivative for $x \neq 0$ is given by $f'(x) = 2x \sin(1/x) - \cos(1/x)$ which has no limit as $x \to 0$ and so f' is not continuous at $x = 0$. The derivative exists at

$x = 0$, however, since $(f(x) - f(0)) / (x - 0) = x \sin(1/x) \to 0$ as $x \to 0$ and so $f'(0) = 0$. ♦

(We shall soon justify the rules for differentiation of products and of trigonometric functions – for now we take them for granted. As usual, the reader who finds this illogical may refer to the entertaining account in [14], or simply pass over this example until we have put in the necessary machinery.)

We shall show that a function may be continuous at *all* points of an interval, but differentiable *nowhere*; however this requires a little more technique, and we postpone it to Section 6.6. It is also true that if a function is differentiable on an interval (a, b) then its derivative must have *some* points of continuity, indeed the set of such points must be dense in (a, b). This uses ideas beyond our scope – a proof is in [5, Exercise (6.92)].

Fortunately for the most commonly encountered functions, the (first) derivative, f' is again differentiable, possibly more than once, and we obtain in this way a sequence of functions $f, f', (f')', ((f')')'$, etc., the repeated derivatives of f. Counting the number of dashes is inconvenient, so for derivatives beyond the second it is customary to write $f^{(n)}$ or $d^n f/dx^n$ for the nth derivative.

Example 3.5

(i) Let $f(x) = ax^2 + bx + d$. Then $f'(x) = 2ax + b$, $f''(x) = 2a$, $f^{(n)}(x) = 0$ for $n \geq 3$. More generally, if f is a polynomial of degree k, then f has derivatives of all orders and $f^{(n)}(x) = 0$ for $n > k$.

(ii) Let $f(x) = x^\alpha$ for $x > 0$. Then f has derivatives of all orders with

$$f^{(n)}(x) = \alpha(\alpha - 1)\ldots(\alpha - n + 1)x^{\alpha - n} \qquad \text{for } x > 0,\ n = 1, 2, 3, \ldots. ♦$$

3.2 The Significance of the Derivative

The previous section introduced the derivative; now we find out what it tells us about the behaviour of the function. To begin with we look at the relation between the sign of the derivative and monotonicity. Clearly there should be some such relation since for instance we feel that a smooth increasing function should have tangents with positive slope, and conversely.

Proposition 3.6

If f is increasing (respectively decreasing) on an interval I, then $f'(x) \geq 0$

(respectively ≤ 0) on the interval.

Proof

Since f is increasing, $f(x) - f(y)$ must have the same sign as $x - y$ for $x, y \in I$, and so $(f(x) - f(y)) / (x - y)$ and its limit as $x - y \to 0$ must be ≥ 0. ■

Notice that we cannot deduce that $f'(x) > 0$, even when f is strictly increasing, as is shown by the example $f(x) = x^3$ at $x = 0$. Conversely, if there is a positive derivative at a point, then the function values to the right (left) of c must be greater (less) than $f(c)$.

Proposition 3.7

If $f'(c) > 0$, then there is some $h > 0$ such that $f(x) > f(c)$ for $c < x < c + h$, and $f(x) < f(c)$ for $c - h < x < c$. If $f'(c) < 0$ then similarly there is some $h > 0$ such that $f(x) < f(c)$ for $c < x < c + h$, and $f(x) > f(c)$ for $c - h < x < c$.

Proof

Since

$$\lim_{x \to c} \frac{f(x) - f(c)}{x - c} = f'(c) > 0$$

the ratio $(f(x) - f(c)) / (x - c)$ must be positive for x near c, and hence $f(x) - f(c)$ must have the same sign as $x - c$ as required. ■

Notice that it does not necessarily follow from $f'(c) > 0$ that f is increasing on $(c - h, c + h)$. This follows by considering $x^2 \sin(1/x) + x/2$ which has $f'(0) = 1/2$, but f is not increasing on any interval which contains zero; compare Example 3.4. To show that f is increasing on an interval, we need to strengthen Proposition 3.7 and assume that the derivative is positive on an interval. We need a preliminary result known as Rolle's theorem, since the special case in which f is assumed to be a polynomial was given by Michel Rolle in the 17th century.

Theorem 3.8 (Rolle)

Let f be differentiable on (a, b) and continuous on $[a, b]$ with $f(a) = f(b)$. Then for some $c \in (a, b)$, $f'(c) = 0$.

Proof

If f is constant on $[a,b]$ then $f'(x)$ is identically zero and the result is trivial; otherwise, since $f(a) = f(b)$, we can assume that there is either some point of (a,b) at which $f(x) > f(a)$ or a point where $f(x) < f(a)$. We consider the first possibility – the argument in the second case is similar. Since f is sometimes greater than its value at a, it has, by Theorem 2.19 and its corollary, a maximum value which is attained at some point $x_1 \in (a,b)$ at which f is differentiable. If $f'(x_1) > 0$ then by Proposition 3.7 there would be points y to the right of x_1 at which $f(y) > f(x_1)$ so there would be no maximum at x_1 contrary to our supposition; similarly $f'(x_1) < 0$ is impossible, and it follows that $f'(x_1) = 0$. ■

Rolle's theorem justifies the obvious observation that if a differentiable function returns to its starting value, then there must be some interior point of the interval at which the derivative is zero.

Example 3.9

Let $f(x) = (\sin x)/x$ which is continuous and differentiable on $(0,\infty)$ with $f(\pi) = f(2\pi) = 0$. Then there is some point $c \in (\pi, 2\pi)$ with $f'(c) = 0$.

The reader may check that this point is unique, but is *not* $3\pi/2$! ♦

Theorem 3.10 (Mean Value Theorem)

Let f be continuous on $[a,b]$ and differentiable on (a,b). Then for some $c \in (a,b)$,

$$f'(c) = \frac{f(b) - f(a)}{b-a}. \tag{3.2}$$

Proof

Let $m = (f(b) - f(a))/(b-a)$ and consider $g(x) = f(x) - m(x-a)$. Then

$$g(b) - g(a) = f(b) - f(a) - m(b-a) = 0,$$

so that by Rolle's theorem there is some c with $g'(c) = 0$. But $g'(c) = f'(c) - m$ and the result follows. ■

Comparing this with Rolle's theorem, the Mean Value Theorem says that there must be some interior point at which the tangent is parallel to the chord joining the end points. of the graph (Figure 3.2).

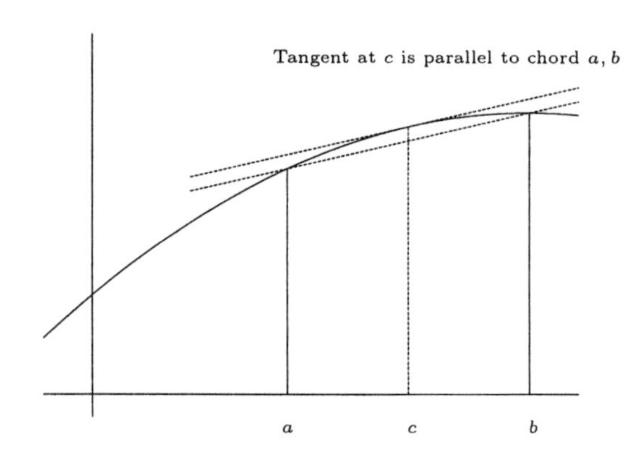

Figure 3.2 Mean Value Theorem.

In other words, if you travel a distance of 40 miles in one hour, then there must have been some time during the journey at which your speed was exactly 40 m.p.h.

The Mean Value Theorem fails if we have only one-sided derivatives – consider $|x|$ on $[-1,1]$. However Exercise 4(ii) gives a partial result.

Example 3.11

Apply the Mean Value Theorem to (i) $f(x) = x^\alpha$, $(\alpha > 0)$ on $[0,1]$, and (ii) $f(x) = \sin x$ on $[0, \pi/2]$.

(i) With $f(x) = x^\alpha$, $f'(x) = \alpha x^{\alpha-1}$, and Equation (3.2) gives

$$\frac{1-0}{1-0} = \alpha z^{\alpha-1},$$

which is satisfied for $z = (1/\alpha)^{1/(\alpha-1)}$.

(ii) With $f(x) = \sin x$, (3.2) gives

$$\frac{1-0}{\pi/2-0} = \cos z,$$

which is satisfied when $\cos z = 2/\pi$. (Since $0 < 2/\pi < 1$, this transcendental equation has a unique solution in $[0, \pi/2]$. However the solution is not a rational multiple of π.) See also Exercise 6. ♦

The result which says that a function with a positive derivative must be increasing now follows from the Mean Value Theorem.

Corollary 3.12

Let f be differentiable on (a,b) and let $f'(x) \geq 0$ (respectively > 0) on (a,b). Then f is increasing (respectively strictly increasing) on (a,b). Similarly if $f' \leq 0$ then f is decreasing, and if $f' = 0$ on (a,b) then f is constant on (a,b).

Proof

Consider any $x, y \in (a,b)$ with $x < y$. By the Mean Value Theorem there is some z with $x < z < y$ and

$$\frac{f(y) - f(x)}{y - x} = f'(z)$$

which is (strictly) positive by our supposition. Hence $f(y) - f(x) \geq (>)\, 0$ as required. ■

This result can be weakened to require only that $f' > 0$ (even $f'_+ > 0$) except on a countable set – an example of this in a very simple case is given by $f(x) = x^3$ where $f' > 0$ except at $x = 0$ and the function is strictly increasing. This is explored in Exercise 10.

Example 3.13

Let $f(x) = x + x^3$ on $\mathbb{R}$.

Since $f'(x) = 1 + 3x^2$ is strictly positive this function is strictly increasing and has a well-defined inverse function on $\mathbb{R}$. We shall find the derivative of the inverse in the next section. ♦

3.3 Rules for Differentiation

Since we shall use differentiation extensively, we must become good at calculating derivatives – fortunately there are (familiar) rules for differentiating the most commonly encountered functions. (The situation is quite different when we come to integration in Chapter 4.) We must find formulae for the derivatives of sums, products, quotients, composites and inverses.

Theorem 3.14

If f, g are differentiable at a point c (or on an interval (a,b)), then $f \pm g$ are differentiable at c (or on (a,b)) and

$$(f \pm g)'(c) = f'(c) \pm g'(c)$$

(at all points of (a,b)).

Similarly

$$\begin{aligned}(f.g)'(c) &= f'(c)\,g(c) + f(c)\,g'(c),\\ \left(\frac{f}{g}\right)'(c) &= \frac{f'(c)\,g(c) - f(c)\,g'(c)}{(g(c))^2}\end{aligned}$$

where the last result requires also $g(c) \neq 0$.

Proof

The result for sums follows at once from Definition 3.1, as the reader will immediately verify. The second requires the simple rearrangement

$$f(x)\,g(x) - f(c)\,g(c) = (f(x) - f(c))\,g(x) + f(c)\,(g(x) - g(c))$$

before dividing by $x - c$ and taking a limit. Notice that g, being differentiable at c must also be continuous there (Proposition 3.3), so $g(x) \to g(c)$ as $x \to c$. The result for quotients requires a similar rearrangement which is left to the reader for practice. ■

Notice the special case of the result for quotients in which f is identically 1:

$$\left(\frac{1}{g}\right)'(c) = \frac{-g'(c)}{(g(c))^2}. \tag{3.3}$$

The following examples follow at once.

Example 3.15

(i) $(x + \sin x)' = 1 + \cos x$,

(ii) $(x \sin x)' = \sin x + x \cos x$,

(iii) $\left(\frac{\sin x}{x}\right)' = \frac{x \cos x - \sin x}{x^2}$ for $x \neq 0$,

(iv) $\left(\frac{1}{\sin x}\right)' = \frac{-\cos x}{(\sin x)^2}$ for $\sin x \neq 0$. ♦

The result for composites is like part (ii) of Theorem 2.13.

Theorem 3.16 (Chain Rule)

Let g be differentiable. at c and f be differentiable at $g(c)$. Then $f \circ g$ is differentiable at c and

$$(f \circ g)'(c) = (f' \circ g)(c)\, g'(c). \tag{3.4}$$

Proof

Write $d = g(c)$ and $m, n = g'(c), f'(d)$ respectively. Since f, g are differentiable at d, c we have

$$\begin{aligned} g(x) - g(c) &= (m+h)(x-c), \\ f(y) - f(d) &= (n+k)(y-d) \end{aligned}$$

where $h \to 0$ as $x \to c$ and $k \to 0$ as $y \to d$. Putting $y = g(x)$, $d = g(c)$ in the second of these gives

$$\begin{aligned} f(g(x)) - f(g(c)) &= (n+k)(g(x) - g(c)) \\ &= (n+k)(m+h)(x-c) \\ &= (nm + (km + hn + kh))(x-c). \end{aligned}$$

Now the required result is that $(f \circ g)'(c) = nm$, and this will follow if we can show that $km + hn + kh \to 0$ as $x \to c$. We already know that $h \to 0$ as $x \to c$. Since g is differentiable at c it must be continuous there, so $y - d = g(x) - g(c) \to 0$ as $x \to c$, and it follows that also $k \to 0$ as $x \to c$. Hence $km + hn + kh \to 0$ as $x \to c$ and we are done. ■

In a first calculus course, when we write $y = f(x)$ and dy/dx for $f'(x)$, we may also have $z = g(y)$, when the chain rule $(g \circ f)'(x) = g'(f(x))\, f'(x)$ appears in the form

$$\frac{dz}{dx} = \frac{dz}{dy}\frac{dy}{dx}. \tag{3.5}$$

This then leads on to a spurious 'proof by cancellation' which is very mystifying to someone who has just been told that dy/dx is not a fraction but a compound symbol for the derivative. The student is liable to say 'well the result is true (as we see from examples which can be verified independently) so the argument must be correct too' – to which well-meaning argument one can only observe that

$$\frac{19}{95} = \frac{1}{5}$$

is correct, but not justified by cancelling the 9's. Of course (3.5) is a convenient way of remembering the chain rule, as the next examples illustrate.

Example 3.17

(i) $\left(\sin\left(x^2\right)\right)' = 2x\cos\left(x^2\right)$, (ii) $\left((\sin x)^2\right)' = 2\sin x\cos x$,

$$\text{(iii)} \quad \left(\frac{x}{\sqrt{x^2+1}}\right)' = \frac{1}{\left(x^2+1\right)^{3/2}}.$$

In (i) we have $f(x) = x^2$, $g(x) = \sin x$, so (3.4) gives $(g \circ f)'(x) = g'(f(x))\, f'(x) = \cos\left(x^2\right) 2x$ as required, and (ii) is similar.

(If we want to use the alternative notation (3.5) then we put $y = x^2$, $z = \sin y = \sin\left(x^2\right)$ and we get

$$\frac{dz}{dx} = \frac{dz}{dy}\frac{dy}{dx} = (\cos y)2x = \left(\cos x^2\right) 2x$$

as before. There is naturally nothing wrong with finding the result this way; in calculation it is a matter of personal taste. The point is that one should not think that (3.5) *proves* anything.)

For (iii) we combine the chain rule with the result for quotients in Theorem 3.14:

$$\left(\frac{x}{\sqrt{x^2+1}}\right)' = \frac{\sqrt{x^2+1} - x\left(2x/2\sqrt{x^2+1}\right)}{x^2+1} = \frac{\left(x^2+1\right) - x^2}{\left(x^2+1\right)^{3/2}}$$

as required. ♦

A special case of the chain rule gives the following result for derivatives of inverses.

Theorem 3.18

Let f be continuous on $[a,b]$, and differentiable on (a,b) with $f'(x) > 0$. Then f has a unique inverse g on $[f(a), f(b)] = [c,d]$ say, which is differentiable on (c,d) with

$$g'(y) = \frac{1}{f'(x)} \qquad \text{when } y = f(x). \tag{3.6}$$

Note that since $f' > 0$, f must be strictly increasing on $[a,b]$ and the the existence and continuity of the inverse g follow from Theorem 2.25. We have to show that g is differentiable and the formula (3.6). There is of course a corresponding theorem for decreasing functions when $f' < 0$.

Proof

Let v, y be distinct points of $[c, d]$ and let $g(v) = u, g(y) = x$ so that

$$\frac{g(v) - g(y)}{v - y} = \frac{u - x}{f(u) - f(x)}. \tag{3.7}$$

As $v \to y$, also $u \to x$, since g is continuous, and so the right side of (3.7) tends to $1/f'(x)$, and the result follows.

(An alternative proof via the Chain Rule, which the reader is invited to complete, says that since $g(f(x)) = x$, then $g'(f(x)) f'(x) = 1$.) ■

Example 3.19

(i) Let $f(x) = x + x^3$ on $\mathbb{R}$. Then f^{-1} is differentiable on $\mathbb{R}$ with

$$\left(f^{-1}\right)'(y) = \frac{1}{1 + 3x^2} \qquad \text{when } y = x + x^3,$$

so e.g. $\left(f^{-1}\right)'(10) = 1/13$.

(ii) Let $g(x) = \sin x$ on $[0, \pi/2]$. Then g^{-1} is differentiable on $(0, 1)$ (notice that $g'(\pi/2) = 0$, so the end point 1 must be excluded) and

$$\left(g^{-1}\right)'(y) = \frac{1}{\sqrt{1 - y^2}},$$

so e.g. $\left(g^{-1}\right)'(1/2) = 2/\sqrt{3}$.

All we need to verify here is that both functions have positive derivatives, and in the second example, that

$$\frac{1}{g'(x)} = \frac{1}{\cos x} = \frac{1}{\sqrt{1 - y^2}} \qquad \text{for } 0 \leq y < 1. \ ♦$$

We shall make a similar study of the sign of the second derivative in Section 3.6 and Exercise 9.

3.4 Mean Value Theorems and Estimation

The Mean Value Theorem 3.10 can be used as follows to give approximate values for functions.

Example 3.20

$6 < \sqrt{37} < 6\frac{1}{12}$.

Take $f(x) = \sqrt{x}$, $f'(x) = 1/(2\sqrt{x})$. Then $1/\left(2\sqrt{37}\right) < f'(x) < 1/12$ on $[36, 37]$ and the Mean Value Theorem does the rest. ♦

When we write the Mean Value Theorem in the form

$$f(b) = f(a) + (b-a)\,f'(c)$$

then, as in the above example, the term $(b-a)\,f'(c)$ gives an estimate of how near the value of f at b is to its value at a, and it is interesting to notice that the formula is equally correct when $b < a$; all that is required is that c is *between* a and b.

For a better approximation, we could try to find terms in $(b-a)^2, (b-a)^3$, etc. which would give greater accuracy near a. Suppose that we write

$$f(x) = f(a) + k_1(x-a) + k_2(x-a)^2 + \cdots + k_n(x-a)^n + r_n(x) \tag{3.8}$$

where the coefficients have to be found, and the final term $r_n(x)$, which represents the error of the approximation, has to be investigated. Write

$$p_n(x) = f(a) + k_1(x-a) + k_2(x-a)^2 + \cdots + k_n(x-a)^n$$

and choose the k's so that f and p_n have the same derivatives of order $1, 2, \ldots, n$ at $x = a$. This gives

$$\begin{aligned} p_n'(a) &= k_1 = f'(a), \\ p_n''(a) &= 2k_2 = f''(a), \\ &\cdots \\ p_n^{(j)}(a) &= j!k_j = f^{(j)}(a), \end{aligned}$$

for $1 \le j \le n$.

Definition 3.21

The polynomial given by

$$p_n(x) = f(a) + (x-a)\,f'(a) + \frac{(x-a)^2}{2!}f''(a) + \cdots + \frac{(x-a)^n}{n!}f^{(n)}(a) \tag{3.9}$$

is called the nth order Taylor polynomial for f at a. If we want to draw attention to the point a at which the approximation occurs, we shall write $p_n(x, a)$ for $p_n(x)$, and similarly for r_n.

Example 3.22

(i) Let $f(x) = 1 + x + x^3$. Then

$$p_1(x,0) = p_2(x,0) = 1 + x, \qquad r_1(x,0) = r_2(x,0) = x^3,$$
$$p_3(x,0) = 1 + x + x^3, \qquad r_3(x,0) = 0,$$
$$p_2(x,1) = 3 + 4(x-1) + 3(x-1)^2,$$
$$p_3(x,1) = 3 + 4(x-1) + 3(x-1)^2 + (x-1)^3.$$

(ii) Let $f(x) = \sin x$. Then

$$p_{2k-1}(x,0) = x - \frac{x^3}{3!} + \cdots + (-1)^{k-1}\frac{x^{2k-1}}{(2k-1)!} = p_{2k}(x,0).$$

Both of these results are immediate on calculating derivatives. ♦

It is tempting in (ii) of the example, and in (3.9), to replace the finite sum by an infinite series, hoping that this will converge, and that the sum will be related to $f(x)$; we shall investigate these questions in Sections 6.4 and 6.6. For now we emphasise that n is finite and that p_n is a polynomial of degree n. Our concern will be to find an explicit expression for the remainder r_n from which we can estimate the accuracy of the approximation implied by (3.8). In this section the remainder will be found in terms of the derivatives of f, while in Section 4.5 it will be in terms of an integral.

Theorem 3.23 (Taylor)

Let f have $n-1$ continuous derivatives on $[a,b]$ and an nth derivative which exists on (a,b). Then for some $c \in (a,b)$

$$f(b) = f(a) + (b-a)f'(a) + \frac{(b-a)^2}{2}f''(a) + \cdots + \frac{(b-a)^{n-1}}{(n-1)!}f^{(n-1)}(a) + r_n(c),$$

where

$$r_n(c) = \frac{(b-a)^n}{n!}f^{(n)}(c).$$

Proof

Consider

$$F_n(x) = f(b) - f(x) - (b-x)f'(x) - \frac{(b-x)^2}{2}f''(x) - \cdots - \frac{(b-x)^{n-1}}{(n-1)!}f^{(n-1)}(x). \tag{3.10}$$

(See the note following the proof for a comment on this unobvious choice.) If we differentiate we find that most terms cancel, leaving only

$$F_n'(x) = -\frac{(b-x)^{n-1}}{(n-1)!} f^{(n)}(x). \tag{3.11}$$

Now look at

$$G_n(x) = F_n(x) - \left(\frac{b-x}{b-a}\right)^n F_n(a)$$

(another unobvious choice). This satisfies $G_n(a) = G_n(b) = F_n(b) = 0$, and hence by Rolle's theorem applied to G_n, there is some $c \in (a,b)$ with $G_n'(c) = 0$. But

$$\begin{aligned} G_n'(c) &= F_n'(c) + n\frac{(b-c)^{n-1}}{(b-a)^n} F_n(a) \\ &= (b-c)^{n-1}\left[\frac{-1}{(n-1)!} f^{(n)}(c) + \frac{n}{(b-a)^n} F_n(a)\right], \text{ or} \\ F_n(a) &= \frac{(b-a)^n}{n!} f^{(n)}(c) \end{aligned}$$

which gives the result. ■

The reader is entitled to ask at this point, how the choice of F_n in this proof was arrived at, to which the only answer is that the proof evolved through various stages, each giving different forms of the remainder term, and that the above is simply one of a number of variants – the one which gives the remainder in the easiest and most natural form. Indeed Brook Taylor (a contemporary of Isaac Newton) is flattered by having his name attached to the theorem, his contribution being little more that to determine the form of the coefficients in (3.9). Consideration of the remainder, in particular the conditions under which $r_n \to 0$, did not come for another century.

We shall find it useful to have the result in Theorem 3.23 in the alternative form

$$f(a+h) = f(a) + hf'(a) + \frac{h^2}{2} f''(a) + \cdots + \frac{h^{n-1}}{(n-1)!} f^{(n-1)}(a) + r_n, \tag{3.12}$$

in which f is supposed n times differentiable on an interval $(a-\delta, a+\delta)$, h is any element (positive or negative!) of $(-\delta, \delta)$, and $r_n = h^n f^{(n)}(c)/n!$ where c is between a and $a+h$ (equivalently $c = a + \theta h$ where $0 < \theta < 1$). This follows from the theorem by putting $b = a + h$ (and interchanging a, b if $h < 0$).

Example 3.24

(i) Let $f(x) = \sqrt{x}$ on $(0, \infty)$ and let $a > 0$. Then

$$\sqrt{a+h} = \sqrt{a} + \frac{1}{2\sqrt{a}}h - \frac{1}{8a^{3/2}}h^2 + \frac{1}{16c^{5/2}}h^3$$

for $-a < h < \infty$ and c between a and $a + h$. In particular if a, h are both positive then

$$\sqrt{a} + \frac{1}{2\sqrt{a}}h > \sqrt{a+h} > \sqrt{a} + \frac{1}{2\sqrt{a}}h - \frac{1}{8a^{3/2}}h^2.$$

Applied to $\sqrt{37}$ which we considered in Example 3.20 this gives $6\frac{1}{12} > \sqrt{37} > 6\frac{1}{12} - \frac{1}{1728}$ or $6.0833 > 6.0827625 > 6.0827546$.

(ii) Let $f(x) = \sin x$. Then $|r_{2k+1}(x)| \le |x|^{2k+1} / (2k+1)!$

The result of (i) is immediate from calculating the derivatives of $\sqrt{x}$. For $\sin x$ we have already found thc Taylor polynomial about $x = 0$ in Example 3.22. The remainder is given by

$$r_{2k+1}(x) = (-1)^k \frac{x^{2k+1}}{(2k+1)!} \sin c$$

and so $|r_{2k+1}(x)| \le |x|^{2k+1} / (2k+1)!$ With an eye to future developments in Section 4.5 note that for any real x, $r_{2k+1}(x) \to 0$ as $n \to \infty$. ♦

The next application of Taylor's theorem is useful for finding limits of the form

$$\lim_{x \to a} \frac{f(x)}{g(x)}$$

where both f, g tend either to zero or infinity as $x \to a$.

Theorem 3.25 (L' Hôpital's Rule)

(i) Let f, g be differentiable in some interval $(a - \delta, a + \delta)$, and suppose that the limit $l = \lim_{x \to a} f'(x) / g'(x)$ exists. Then also

$$\lim_{x \to a} (f(x) - f(a)) / (g(x) - g(a))$$

exists and equals l.

(ii) Let f, g be differentiable in some interval $(a - \delta, a + \delta)$, and suppose that $\lim_{x \to a} f'(x) = l_1$, $\lim_{x \to a} g'(x) = l_2 \neq 0$. Then the limit

$$\lim_{x \to a} \frac{f(x) - f(a)}{g(x) - g(a)}$$

exists and equals l_1 / l_2.

Proof

It is enough to show (i), since (ii) is a special case. Notice that we cannot use the Mean Value Theorem on f, g separately, since this would give $f(x) - f(a) = f'(c_1)(x-a)$ and $g(x) - g(a) = g'(c_2)(x-a)$ for different values of c_1, c_2. The idea is to consider instead

$$h(t) = (f(x) - f(a))(g(t) - g(a)) - (f(t) - f(a))(g(x) - g(a))$$

which has clearly $h(a) = h(x) = 0$ and so by Rolle's theorem there is some c between a and x with

$$0 = h'(c) = (f(x) - f(a))\, g'(c) - f'(c)(g(x) - g(a)).$$

The existence of the limit of f'/g' shows that $g' \neq 0$ at points near a, and so we may rearrange to get

$$\frac{f(x) - f(a)}{g(x) - g(a)} = \frac{f'(c)}{g'(c)}$$

and the result follows since $c \to a$ when $x \to a$. ■

Example 3.26

(i) $\lim_{x\to 0} (\sin cx)/x = c$ for any real c.

(ii) $\lim_{x\to 1} \left(\sqrt{x+3} - 2x\right)/\ln x = -7/4$.

(iii) $\lim_{x\to 0} (x - \sin x)/(\tan x - x) = 1/2$.

For (i) take $f(x) = \sin(cx)$ and $g(x) = x$ in part (ii) of the theorem. Results like this which have the special form $\lim_{x\to a} (f(x) - f(a))/(x-a) = f'(a)$ are not really using L'Hôpital's rule at all, but simply stating the definition of the derivative at a.

For (ii) take $f(x) = \sqrt{x+3} - 2x$ and $g(x) = \ln x$, when $f(1) = g(1) = 0$, $f'(x) = 1/\left(2\sqrt{x+3}\right) - 2$, $g'(x) = 1/x$ and $f'(1) = -7/4$, $g'(1) = 1$.

Part (iii) illustrates the need for the more general first part of the theorem, since $f(x) = x - \sin x$ and $g(x) = \tan x - x$ have $f'(0) = g'(0) = 0$. Then $f'(x) = 1 - \cos x$ and $g'(x) = \sec^2 x - 1$ which are both zero at $x = 0$, and the limit of f'/g' is again indeterminate. In this case $f''(x) = \sin x$ and $g''(x) = 2\sec^2 x \tan x = 2\sec^3 x \sin x$ so

$$\frac{f''(x)}{g''(x)} = \frac{1}{2\sec^3 x} \to \frac{1}{2} \qquad \text{as } x \to 0.$$

Hence using (ii) of the theorem twice we have

$$\lim_{x\to 0} \frac{f(x)}{g(x)} = \lim_{x\to 0} \frac{f'(x)}{g'(x)} = \lim_{x\to 0} \frac{f''(x)}{g''(x)} = \frac{1}{2}.$$

(Actually a more alert reader might notice that f' and g' have a common factor of $1-\cos x$ and so the second differentiation was not needed, but it is easy to devise examples where such short cuts don't work.) ♦

L'Hôpital's rule is so easy to remember and to use (it always helps to have a catchy name!) that one sometimes gets into the careless habit of using it when the numerator and denominator do not tend to zero, resulting in such absurdities as $\lim_{x\to 0}(1+2x)/(2+x)=2$ when the correct answer is of course $1/2$.

There are versions of the result in which $f,g\to\infty$ which we look at next, and also when $x\to\infty$ which are left to the reader as Exercise 19.

Theorem 3.27

Let f,g be differentiable in some interval $(a-\delta,a+\delta)$, let $f(x),g(x)\to\infty$ as $x\to a$, and suppose that the limit $l=\lim_{x\to a}f'(x)/g'(x)$ exists. Then also $\lim_{x\to a}f(x)/g(x)$ exists and equals l.

Proof

Given $\varepsilon>0$, choose $\delta>0$ such that if $0<|x-a|<\delta$ then

$$\left|\frac{f'(x)}{g'(x)}-l\right|<\varepsilon \tag{3.13}$$

and in particular, $g'(x)\neq 0$ for $0<|x-a|\leq\delta$. Consider $a<x<b=a+\delta$ (the proof if $a-\delta<x<a$ is similar). Then as in the proof of Theorem 3.25 there is some c with $x<c<b$ and

$$\frac{f(x)-f(b)}{g(x)-g(b)}=\frac{f'(c)}{g'(c)}. \tag{3.14}$$

We can write the left side of (3.14) as

$$\frac{f(x)}{g(x)}\left[\frac{1-f(b)/f(x)}{1-g(b)/g(x)}\right]$$

where the term in square brackets tends to 1 as $x\to a$ for fixed b since $f(x),g(x)\to\infty$. Hence for x near enough to a, $f(x)/g(x)$ is close to $f'(c)/g'(c)$ which in turn is within ε of l by (3.13) so that in fact $f(x)/g(x)\to l$ as $x\to a$ as required. ■

Example 3.28

Show that $\lim_{x\to 0+}x\ln x=0$.

Let $f(x) = -\ln x$, $g(x) = 1/x$ for $x > 0$. Then both $f, g \to \infty$ as $x \to 0^+$. Also $f'(x) = -1/x$, $g'(x) = -1/x^2$, so $f'(x)/g'(x) = x \to 0$ as $x \to 0^+$. Hence $f(x)/g(x) = -x \ln x \to 0$ as required. ♦

3.5 More on Iteration

In Section 2.6 we started to look at the possibility of solving equations $f(x) = 0$ by iteration, i.e. by rearranging the equation as $x = g(x)$ for some well-chosen g, and defining a sequence by $x_{n+1} = g(x_n)$ in the hope that it would converge to the desired root. The discussion at the end of that section shows that this may or may not succeed; our job now is to find some reasonable conditions which will make it work. The answer in the simplest case follows from the Mean Value Theorem.

Theorem 3.29

Let g be a differentiable function on $[a, b]$ into itself, and suppose that g is a contraction, i.e. that there is some $c < 1$ such that $|g'(x)| \le c$ on $[a, b]$. Then for any $x_0 \in [a, b]$, the sequence (x_n) defined by $x_{n+1} = g(x_n)$ is convergent with a limit $d \in [a, b]$, and

$$|x_n - d| \le \frac{c^n}{1-c} |x_0 - x_1|. \tag{3.15}$$

Proof

Once we show that g is a contraction, the rest will follow from Theorem 2.29. But for $x, y \in [a, b]$ we have $|g(y) - g(x)| = |(y - x) g'(t)|$ for some t between x and y, by the Mean Value Theorem. The given condition on g' now gives $|g(y) - g(x)| \le c|(y - x)|$ as required. ■

Thus to solve $f(x) = 0$, we should try to rearrange the equation as $x = g(x)$ for a contraction g, as in the following examples.

Example 3.30

Solve the equations (i) $x + x^5 = 3$ on $[1, 2]$, (ii) $\cos x = x$ on $[0, \pi/3]$, (iii) $\pi - \sin x - (\pi - x)\cos x = \pi/2$ on $[0, \pi/2]$.

For (i), notice that $x = g(x) = 3 - x^5$ has $g'(x) \le -2$ on $[1, 2]$ so g is surely not a contraction. Instead we rearrange to get $x = h(x) = (3-x)^{1/5}$ where h is a contraction on $[1,2]$ since it is decreasing with $h(1) = 2^{1/5} < 2$, $h(2) = 1$ and $|h'(x)| \leq 1/5$. Starting with $x_0 = 1.5$, our calculator gives us $x_1 = 1.08$, $x_2 = 1.13$, $x_3 = 1.132$, $x_4 = 1.1330$, $x_5 = 1.13298$, etc. Then (3.15) gives the estimate $|x_n - d| \leq (1.25)(0.2)^n(0.5)$ for the accuracy of our iteration.

For (ii), notice that cos is already a contraction on $[0, \pi/3]$ since $\pi/3 > \cos 0 = 1 > 1/2 = \cos(\pi/3) > 0$. Repeated use of the cosine key (in radians mode!) shows that the iteration settles down rather slowly to around 0.739..., the relative slowness being caused by the fact that the derivative $g'(x)$ is bounded by $\sin(\pi/3) = \sqrt{3}/2 = 0.866$ which is too near 1 to be of much practical value. In both (i) and (ii) it is interesting to compare the successive values of $|x_{n+1} - x_n| / |x_n - x_{n-1}|$ with $g'(x)$ near the solution. An appropriate rearrangement of (iii) (this is Example 2.22, the problem of the farmer and the goats) is left to the reader. ♦

These examples should have convinced the reader that iteration, while robust and easy to spot, may not be very rapid in finding roots to high accuracy. Many other techniques are known, of which the most interesting is Newton's method which from our point of view may be introduced as follows. (See below for a graphical interpretation.) We want to solve $f(x) = 0$, which we rearrange in the form $x = x - h(x) f(x)$ where h is to be chosen. Writing $g(x) = x - h(x) f(x)$, we want $g'(x)$ as small as possible for rapid converge of $x_{n+1} = g(x_n)$ near the supposed root d. But $g'(x) = 1 - h'(x) f(x) - h(x) f'(x)$, and since we suppose that f is small near d, we neglect the $h'(x) f(x)$ term. This leaves $g'(x) = 1 - h(x) f'(x)$, and to get g' near zero, we choose $h(x) = 1/f'(x)$, and $g(x) = x - f(x)/f'(x)$. This motivates the following definition.

Definition 3.31

Given a function f on $[a,b]$ and a point $x_0 \in [a,b]$, the iterative sequence (x_n) given by

$$x_{n+1} = x_n - \frac{f(x_n)}{f'(x_n)}, \qquad n \geq 0, \tag{3.16}$$

determines Newton's method (or Newton's iteration) with initial value x_0.

From the discussion preceding the definition, we may hope to get rapid convergence, but there are a number of difficulties of which perhaps the most fundamental is to ensure that the values x_n remain in the interval $[a,b]$ of definition. If for some x_n the value of x_{n+1} given by (3.16) is not in $[a,b]$ then the iteration terminates at this point, and the method fails; the method may also fail if the sequence does not converge, as examples below will illustrate. Division by f' may also cause difficulties if there are zeros of the derivative in the interval.

From a geometrical point of view, Newton's method determines x_{n+1} as the point of intersection of the tangent at $(x_n, f(x_n))$ with the x-axis (Figure 3.3).

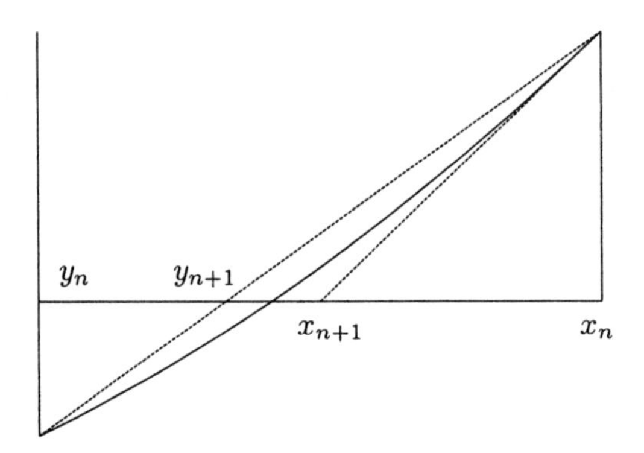

Figure 3.3 Newton's method.

Before going into any theoretical details we look again at Example 3.30 above, using Newton's method in a more or less uncritical way. For (i) we have $f(x) = x + x^5 - 3$ and the scheme is given by

$$x_{n+1} = x_n - \frac{x_n + x_n^5 - 3}{1 + 5x_n^4}.$$

Beginning with $x_0 = 2$ we have successively

$$2, 1.617, 1.342, 1.1856, 1.1370,\ 1.13302, 1.13299756693, 1.13299756589,$$

and after a slow start we have a sudden increase in the number of figures to which the terms agree. Beginning with $x_0 = 1$ which is nearer to (but less than) the root, we find $(1, 1.166, 1.134, 1.13300, 1.13299756592)$ which gives agreement after an initial overshoot as could be expected from the figure.

For (ii) we have

$$x_{n+1} = x_n - \frac{x_n - \cos x_n}{1 + \sin x_n}$$

and starting with $x_0 = 1$, we find

$$1, 0.75, 0.739112, 0.739085133386, 0.739085133215.$$

For (iii) we have

$$x_{n+1} = x_n - \frac{\pi/2 - \sin x_n - (\pi - x_n)\cos x_n}{(\pi - x_n)\sin x_n}$$

and starting with $x_0 = 1$ we get $1, 1.237, 1.2358967, 1.23589692428$ which is a spectacular improvement on the method of bisection in Section 2.4. ♦

These examples should convince the reader that Newton's method is worth the effort of calculating derivatives, provided that we can find some simple conditions on the function which will give convergence, and allow us to estimate the position of the required root. Notice that in the examples we carefully avoided saying that the calculated digits actually agreed with those of the root, and indeed at this point we have no way of knowing this. (The fact that successive terms of a sequence agree to a certain number of decimal places does *not* imply that these figures agree with those of the limit, as is shown by examples such as $x_n = \sqrt{n}$ in which $x_{n+1} - x_n \to 0$ while $x_n \to \infty$.) The following result gives some sufficient conditions.

Theorem 3.32

Let f be twice continuously differentiable (i.e. f'' exists and is continuous) on $[a, b]$ with $f(a) f(b) < 0$ and suppose that both f', f'' have fixed sign (not necessarily the same) on $[a, b]$. Then taking x_0 as whichever of a, b has the larger value of $|f'|$, the sequence given by Newton's method will converge monotonically to the root d say, with

$$k|x_n - d| \leq (k|x_0 - d|)^{2^n} \tag{3.17}$$

where

$$k = \frac{1}{2} \sup \left\{ \left| \frac{f''(x)}{f'(x)} \right|, a \leq x \leq b \right\}.$$

Before we begin the proof, several comments are in order. The facts that f' has fixed sign and f'' is continuous ensure that k is finite. The very rapid convergence is a result of the 2^n in the exponent of (3.17), but this is of no use unless the base $k|x_0 - d|$ is < 1. If this is not satisfied then we have to begin the iteration and wait until the value of $k|x_j - d|$ becomes < 1 before applying the estimate (3.17) with this value of x_j in place of x_0 :

$$k|x_n - d| \leq (k|x_j - d|)^{2^{n-j}}. \tag{3.18}$$

A stronger objection is that initially we have no idea of the value of d, and hence the only quantity which can be used on the right side of (3.17) is $k(b - a)$ which is unlikely to be < 1. We will deal with this second point after we have proved the theorem as stated.

Proof

We can suppose for the proof that both f', f'' are positive – if not then we consider $-f(x)$ or $f(a+b-x)$ or $-f(a+b-x)$, one of which will have the given signs (as the reader will easily verify). Then both f and f' are strictly increasing on $[a,b]$ so there is only one root d with $f(d) = 0$, and we take $x_0 = b$. The sequence (x_n) is decreasing from (3.16) since f, f' are positive. Also, by induction, the sequence. is bounded below by d since (i) $x_0 = b > d$, and (ii) if $x_n > d$, then we have

$$\frac{f(x_n)}{x_n - d} = \frac{f(x_n) - f(d)}{x_n - d} = f'(u)$$

for some $u \in (d, x_n)$ by the Mean Value Theorem, and from (3.16) that

$$\frac{f(x_n)}{x_n - x_{n+1}} = f'(x_n).$$

But f' is increasing, so $f'(u) < f'(x_n)$, and hence

$$\frac{f(x_n)}{x_n - d} = f'(u) < f'(x_n) = \frac{f(x_n)}{x_n - x_{n+1}}$$

which gives $x_{n+1} > d$.

So far we have proved that the sequence (x_n) is decreasing and bounded below by d, and so has a limit $d' \geq d$. But since f, f' are continuous, it follows from (3.16) that if $x_n \to d'$ then $d' = d' - f(d')/f'(d')$ and so $f(d') = 0$. Hence $d' = d$ since there is only one root in the interval. It remains to prove the estimate (3.18).

Taylor's theorem with $n = 2$ gives

$$0 = f(d) = f(x_n) + (d - x_n) f'(x_n) + \frac{(d - x_n)^2}{2} f''(v), \qquad d < v < x_n,$$

or, on rearranging,

$$\begin{aligned} x_{n+1} - d &= x_n - d - \frac{f(x_n)}{f'(x_n)} = \frac{(x_n - d)^2}{2} \frac{f''(v)}{f'(x_n)} \\ &< \frac{(x_n - d)^2}{2} \frac{f''(v)}{f'(v)} \leq k(x_n - d)^2, \end{aligned}$$

where we used the fact that f' is increasing and hence $f'(x_n) > f'(v)$. Thus $x_{n+1} - d \leq k(x_n - d)^2$, and the result follows by applying this repeatedly. ■

Example 3.33

Find k for $f(x) = x + x^5 - 3$ on $[1, 2]$.

We have $f''(x) / f'(x) = 20x^3 / (1 + 5x^4)$ which is positive and decreasing on $[1, 2]$ (its derivative is negative). Hence its maximum is at $x = 1$, and we have $k = 10/6$. Notice that this is much more efficient than calculating the maximum of f'' and the minimum of f' separately – this gives only $k \leq 10$. ♦

We continue our investigation of Newton's method, by showing how the process can actually determine the root d to a given accuracy. One rather direct method is to look at the sequence at the point where the rapid increase in agreement in digits becomes apparent, and truncate the value obtained, hoping in this way to find a bound for the root in the opposite direction to that given by the Newton iteration. For example in the calculations above for the root of $f(x) = x + x^5 - 3$, we could take $x = 1.13$ which has $f(x) < 0$, and so the root is now known to be in the interval $[1.13, x_n]$ for any of the calculated values of x_n, and we can use $x_n - 1.13$ in (3.18). However this is a little hit-and-miss, and a more systematic method is to combine Newton's method with a different method which gives an estimate in the opposite direction. An alternative method for finding a lower bound for the root is given in Exercise 14.

Suppose for simplicity that, as in the proof above, we have $f(a) < 0 < f(b)$ and both $f', f'' > 0$ on $[a, b]$. Then it follows easily from the Mean Value Theorem that on any subinterval, the graph of f lies below the chord joining the end points of the subinterval (the graph is concave upwards in the sense of Exercise 9 below); in particular, if $f(x) < 0 < f(y)$ then the root must be to the right of the point where the chord joining $(x, f(x))$ to $(y, f(y))$ meets the axis. (Used on its own, this forms the basis for the *secant method.*) We consider a new sequence (y_n) with $y_0 = a$ (in general y_0 is the opposite end of the interval from where we started the Newton iteration), and for $n \geq 0$, we define y_{n+1} by

$$y_{n+1} = \frac{y_n f(x_n) - x_n f(y_n)}{f(x_n) - f(y_n)} \tag{3.19}$$

so that y_{n+1} is the point where the chord from $(y_n, f(y_n))$ to $(x_n, f(x_n))$ meets the axis. The new sequence (y_n) approaches the root from the opposite side at almost the same rate as does (x_n), giving a sequence of intervals (y_n, x_n) each of which contains the root, and thus determining it to any required accuracy (within the limits of the calculating machinery employed!). We leave the proof of these statements as exercises, and return to consider the examples from this new viewpoint.

Consider for a change the example in which $f(x) = x - \cos x$ with $y_0 = a = 0$, $x_0 = b = \pi/2$. Then $y_1 = (af(b) - bf(a)) / (f(b) - f(a)) = \pi / (\pi + 2) = 0.611$. Continuing the scheme gives

$$\begin{array}{rr} y_1 = 0.611 & x_1 = 0.785 \\ y_2 = 0.7377 & x_2 = 0.7395 \\ y_3 = 0.7390849 & x_3 = 0.7390851 \\ y_4 = 0.739085133213 & x_4 = 0.739085133215 \end{array}$$

demonstrating that for instance the value of the root, which lies in the interval (y_4, x_4), is 0.7390851332 to 10 decimal places.

This discussion is summarised in the following theorem.

Theorem 3.34

Let $f, a, b, (x_n)$ be as in Theorem 3.32, let y_0 be the end point of $[a, b]$ opposite to x_0, and let (y_n) be given by (3.19). Then for all n, the root of the equation $f(x) = 0$ lies between y_n and x_n.

We have not yet considered how to find starting values. Once a root has been isolated (by finding a sign change of the function for instance) it is relatively easy to determine a subinterval on which f' and f'' have fixed signs, as in the above examples. The only theoretical difficulty is that the function might have a zero of f'' at the same point as the required zero of f. If $f' \neq 0$ at the root, then this serves only to accelerate the convergence further, since the graph is then approximately straight near the zero, and the tangent gives an even better approximation; the example of $f(x) = \sin x$ at $x = 0$ illustrates this. In the rare case in which both f' and f'' are zero at the root, then Newton's method can be slow to converge, and it is time to go back to simple iteration of f, as the example $f(x) = x - \sin x$ shows.

The last example of this section illustrates how, if we stray away from the intervals on which f', f'' have fixed sign, the behaviour of Newton's method can be highly irregular.

Example 3.35

Let $f(x) = x^3 - 7x + 6 = (x - 1)(x - 2)(x + 3)$.

The function has a (local) maximum at $a = -\sqrt{7/3} = -1.5275$ and a minimum at $b = \sqrt{7/3} = 1.5275$, and from the graph (Figure 3.4) it is clear that Newton's method will converge to 2 if $x_0 > b$, and to -3 if $x_0 < a$. The zero of f'' is at 0 so Theorem 3.32 gives convergence to 1 for $0 < x_0 < 1$. With care we can extend this region by finding the points $p = 1.47402$ and

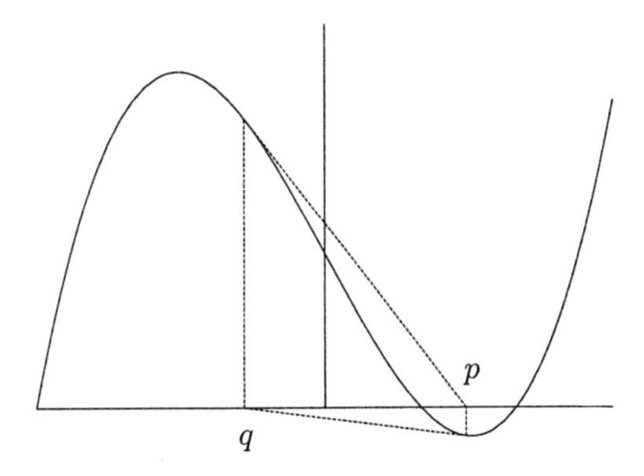

Figure 3.4 Example 3.35.

$q = -0.841071$ such that the tangent at $(p, f(p))$ passes through $(q, 0)$ and vice versa. This gives convergence (by looking at the graph!) to the root at 1 for $q < x_0 < p$. Outside these intervals, chaos reigns. For instance $x_0 = 1.48$ gives convergence to 2 and $x_0 = 1.49$ gives convergence to -3. For complex starting values the situation is even more complicated, and the regions for which convergence occurs are fractal sets. ♦

A final reassuring comment, is that Newton's method is 'self-correcting' (locally, at any rate) in the sense that an error of calculation is not fatal, provided it does not take the iteration outside the region of convergence – the effect is simply that of taking a new starting value.

3.6 Optimisation

One of the most interesting uses of the derivative is in finding the greatest and least values of functions which vary continuously; all functions in this section will be assumed continuous unless explicitly stated otherwise. The existence of such extreme values is guaranteed by Theorem 2.19 and its corollary, but those results do not give any method for their location. On the other hand, the proof of Rolle's theorem shows that at a maximum or minimum point, a function must, if differentiable, have derivative equal to zero. This motivates our first definition.

Definition 3.36

(i) A point at which a function is differentiable, and has derivative equal to zero is called a stationary point of the function.

(ii) A point which is either a stationary point, or a point where f is not differentiable, or an end point of the interval of definition, is called a critical point of the function.

Example 3.37

(i) $f(x) = x^3 - 3x$ has stationary points at $x = \pm 1$.

(ii) $f(x) = |x|$ for $-1 \leq x \leq 1$ has no stationary points. It has critical points at 0 and ± 1.

Notice that in example (ii) the function has a minimum at $x = 0$ where it is not differentiable, and maxima (relative to its domain $[-1, 1]$ of definition) at ± 1, the end points of the interval. ♦

This example illustrates how maximum or minimum points must lie in the set of critical points of the function; this is stated formally below as a theorem.

To clarify the discussion, we have to distinguish between those points at which a function takes its largest (or smallest) value on its domain, and those points at which the function takes its largest value relative to some subinterval of the domain of definition.

Definition 3.38

Let f be defined on an interval I, which may be open or closed, bounded or unbounded, and let $a \in I$. We say that f has a global maximum (relative to I) at a if $f(a) \geq f(x)$ for all $x \in I$. We say f has a local maximum (relative to I) at a if for some $\delta > 0$, $f(a) \geq f(x)$ for all $x \in I \cap (a - \delta, a + \delta)$. Global and local minima are defined similarly, with the direction of the inequality reversed. See Figure 3.5.

Example 3.39

Relative to the interval $[-1, 1]$ the function given by $f(x) = x^3 - 3x$ has a global maximum at -1 and a global minimum at 1. Relative to the interval $[-3, 3]$ it has local maxima/minima at $-1, +1$ respectively, and global maxima/minima at $+3, -3$ respectively. Relative to $\mathbb{R}$ it has local maxima/minima at $-1, +1$

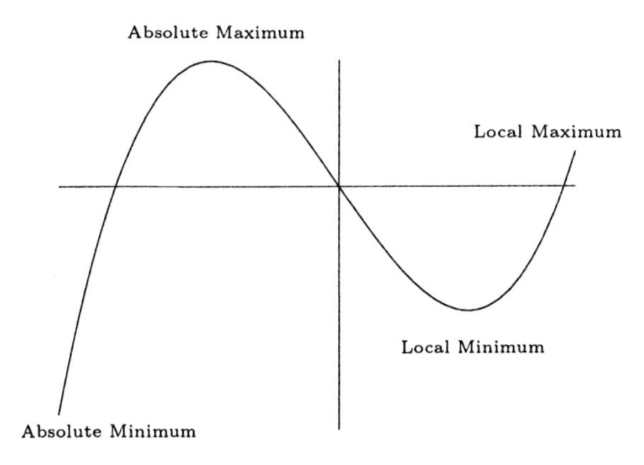

Figure 3.5 Maxima and minima.

respectively, and no global maxima/minima.

These statements are evident from the well-known shape of the graph, though we should really wait until after Theorem 3.41 to justify them properly. ♦

Theorem 3.40

Every local or global maximum or minimum point of a continuous function must lie in the set of critical points.

Proof

Proposition 3.7 shows that if f is differentiable at a point c which is not an end point of the interval of definition, then f does not have a maximum or minimum point at c. ■

The key to distinguishing these different kinds of behaviour comes from Corollary 3.12 which says that if a function f has $f'(x) > 0$ on an interval then f is strictly increasing there; similarly if the derivative is negative then it is strictly decreasing. The gives the most useful test for classifying critical points.

Theorem 3.41 (First Derivative Test)

Let f be continuous on an interval I.

(i) Let a be an interior point of I so that for some $\delta > 0$, $(a - \delta, a + \delta) \subset I$ and suppose that $f'(x) > 0$ on $(a - \delta, a)$ and $f'(x) < 0$ on $(a, a + \delta)$ Then f has a local maximum at a. Similarly if $f'(x) < 0$ on $(a - \delta, a)$ and $f'(x) > 0$ on $(a, a + \delta)$ then f has a local minimum at a.

(ii) If a is an end point of the interval, say the left, then if $f'(x) > 0$ on $(a, a+\delta)$ then f has a local minimum at a; while if $f'(x) < 0$ on $(a, a + \delta)$ then f has a local maximum at a (and there is an obviously similar statement at the right end point).

Notice that the theorem requires nothing of the function at a except continuity, and thus it applies equally to stationary points where f' is zero and to critical points where it is undefined.

Proof

The sign of f' implies that f is increasing on $(a - \delta, a)$ and decreasing on $(a, a + \delta)$ and hence by continuity there is a local maximum at a. The other statements are similar. ■

The results of Example 3.39 can now be justified by observing that $f'(x) = 3(x^2 - 1)$ is positive for $x > 1$ or $x < -1$, and negative for $-1 < x < 1$.

Example 3.42

(i) Let $f(x) = x^3(x + 4)$. Then f has a stationary point at 0 and a global minimum at -3.

(ii) Let $f(x) = (x - 2)/(x^2 + 5)$. Then f has a global maximum at -1 and a global minimum at 5.

(i) If $f(x) = x^3(x + 4) = x^4 + 4x^3$ then $f'(x) = 4x^3 + 12x^2 = 4x^2(x + 3)$. Thus f' is negative on $(-\infty, -3)$, and positive on $(-3, \infty)$ except at $x = 0$, so f is decreasing on $(-\infty, -3)$ and increasing on $(-3, \infty)$. It follows that $x = -3$ gives a global minimum and that $x = 0$ gives a stationary point in the neighbourhood of which f is strictly increasing and so is neither a maximum nor a minimum point.

(ii) With $f(x) = (x - 2)/(x^2 + 5)$, f is positive for $x > 2$, negative for $x < 2$, and tends to zero as $x \to \pm\infty$. The derivative is

$$f'(x) = \frac{(x^2 + 5) - 2x(x - 2)}{(x^2 + 5)^2} = \frac{(5 - x)(1 + x)}{(x^2 + 5)^2}$$

so f is decreasing on $(-\infty, -1)$ and $(5, \infty)$, and increasing on $(-1, 5)$. Thus -1 is a local minimum and $x = 5$ is a local maximum, and since $f(-1) < 0 = f(\pm\infty) < f(5)$ these are also global extreme values. (The superiority of the first derivative test, Theorem 3.41 above, to the second derivative test below is plain from this example, where calculating the second derivative is both messy and unnecessary.) ♦

Part (i) of this example shows that stationary points need not be maxima or minima. To investigate this a little further, we make the following definition.

Definition 3.43

A point at which f has a local maximum or minimum of its derivative is called a point of inflexion.

Applying the first derivative test to f' we see that f has a point of inflexion at a when the second derivative (if it exists) changes sign at a. Looking back at Example 3.42(i) this shows that $f(x) = x^4 + 4x^3$ has $f''(x) = 12x(x+2)$ and so has points of inflection at $x = 0$ (a minimum of f') which is the stationary point already found, and at $x = -2$ (a maximum of f') which is not a stationary point of f.

Consideration of the sign of f'' gives another familiar test for extreme values.

Theorem 3.44 (Second Derivative Test)

Let f have a stationary point at a, and suppose f'' exists at a. Then if $f''(a) > 0$ the stationary point is a minimum, while if $f''(a) < 0$ it is a maximum. No conclusion is possible if $f''(a) = 0$.

This is the test which most beginners use to tell whether a given stationary point is a maximum or minimum. It suffers from two disadvantages, (a) it requires that the second derivative exists, and is found (or at least the sign is determined) and (b) it gives no information in case $f''(a) = 0$ (though this is another place where generations of students get it wrong: saying that $f'' = 0$ determines a point of inflection is the commonest of errors, despite the example of $f(x) = x^4$ which has a minimum at $x = 0$). For a proof we leave it to the reader to combine Proposition 3.7 (applied to f') with the first derivative test. In fact calculation of second derivatives is unnecessary, as Example 3.42(ii) shows, unless all (non-stationary) points of inflexion are specifically required.

The kind of geometrical questions to which calculus techniques can be applied are illustrated in the next examples.

Example 3.45

(i) Find the shape of a cylindrical can which encloses the maximum volume for a given surface area.

(ii) Find the shape of an open-topped cone which encloses the maximum volume for a given surface area.

(i) The volume of a cylinder with base radius r and height h is given by $V = \pi r^2 h$ and its surface area is $A = 2\pi r^2 + 2\pi rh$. We want the maximum volume for a given surface area; equivalently we shall find the minimum surface area for a given volume. Suppose then that V is fixed, so that $h = V/\pi r^2$ and $A = 2\pi r^2 + 2V/r$. Then

$$\frac{dA}{dr} = 4\pi r - \frac{2V}{r^2}$$

which is negative when $r^3 < V/2\pi$ and positive when $r^3 > V/2\pi$. Thus A has a minimum when $r^3 = V/2\pi = r^2h/2$, so $h = 2r = (2V/\pi)^{1/3}$ and the height is equal to the diameter.

(A visit to the local supermarket will soon show that most cans have height greater than their diameter. A possible explanation for this is that the circular ends for the can are punched from steel plate in a square pattern, and the above calculation takes no account of the waste involved. Show that if this is taken into account (i.e. the ends are regarded as squares of side $2r$ from which the corners are rounded off), the optimal ratio of height to diameter is $4/\pi$, which is a little closer to observed values. All kinds of variation on this argument are possible and plausible.)

(ii) Let the cone have height h and slant length l so that $h = l\cos\theta$ where θ is the angle between the side and the axis of the cone. When opened out, the cone forms the sector of a circle with angle $\phi = 2\pi\sin\theta$. Thus the volume is $V = \pi r^2 h/3 = \pi (l\sin\theta)^2 l\cos\theta/3$, and its surface area is $A = \phi l^2/2 = \pi\sin\theta l^2$. As before we minimise A for given V. Elimination of l gives

$$\begin{aligned} A^3 &= (\pi\sin\theta)^3\left(\frac{3V}{\pi\sin^2\theta\cos\theta}\right)^2 \\ &= \frac{9\pi V^2}{\sin\theta\cos^2\theta} \end{aligned}$$

which is easily seen to be a minimum when $\tan\theta = 1/\sqrt{2}$. This gives a value of θ of around 35° which is not too far from observed values, though further research into the consumption of ice cream is obviously needed. ♦

EXERCISES

1. At what points are the functions given by (i) $\sqrt{|x|}$, (ii) $|\sin x|$, (iii) $\cos(\sqrt{x})$, (iv) $\sqrt{1-x^2}$ (a) continuous, (b) differentiable?

2. Show that although a derivative may not be continuous, it must still have the intermediate value property; in other words, if f is differentiable on (a,b), $c,d \in (a,b)$ and t lies between $f'(c)$ and $f'(d)$ then there is some $e \in (c,d)$ with $f'(e) = t$.

3. (i) Show that if f is differentiable at t and if $x < t < y$ then

$$\frac{f(y) - f(x)}{y - x} \to f'(t) \qquad \text{as } x, y \to t.$$

   (ii) Given $f(x) = x^2 \cos(1/x)$, find (x_n) with $0 < x_{n+1} < x_n$ and $x_n \to 0$ as $n \to \infty$, such that $(f(x_n) - f(x_{n+1}))/(x_n - x_{n+1})$ has no limit as $n \to \infty$. This shows that the result of (i) fails without the condition $x < t < y$.

4. (i) Show that if f is continuous on $[a,b]$, differentiable on (a,b) and $f'(x) \to l$ as $x \to a+$ then f is differentiable on the right at a and $f'_+(a) = l$. (In other words, $f'(a+) = f'_+(a)$.)

   (ii) Show that if f is continuous on $[a,b]$ and f'_+ exists and satisfies $|f'_+(x)| \le m$ on $[a,b)$ then $|f(b) - f(a)| \le m(b-a)$.

   (iii) Use (ii) to show that (i) is still true if only $f'_+(x) \to l$ as $x \to a+$ is assumed.

5. (i) Find the points (if any) on the parabola $y = x^2$ at which the tangent passes through (a) $(0,-1)$, (b) $(0,1)$.

   (ii) Find the points (if any) on the parabola $y = x^2$ at which the normal (the line at right angles to the tangent) passes through (a) $(0,-1)$, (b) $(0,1/2)$, (c) $(0,4)$.

   (iii) Find the point (or points) on the curve $y = e^x$ at which the tangent passes through the origin.

6. Show that Rolle's theorem applies to $f(x) = x^\alpha (1-x)^\beta$ on $[0,1]$ whenever both $\alpha, \beta > 0$.

7. Let c be the value given by the Mean Value Theorem

$$f'(c) = \frac{f(b) - f(a)}{b - a}$$

   on $[a,b]$ when $f(x) = x^2$. Find $\lim_{b \to a} (c-a)/(b-a)$. Repeat with $f(x) = x^n$ for some other values of n.

8. Show that if $f(x) \to \infty$ as $x \to a+$ then f' (if it exists) cannot be bounded for $x > a$.

9.* Convexity: Suppose that f is defined on (a, b) and satisfies, for all $x, y \in (a, b)$ and $t \in (0, 1)$,

$$f(tx + (1-t)y) \leq tf(x) + (1-t)f(y). \tag{3.20}$$

(i) Show that the condition implies that for any $x, y \in (a, b)$, (a) the graph of the function on (x, y) is below the chord joining $(x, f(x))$ to $(y, f(y))$, and (b) if $x < t < y$ then the graph of the function on (x, t) is above the (extended) chord joining $(t, f(t))$ to $(y, f(y))$, and similarly on (t, y).

(ii) Deduce that f must be continuous on (a, b) and that f has right- and left-hand derivatives everywhere which are increasing on (a, b).

(iii) Show that if $x < y$ then $f'_+(x) \leq f'_-(y)$, and hence that f is differentiable except for at most a countable number of points.

(iv) Show that conversely if f is differentiable and f' is increasing on (a, b) then f is convex on (a, b).

10.* (i) Show that if f is continuous on (a, b) and $f'_+ > 0$ on $(a, b) - D$, where D is a countable subset of (a, b), then f is strictly increasing on (a, b).

(ii) Deduce that if f is continuous on (a, b) and $f'_+ \geq 0$ on $(a, b) - D$ then f is increasing on (a, b), and hence that if f is continuous on (a, b) and $f'_+ = 0$ on $(a, b) - D$ then f is constant on (a, b).

11. Let $f(x) = x^3 - 13x^2 + 40x$. Show that f has a local minimum at $20/3$ and a local maximum at 2. Modify this example (or give another) to find a polynomial of degree 3 such that both it and its derivative have distinct integer roots.

12.** Find a polynomial f of degree 4 such that f, f', f'' all have distinct integer roots.

13. Justify the statement made in the discussion preceding Theorem 3.34, that the sequence given by (3.19) increases to the required root of the equation at the same rate as (x_n).

14.* Show that if f, (x_n), (y_n) are as in Theorem 3.34, then $x_{n+1} - d \leq f'(x_n)(x_n - x_{n+1}) / f'(y_n)$. This gives an alternative to Theorem 3.34 for finding an estimate of the error at the nth stage in Newton's method. Apply this to some of the numerical exercises in the text.

15. Give an example of a sequence (f_n) of continuous functions with $f_n(x) \to 0$ for each $x \in [0,1]$ but for which m_n, the least upper bound of f_n, does not tend to zero. Compare Exercise 27, Chapter 6.

16. Find and classify the stationary points of $f(x) = x^2(x-5)(x-9)$.

17. Show that if a, b have the same sign then $x^a(1-x)^b$ has a global maximum or minimum at $x = a/(a+b)$ depending on whether a, b are both positive or both negative. Show that if a, b have opposite sign the function is monotonic on $(0,1)$.

18. Locate the positions of the stationary and inflexion points of $(\sin x)/x$, in relation to $(n+1/2)\pi$, $n\pi$, which are the stationary and inflexion points of $\sin x$.

19. Formulate (and prove) a version of Theorem 3.27 which holds when $f, g \to \infty$ as $x \to \infty$.

20.* The limit $\lim_{x\to\infty}(f(x+h) - f(x))/h$, if it exists and is independent of h, could be called the 'derivative at infinity' of f. Find some (linear and non-linear) examples for which the limit exists. To what extent is the name 'derivative at infinity' appropriate?

21. Let $g = f_1 f_2 \cdots f_n$ where each f_j is a differentiable function. Show that g is differentiable, and

$$\frac{g'}{g} = \frac{f_1'}{f_1} + \frac{f_2'}{f_2} + \cdots + \frac{f_n'}{f_n}$$

at all points where $g \neq 0$.

4
Constructive Integration

In a first course in calculus, the integral is usually introduced as an 'anti-derivative', i.e. given f, we look for a function F such that $F' = f$. This works well enough when f is given by a simple formula, but the class of functions for which it succeeds is difficult to identify – who knows whether there is a function F with $F'(x) = \sin x/\sqrt{1-x^7}$?

A more satisfactory approach is to define at the outset the class of functions to be integrated and to construct the integral for such functions. Then the process is related to differentiation at a later stage. The class of functions may be chosen in many ways – the class of continuous functions gives one obvious example. We choose the wider class of regulated functions; these were introduced in Section 2.7 as those which have both right- and left- hand limits at each point of an interval, though the limits need not be equal either to each other or to the value of the function at the point (Definition 2.30(iii)). This class is sufficient for our needs since it contains all continuous and all monotone functions, and is closed under uniform convergence (Section 6.6 below).

This constructive approach ensures that the integral of a given function on an interval has a definite value, and allows us to define new functions (for instance the logarithm) by integration of existing ones. The relation with differentiation is established in Section 4.3, where all the familiar rules and methods are established.

The familiar intuitive notion of a definite integral is that for a given function f on an interval $[a, b]$, the integral measures the area between the graph of the function and the x-axis. In particular we should require that if the function is constant, say $f(x) = k$ for all $x \in [a, b]$, then $I(f) = k(b-a)$ since this is simply the elementary formula for the area of a rectangle. In addition, denoting the integral of f (temporarily) by $I(f)$, we should have

(i) $I(f) \geq 0$ when $f \geq 0$ (positivity)

(ii) $I(f+g) = I(f) + I(g)$ (linearity), and

(iii) $I(cf) = cI(f)$, for $c \in \mathbb{R}$ (scalar multiplicativity).

Theorem 4.12 below shows that these are correct.

In this chapter all integrals will be over a finite interval $[a,b] \subset \mathbb{R}$; for integration over unbounded intervals see Chapter 5.

4.1 Step Functions

We decided in the introduction that the integral of a constant function $f(x) = k$ for all $x \in [a,b]$ is $k(b-a)$. This unsurprising fact will be the basis for our method of integration – all we have to do is to show how more general functions can be approximated by functions which are constant on subintervals. Such locally constant functions are called step functions – the formal definition follows. Write χ_A for the characteristic function of a set A, i.e. the function which is 1 at points of A, 0 elsewhere.

Definition 4.1

(i) A finite set $D = \{x_j\}_0^n = \{x_0, x_1, \ldots, x_j, \ldots, x_{n-1}, x_n\}$ is called a dissection of $[a,b]$ if

$$a = x_0 < x_1 < \ldots < x_j < \ldots < x_{n-1} < x_n = b.$$

If D, E are two dissections with $D \subset E$ so that every point of D is also a point of E, we say E is a refinement of D.

(ii) A function which is defined on $[a,b]$ and constant on each interval (x_{j-1}, x_j) of a dissection is called a step function on $[a,b]$. The values of the function at the points x_j may be any real numbers. If $f(x) = c_j$ on (x_{j-1}, x_j), we shall write

$$s = \sum_{j=1}^{n} c_j \chi_{(x_{j-1}, x_j)}, \tag{4.1}$$

ignoring the values at x_j.

(iii) The integral of a step function of the form (4.1) is

$$I(s) = \sum_{j=1}^{n} c_j (x_j - x_{j-1}). \tag{4.2}$$

If it is necessary (as in Theorem 4.5(i) below) to draw attention to the interval $[a, b]$ over which the integral is taken, we shall write $I(s)$ in the extended form

$$I_a^b(s)$$

which corresponds to the conventional notation for integrals (Figure 4.1).

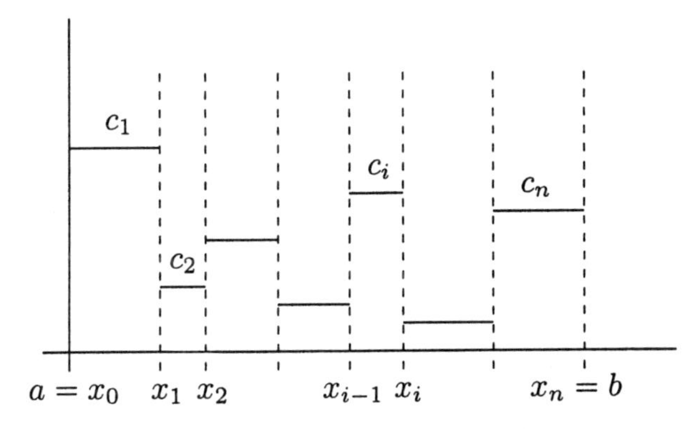

Figure 4.1 Step function. From *Handbook of Applicable Mathematics*, Volume 4: Analysis, W. Ledermann and S. Vajda (Eds), 1982. ©John Wiley & Sons Limited. Reproduced with permission.

Example 4.2

Let $s(x) = 1$ on $(0,1)$, 3 on $(1,3)$ and -2 on $(3,4)$. Then $s = 1\chi_{(0,1)} + 3\chi_{(1,3)} - 2\chi_{(3,4)}$ and $I_0^4(s) = 1.1 + 3.2 - 2.1 = 5$. ♦

The definition calls for several comments. Note again that we are not interested in the value of the step function at x_j; this may be any real number and plays no part in defining $I(s)$. In particular we do *not* require that s is discontinuous at each x_j, although any discontinuities of s must be included among the x_j. It follows that a step function can be written in more than one way – for instance $\chi_{(0,1)} = \chi_{(0,1/2)} + \chi_{(1/2,1)}$ (except at $1/2$). Consequently we must check that the different ways of writing a step function in the form

(4.1) have no effect on the sum given by (4.2), otherwise the definition would be faulty. The next result shows this.

Proposition 4.3

Given two step functions, say

$$s = \sum_{j=1}^{n} c_j \chi_{(s_{j-1}, s_j)}, \text{ and } t = \sum_{k=1}^{p} d_k \chi_{(t_{k-1} t_k)},$$

which are equal except possibly at s_j or t_k, then $I(s) = I(t)$, i.e.

$$\sum_{j=1}^{n} c_j (s_j - s_{j-1}) = \sum_{k=1}^{p} d_k (t_k - t_{k-1}).$$

Proof

Suppose first that $\{t_k\}$ is a refinement of $\{s_j\}$, i.e. each s_j belongs to $\{t_k\}$. If (s_{j-1}, s_j) is some interval of S then the value c_j of s will equal the values of t on the corresponding intervals of T; more precisely, if say $s_{j-1} = t_{q-1}$ and $s_j = t_r$, then $c_j = d_q = \cdots = d_r$ and consequently $c_j (s_j - s_{j-1}) = \sum_{k=q}^{r} d_k (t_k - t_{k-1})$. Adding these terms over all values of j gives $I(s) = I(t)$ in this case. In general when neither dissection is a refinement of the other, consider the common refinement obtained by taking the union of the dissections. This produces a new step function u say with $I(s) = I(u)$ and $I(u) = I(t)$ by what we just proved, and the result follows. ■

Example 4.4

We already noticed that $\chi_{(0,1)} = \chi_{(0,1/2)} + \chi_{(1/2,1)}$, and in accordance with the proposition, $1(1-0) = 1(1-1/2) + 1(1/2-0)$. ♦

The next result is a necessary technicality which shows that the properties listed in the introduction are satisfied, at least for step functions.

Theorem 4.5

Let s, t be step functions on $[a, b]$ and let $|s(x)| \leq M$ on $[a, b]$. Then

(i) for $a < c < b$, $I_a^b(s) = I_a^c(s) + I_c^b(s)$.

(ii) $I(s \pm t) = I(s) \pm I(t)$ and $I(ks) = kI(s)$ for any real number k.

(iii) $|I(s)| \leq M(b-a)$ and $|I(s.t)| \leq MI(|t|)$.

(iv) If $s(x) \geq 0$ on $[a,b]$ then $I(s) \geq 0$.

(v) If $s(x) \leq t(x)$ on $[a,b]$ then $I(s) \leq I(t)$.

Proof

For (i) simply introduce c (if necessary) as a new point in the dissection determined by s, and use Proposition 4.3.

For (ii), notice that the result is immediate if s, t are constant on $[a,b]$. In the general case, consider the common refinement which was introduced in the proof of Proposition 4.3, and observe that s, t are constant on each of its subintervals. The proofs of the second statement, and of (iii) and (iv) are left as exercises. For (v), combine (ii) and (iv). ■

Notice that these results would be quite messy to prove if we required that each step function was discontinuous only at points of the dissection, since for instance a discontinuity of f, g need not be a discontinuity of $f+g$. (The reader may also have noticed that we have assumed such results as 'if f is a step function on $[a,b]$ then it is a step function on any subinterval of $[a,b]$', and 'if f, g are step functions on $[a,b]$, then so are $f \pm g$ and kf for any real k' – these are obvious as soon as they have been recognised.)

4.2 The Integral of a Regulated Function

Now that we have defined the integral for step functions we can extend the definition to regulated functions since for any regulated function there is a step function nearby. More precisely

Theorem 4.6

Given a regulated function f on $[a,b]$ and any $\varepsilon > 0$, there is a step function s such that

$$|f(x) - s(x)| < \varepsilon$$

for all $x \in [a,b]$. If $f(x) \geq 0$ on $[a,b]$ then we can take $s(x) \geq 0$.

Proof

We construct the required step function by starting from a and moving to the right. Consider the set

$$E = \{t : \text{there is some step function } s \text{ with } |f(x) - s(x)| < \varepsilon \text{ for all } x \in [a, t]\}.$$

(We include in the definition of E that $s \geq 0$ if $f \geq 0$ on $[a, t]$.)

The proof consists of showing the following facts about E: (i) $a \in E$, so it is non-empty, (ii) E is a closed interval, $E = [a, d]$ say, and (iii) $d = b$ so that $E = [a, b]$ as required.

For (i) define $s(a) = f(a)$; this gives the required function on $[a, a]$.

For (ii) notice that if $t \in E$ then a suitable function s exists on $[a, t]$. Then if $a \leq t' < t$, the restriction of s to $[a, t']$ shows that $t' \in E$, so E must be an interval, with either $E = [a, d]$ or $E = [a, d)$ where d is the least upper bound of E. We show that the second possibility does not occur.

If $d = a$ (which may still be the case at this stage) then $E = [a, a]$ is immediate. If $d > a$ then f has a left-hand limit, l_1 say, at d, so that for some $0 < \delta < d - a$, $|f(x) - l_1| < \varepsilon$ on $(d - \delta, d)$. But since E is an interval, $d - \delta \in E$ and so there is a step function s_1 of the required form on $[a, d - \delta]$. If we define a new step function s_2 equal to s_1 on $[a, d - \delta]$, to l_1 on $(d - \delta, d)$ and to $f(d)$ at d, then $|s_2 - f| < \varepsilon$ on $[a, d]$ and so $d \in E$ as required for (ii).

Finally for (iii), suppose that $d < b$ (including $d = a$ which is still possible!). Then f has a right-hand limit l_2 say at d, so that there is $\delta' > 0$ such that $|f(x) - l_2| < \varepsilon$ on $(d, d + \delta')$. Define s_3 equal to s_2 on $[a, d]$, to l_2 on $(d, d + \delta')$, and to $f(d + \delta')$ at $d + \delta'$. This gives a step function of the required form on $[a, d + \delta']$ which contradicts the definition of d as the least upper bound of E, and (iii) is proved.

Notice that the construction uses only values (and limiting values) of f, so that if $f \geq 0$ then $s \geq 0$ also. ■

Example 4.7

(i) Let $f(x) = x^2$ on $[0, 1]$ and let $\varepsilon > 0$ be given. A step function of the form required by the theorem is given for $j = 1, 2, \ldots$ by

$$s(x) = \begin{cases} 0 & \text{for } 0 \leq x^2 < \varepsilon, \\ \varepsilon & \text{for } \varepsilon \leq x^2 < 2\varepsilon, \\ \cdots & \\ j\varepsilon & \text{for } j\varepsilon \leq x^2 < (j+1)\varepsilon, \end{cases}$$

until $x = 1$ is reached. More generally for any increasing function, a step function of the form required by the theorem is given by $s(x) = j\varepsilon$ when $j\varepsilon \leq f(x) < (j+1)\varepsilon$.

(ii) Example 2.1(ix) in which $f(p/q) = 1/q$ and otherwise zero, was found to be regulated in Example 2.32. A step function of the required form is given by $s(x) = f(x)$ if $f(x) > \varepsilon$, otherwise $s(x) = 0$. ♦

Given Theorem 4.6, it is natural to use step functions to define the integral of f. Informally $I(f)$, the integral of f over $[a, b]$, is defined to be the limit of $I(s_n)$ for a sequence of step functions which approximate f with increasing accuracy; however we need to do some work to see how the idea works out in detail. The step function s which is constructed in Theorem 4.6 has the property that its graph is close to that of f with an error of less than ε *at all points of the interval.* This motivates the following definition.

Definition 4.8

A sequence (f_n) of functions such that for each $\varepsilon > 0$ there is some integer N with $|f_n(x) - f(x)| < \varepsilon$ for all $x \in [a, b]$ and $n \geq N$ is said to be uniformly convergent with limit f on $[a, b]$.

We shall consider this mode of convergence in more detail in Section 6.6. The next theorem shows that the integral of a regulated function can be defined as the limit of the integrals of a uniformly convergent sequence of step functions whose limit is f.

Theorem 4.9

Given a regulated function f, choose a sequence (ε_n) with $\varepsilon_n \to 0$ as $n \to \infty$, and for each n, use Theorem 4.6 to choose a step function s_n with $|f(x) - s_n(x)| < \varepsilon_n$ for all $x \in [a, b]$. Then the sequence $(I(s_n))$ is a Cauchy sequence whose limit is independent of the choices of ε_n, s_n.

Proof

Given $\varepsilon > 0$, there is some N with $\varepsilon_n < \varepsilon$ if $n \geq N$ and so $|f(x) - s_n(x)| < \varepsilon$ for $n \geq N$ and all $x \in [a, b]$. It follows that if $m, n \geq N$ then $|s_m(x) - s_n(x)| \leq |s_m(x) - f(x)| + |f(x) - s_n(x)| < 2\varepsilon$ for all $x \in [a, b]$. Hence

$$|I(s_m) - I(s_n)| \leq 2\varepsilon(b - a) \qquad \text{for } m, n \geq N$$

by Theorem 4.5 (iii) and $(I(s_n))$ is a Cauchy sequence as required.

To show that the limit is independent of the choice of ε_n, s_n, suppose also that $\delta_n \to 0$ and let (t_n) be a sequence of step functions with $|f(x) - t_n(x)| <$

δ_n. Form a new sequence (u_n) by interlacing (s_n) and (t_n), i.e.

$$(u_1, u_2, u_3, u_4, \ldots, u_{2j-1}, u_{2j}, \ldots) = (s_1, t_1, s_2, t_2, \ldots, s_j, t_j, \ldots).$$

Then again $u_n(x) \to f(x)$ uniformly and the argument above shows that $(I(u_n))$ is a Cauchy sequence. Then the limit of $(I(u_n))$ is the same as the limits of its two subsequences $(I(s_n))$ and $(I(t_n))$ which are therefore equal as required. ■

This hard work enables us to make the following definition.

Definition 4.10

Given a regulated function f, its integral $I(f)$ is the limit of $I(s_n)$ where (s_n) is any sequence of step functions which converges uniformly to f on $[a, b]$. We write $I(f)$ in any of the forms

$$I(f) = I_a^b(f) = \int_a^b f = \int_a^b f(x)\, dx,$$

with the understanding that in the last, x is a variable which may be replaced by any other (except a, b or f). The function f to be integrated is referred to as the integrand.

Example 4.11

(i) Let $f(x) = c$ on $[a, b]$. Then $I(f) = c(b-a)$.

(ii) Let

$$f(x) = \begin{cases} 0 & \text{if } x = 0, \\ 1/2^n & \text{if } 1/2^n \le x < 1/2^{n-1} \quad \text{for} \quad n = 1, 2, 3, \ldots, \\ 1 & \text{if } x = 1. \end{cases}$$

which is Example 2.1(vii). Then $I(f) = 1/3$.

(iii) Example 2.1(ix) in which $f(p/q) = 1/q$ and otherwise zero, has a step function of the required form given by $s(x) = f(x)$ if $f(x) > \varepsilon$, otherwise $s(x) = 0$. Thus since s is zero except at a finite number of points, both $I(s)$ and $I(f)$ are zero.

For (i), take $s(x) = c$ on $[a, b]$. For (ii), given $\varepsilon_n = 1/2^n$, define a step function s_n by $s_n(x) = f(x)$ if $f(x) > \varepsilon_n$, $s_n(x) = 0$ otherwise. Then

$$\begin{aligned} I(s_n) &= \frac{1}{2}\cdot\frac{1}{2} + \frac{1}{4}\cdot\frac{1}{4} + \cdots + \frac{1}{2^{n-1}}\cdot\frac{1}{2^{n-1}} \\ &= \frac{1}{4} + \frac{1}{4^2} + \cdots + \frac{1}{4^{n-1}} \\ &= \frac{1}{3}\left(1 - \frac{1}{4^{n-1}}\right) \to \frac{1}{3} \text{ as } n \to \infty \end{aligned}$$

so $I(f) = \int_0^1 f = 1/3$. ♦

The last result of this section confirms that the definition of the integral has the required properties of linearity and monotonicity, as predicted.

Theorem 4.12

Let f, g be regulated functions on $[a, b]$ and let $|f|$ be bounded by M. Then:

(i) For $a < c < b$, $I_a^b(f) = I_a^c(f) + I_c^b(f)$.

(ii) $I(f \pm g) = I(f) \pm I(g)$ and $I(kf) = kI(f)$ for any real number k.

(iii) $|I(f)| \leq M(b-a)$ and $|I(f.g)| \leq MI(|g|)$.

(iv) If $f \geq 0$ on $[a, b]$ then $I(f) \geq 0$.

(v) If $f \leq g$ on $[a, b]$ then $I(f) \leq I(g)$.

Proof

These follow at once from the corresponding results for step functions (Theorem 4.5) by taking limits. For (iv) recall that the construction in Theorem 4.6 gives a positive step function to approximate a positive regulated function. ■

4.3 Integration and Differentiation

The construction of the integral as a limit of integrals of step functions is of limited use in finding its value. In practice what is needed is what is known as the 'Fundamental Theorem of Calculus,' in which the value of an integral is found in terms of an anti-derivative, or indefinite integral. We begin with the reverse situation; the derivative of an integral is the function integrated.

Theorem 4.13

Let f be regulated on $[a,b]$. Then $F(x)=\int_a^x f$ is continuous on $[a,b]$. It has a right derivative at all points of $[a,b)$ with $F'_+(x)=f(x+)$, and a left derivative at all points of $(a,b]$ with $F'_-(x)=f(x-)$. In particular F is differentiable at every point of continuity of f with $F'(x)=f(x)$.

Proof

The continuity follows since $|f|$ is bounded, say by m, (Exercise 6, Chapter 2), and so

$$|F(x+h)-F(x)|=\left|\int_x^{x+h} f\right|\le m|h|.$$

Let x be any point of $[a,b)$. Then f has a right-hand limit, $l=f(x+)$, at x and so for any $\varepsilon>0$, there is a $\delta>0$ such that $l-\varepsilon<f(u)<l+\varepsilon$ for $x<u<x+\delta$. To show that F has a right derivative at x, we take $u>x$ and consider

$$\frac{F(u)-F(x)}{u-x}=\frac{1}{u-x}\left(\int_a^u f-\int_a^x f\right)=\frac{1}{u-x}\int_x^u f$$

and so

$$\left|\frac{F(u)-F(x)}{u-x}-l\right|=\left|\frac{1}{u-x}\int_x^u (f-l)\right|\le\frac{|(u-x)\,\varepsilon|}{|u-x|}=\varepsilon$$

for $x<u<x+\delta$. Thus $F'_+(x)$ exists and is equal to l as required. (Notice the crucial role in the proof played by part (i) of Theorem 4.12.) The argument for left-hand limits is similar, and the result at points of continuity follows by combining the results on the right and the left. ■

Example 4.14

Let $F(x)=\int_0^x(2+\sin t)^{1/5}dt$. Then F is differentiable for all x, with $F'(x)=(2+\sin x)^{1/5}$.

All that is required here is to notice that the integrand is continuous for all x. ♦

The expected 'Fundamental Theorem of Calculus' follows from this.

Theorem 4.15 (Fundamental Theorem of Calculus)

Let f be differentiable on $[a, b]$ with a continuous derivative. Then

$$\int_a^b f' = f(b) - f(a).$$

Proof

Since f' is continuous we know that $\phi(x) = \int_a^x f'$ exists, and from the previous result we have $\phi'(x) = f'(x)$ for all $x \in (a, b)$. Hence ϕ and f can differ at most by a constant (this is the final statement in Corollary 3.12), say $\phi(x) - f(x) = c$ on (a, b), and this is true also at a, b since f, ϕ are continuous. Putting $x = a$ gives $c = \phi(a) - f(a) = -f(a)$ and so $\phi(x) - f(x) = -f(a)$ for all $x \in [a, b]$. In particular $\phi(b) = \int_a^b f' = f(b) - f(a)$ which is what we wanted. ■

The obvious question here is 'why not suppose only that f' is regulated?' The answer is that a regulated derivative is necessarily continuous – this is Exercise 2. On the other hand if we only consider one-sided derivatives the situation again becomes interesting (Exercise 6).

Once we have the Fundamental Theorem of Calculus, all the familiar results and rules for the calculus follow.

Example 4.16

(i) For any integer $n \geq 0$, $\int_0^a x^n dx = a^{n+1}/(n+1)$.

(ii) $\int_0^a (x^2+1)^{-3/2} dx = a(a^2+1)^{-1/2}$.

(iii) $\int_0^a xe^{x^2} dx = \left(e^{a^2} - 1\right)/2$.

These are all immediate on calculating the derivatives of the functions on the right (and perhaps recalling Example 3.17(ii)). Part (iii) reminds us not to commit the elementary error of automatically putting the integrated function equal to zero when $x = 0$. ♦

For examples which cannot be integrated simply by inspection, there are the familiar rules for integration by parts and by substitution. However we have to remember that there is no guarantee of success. Each method works for certain types of integrand, but there will always be others, for instance the function in Example 4.14 above, for which no elementary method will give the exact value of the integral.

Theorem 4.17 (Integration by Parts)

Let f, g have continuous derivatives on $[a, b]$. Then

$$\int_a^b fg' = f(b)\,g(b) - f(a)\,g(a) - \int_a^b f'g.$$

Proof

This is just a disguised form of the product rule for derivatives from Theorem 3.14. For since $(f.g)' = f'g + fg'$ it follows that

$$\int_a^b fg' + \int_a^b f'g = \int_a^b (fg)' = f(b)\,g(b) - f(a)\,g(a) \tag{4.3}$$

as required. ■

The indefinite integral $F = \int f$ is a convenient way of writing an anti-derivative when $f = F'$ (but notice that $\int f$ is defined only up to an additive constant). It is useful in calculation when we are not interested in the limits of integration. Another useful notational convention is that in which we attach limits to square brackets: $[f]_a^b = f(b) - f(a)$.

Example 4.18

(i) $\int_0^\pi x \sin x \, dx = \pi$.

(ii) $\int e^{ax} \sin(bx)\, dx = e^{ax} (a \sin(bx) - b \cos(bx)) / (a^2 + b^2)$,
$\int e^{ax} \cos(bx)\, dx = e^{ax} (a \cos(bx) + b \sin(bx)) / (a^2 + b^2)$.

(iii) $\int_1^a x^n \ln x \, dx = \left((n+1)\, a^{n+1} \ln a - (a^{n+1} - 1)\right) / (n+1)^2$,
$\int_1^a \ln x \, dx = a \ln a - a + 1$.

When doing integration by parts, the first thing to decide is which of the functions is to play the role of f (i.e. is to be differentiated in (4.3)), and which is g. In (i) it is reasonable to take $f(x) = x$, since then $f'(x) = 1$, and the integral is simplified. Then with $g'(x) = \sin x$, $g(x) = -\cos x$ we get

$$\begin{aligned}\int_0^\pi x \sin x \, dx &= [-x \cos x]_0^\pi + \int_0^\pi \cos x \, dx \\ &= -\pi(-1) - 0 + [\sin x]_0^\pi = \pi.\end{aligned}$$

For (ii) it is not clear which is f and which is g – we take $f(x) = \sin(bx)$, $g(x) = e^{ax}$ and hope for the best. (The reader can check that the opposite choice also works.) We get

$$\int e^{ax} \sin(bx)\, dx = \frac{1}{a} e^{ax} \sin(bx) - \frac{b}{a} \int e^{ax} \cos(bx)\, dx. \tag{4.4}$$

At first sight we seem to have made no progress; the new integral is no easier than the first. We press on, hopefully, to integrate by parts a second time, now with $f(x) = \cos(bx)$, $g(x) = e^{ax}$ (the opposite choice here just returns us to the starting point – snakes and ladders in the calculus). This gives

$$\int e^{ax} \sin(bx)\, dx = \frac{1}{a} e^{ax} \sin(bx) - \frac{b}{a}\left(\frac{1}{a} e^{ax} \cos(bx) + \frac{b}{a} \int e^{ax} \sin(bx)\, dx\right).$$

We have recovered the required integral on the right side, so transferring this to the left gives

$$\left(1 + \frac{b^2}{a^2}\right) \int e^{ax} \sin(bx)\, dx = \frac{1}{a^2} e^{ax} \left(a \sin(bx) - b \cos(bx)\right)$$

which is the first result given. The second can be done similarly, or deduced from (4.4). (Or, using complex functions, as in Section 6.5, both can be deduced from $\int e^{(a+ib)x} dx$ by taking real and imaginary parts.)

For (iii) take $f(x) = \ln x$ which has $f'(x) = 1/x$, and we get

$$\begin{aligned} \int_1^a x^n \ln x\, dx &= \left[\frac{1}{n+1} x^{n+1} \ln x\right]_1^a - \int_1^a \frac{x^{n+1}}{(n+1)x} dx \\ &= \frac{1}{n+1} a^{n+1} \ln a - \left[\frac{x^{n+1}}{(n+1)^2}\right]_1^a \end{aligned}$$

as required. ♦

The other standard rule for integration 'by substitution' follows from the chain rule for derivatives, in the same way that integration by parts comes from the product rule.

Theorem 4.19 (Integration by Substitution)

Let g have a continuous derivative on $[a, b]$ and map $[a, b]$ into $[c, d]$, and let f be continuous on $[c, d]$. Then

$$\int_a^b f(g(x))\, g'(x)\, dx = \int_{g(a)}^{g(b)} f(t)\, dt. \tag{4.5}$$

Notice that there is no need for g to be monotone here as is often supposed, and that the interval $[c, d]$ may be strictly larger then $[g(a), g(b)]$; this generality actually makes the result easier, both to prove and to apply. We shall also see that (4.5) can be used in either direction, i.e. either the right side or the left can be taken as the starting point.

Proof

For $u \in [a, b]$, let $F_1(u) = \int_a^u f(g(x)) g'(x) \, dx$, so that $F_1'(u) = f(g(u)) g'(u)$. Also for $v \in [c, d]$ let $F_2(v) = \int_{g(a)}^v f(t) \, dt$, so that $F_2'(v) = f(v)$. Then $F_2'(g(u)) = f(g(u))$ for $u \in [a, b]$, so

$$\begin{aligned} F_2'(g(u)) g'(u) &= f(g(u)) g'(u) = F_1'(u), \text{ and} \\ F_2(g(u)) &= F_1(u) + k \text{ say,} \end{aligned}$$

where $k = 0$ by putting $u = a$.

In particular, with $u = b$,

$$\int_{g(a)}^{g(b)} f(t) \, dt = F_2(g(b)) = F_1(b) = \int_a^b f(g(x)) g'(x) \, dx \qquad (4.6)$$

which is what we want. ■

Example 4.20

(i) $\int_0^{\pi/2} \sin^3 x \cos x \, dx = 1/4$, $\int_0^{\pi} \sin^3 x \cos x \, dx = 0$.

(ii) $\int_0^{1/2} \left(1 - x^2\right)^{-1/2} dx = \pi/6$, $\int_0^{1/2} x \left(1 - x^2\right)^{-1/2} dx = 1 - \sqrt{3}/2$.

For (i), put $g(x) = \sin x$, $f(t) = t^3$ and $a = 0$, $b = \pi/2$, $g(a) = 0$, $g(b) = 1$, clearly satisfying the conditions of the theorem. Then

$$\begin{aligned} \int_0^{\pi/2} \sin^3 x \cos x \, dx &= \int_a^b f(g(x)) g'(x) \, dx \\ &= \int_{g(a)}^{g(b)} f(t) \, dt = \int_0^1 t^3 dt = 1/4. \end{aligned}$$

Similarly with $a = 0$, $b = \pi$, $g(a) = g(b) = 0$, gives $\int_0^{\pi/2} \sin^3 x \cos x \, dx = \int_0^0 t^3 dt = 0$.

For (ii) we use the result in reverse, that is we put $f(x) = 1/\sqrt{1 - x^2}$ and $x = g(t) = \sin t$ and start from the right side of (4.5). Then

$$\int_0^{1/2} \frac{dx}{\sqrt{1 - x^2}} = \int_{g(0)}^{g(\pi/6)} \frac{dx}{\sqrt{1 - x^2}} = \int_0^{\pi/6} \frac{\cos t}{\sqrt{1 - \sin^2 t}} dt = \pi/6$$

since for $0 < t < \pi/2$, $\sqrt{1-\sin^2 t} = \cos t$. The second integral may be done in either direction. With $f(x) = 1/\sqrt{1-x^2}$ and $x = g(t) = \sin t$ we have

$$\begin{aligned}\int_0^{1/2} \frac{xdx}{\sqrt{1-x^2}} &= \int_{g(0)}^{g(\pi/6)} \frac{xdx}{\sqrt{1-x^2}} = \int_0^{\pi/6} \frac{\sin t \cos t}{\sqrt{1-\sin^2 t}} dt \\ &= \int_0^{\pi/6} \sin t dt = [-\cos t]_0^{\pi/6} = 1 - \sqrt{3}/2.\end{aligned}$$

Alternatively with $u = 1 - x^2$,

$$\int_0^{1/2} \frac{xdx}{\sqrt{1-x^2}} = -\int_1^{3/4} \frac{du}{2\sqrt{u}} = \left[\sqrt{u}\right]_{3/4}^{1} = 1 - \sqrt{3}/2.$$

It is a relief when two different methods give the same answer! ♦

To use Leibniz notation in the above examples, for instance $\int_0^1 x(1 - x^2)^{-1/2} dx$, take $x = \sin t$, $dx/dt = \cos t$, and then $t = 0, \pi/2$ when $x = 0, 1$ and the integral becomes $\int_0^{\pi/2} \sin t \, dt = 1$. As we have said before, the choice of notation is purely a matter of taste.

4.4 Applications

To illustrate the process of integration in action, we consider some examples from geometry and mechanics. Any textbook of calculus will give a good selection of such problems – we are choosing from a very wide area.

One obvious use is directly to the calculation of areas. For instance

Example 4.21

(i) Show that the area inside the ellipse $x^2/a^2 + y^2/b^2 = 1$, $a, b > 0$ is πab.

(ii) Show that the area between the line $y = 4x + 2$ and the curve $y = x^3 + x$ is $27/4$.

(iii) Show that the total area between the line $y = 5x$ and the curve $y = x^3 + x$ is 8.

(i) Since the curve is unaltered by putting $-x, -y$ for x, y, it is enough to find the area in the first quadrant where $x, y \geq 0$ and multiply the result by 4.

This results in

$$\begin{aligned} A &= 4\int_0^a \sqrt{b^2\left(1-x^2/a^2\right)}dx \\ &= 4ab\int_0^{\pi/2}\cos^2\theta d\theta \qquad \text{after the substitution } x = a\sin\theta \\ &= 2ab\int_0^{\pi/2}(1+\cos 2\theta)\,d\theta = \pi ab. \end{aligned}$$

Parts (ii) and (iii) illustrate that we have to be careful to understand what the question is asking for. In (ii) the line meets the curve where $4x+2 = x^3+x$, i.e. where $x^3 - 3x - 2 = (x+1)^2(x-2) = 0$. Thus the line is tangent at $x = -1$, and meets the curve again when $x = 2$, and the required area lies above the curve, below the line, between $x = -1$ and $x = 2$.

The formula for the area between curves is

$$A = \int (\text{upper} \; - \; \text{lower}) \tag{4.7}$$

which in this case gives

$$\begin{aligned} A &= \int_{-1}^{2}\left(4x+2-\left(x^3+x\right)\right)dx \\ &= \left[2x+3x^2/2-x^4/4\right]_{-1}^{2} \\ &= 6+3(4-1)/2-(16-1)/4 = 27/4. \end{aligned}$$

In (iii) the line meets the curve when $x^3 = 4x$, i.e. when $x = 0, \pm 2$ and there are two areas, given according to (4.7) by

$$\begin{aligned} A_1 &= \int_0^2\left(5x-\left(x^3+x\right)\right)dx \\ &= \left[2x^2-x^4/4\right]_0^2 = 4 \end{aligned}$$

and

$$A_2 = \int_{-2}^{0}\left(x^3+x-5x\right)dx = 4.$$

Thus the total area required is $A_1+A_2 = 8$. Of course $A_1 = A_2$ is predictable from symmetry, but notice that the total area is not $\int_{-2}^{2}\left(5x-\left(x^3+x\right)\right) = 0$. And writing (4.7) in the form $A = \int |f-g|$ is not much help, since we still need to know which of f, g is greater. ♦

Integration is constantly required for problems in mechanics, for instance in finding centres of gravity.

Example 4.22

(i) Find the centre of gravity of a straight rod whose density is proportional to the distance from one end.

(ii) Find the centre of gravity of a uniform semicircular plate.

(i) Suppose that the length of the rod is a and the density is kx where k is constant, and x is the distance from one end of the rod, $0 \leq x \leq a$. Then the total mass of the rod is $M = \int_0^a kx\,dx = ka^2/2$. The position of the centre of gravity is given by b where $Mb = \int_0^a kx\,xdx = ka^3/3$, and so $b = ka^3/3M = 2a/3$, i.e. the centre of gravity is two-thirds of the way from the lighter end. (This is the answer to the tree-surgeon's problem of where to attach a rope to the branch of a tree, so that the branch will be balanced when it is cut off at the trunk.)

(ii) Let the density be k (a constant this time), and the radius of the semicircle be a. The mass of the plate is $k\pi a^2/2$ and the the distance of the centre of gravity from the diameter of the semicircle is given by b where

$$\begin{aligned} Mb &= \int_0^a 2k\sqrt{a^2 - x^2}x dx \\ &= \left[-2k\left(a^2 - x^2\right)^{3/2}/3\right]_0^a = 2ka^3/3 \end{aligned}$$

so $b = 4a/(3\pi)$. ♦

4.5 Further Mean Value Theorems

In Section 3.4 we found Taylor's theorem with a remainder of the form $r_n(b) = (b-a)^n f^{(n)}(c)/n!$. This form of remainder which depends on an unknown point $c \in [a,b]$ is not always the most useful. This section is concerned with a form of the theorem in which the remainder is in the form of an integral.

Theorem 4.23

Let f be n times continuously differentiable on $[a,b]$. Then

$$\begin{aligned} f(b) &= f(a) + (b-a)f'(a) + \frac{(b-a)^2}{2}f''(a) + \cdots \\ &+ \frac{(b-a)^n}{n!}f^{(n)}(a) + \frac{1}{n!}\int_a^b (b-x)^n f^{(n+1)}(x)\,dx. \end{aligned} \tag{4.8}$$

Proof

The proof begins with the simple observation that $f(b) - f(a) = \int_a^b f'(x)\,dx$ and proceeds by integrating by parts repeatedly on the right. This gives

$$\begin{aligned} f(b) - f(a) &= \int_a^b f'(x)\,dx = -\int_a^b (b-x)'\,f'(x)\,dx \\ &= -\left[(b-x)\,f'(x)\right]_a^b + \int_a^b (b-x)\,f''(x)\,dx \\ &= (b-a)\,f'(a) - \frac{1}{2}\left[(b-x)^2 f''(x)\right]_a^b + \frac{1}{2}\int_a^b (b-x)^2 f^{(3)}(x)\,dx \\ &= \sum_{r=1}^{n} \frac{(b-a)^r}{r!} f^{(r)}(a) + \frac{1}{n!}\int_a^b (b-x)^n f^{(n+1)}(x)\,dx \end{aligned}$$

as required. ■

We shall come across many uses for this result later; for now we use it to deduce the binomial series for functions of the form $f(x) = (1+x)^a$ where a may be any real number. (Such non-integer powers will not be formally defined until Section 6.5, but it is such a nice example that it is worth giving immediately after the theorem.)

Theorem 4.24 (Binomial Theorem)

Let $a \in \mathbb{R}$. Then

$$(1+x)^a = \sum_{r=0}^{n} \frac{a(a-1)\cdots(a-r+1)}{r!} x^r + s_n(x)$$

where $s_n(x) \to 0$ as $n \to \infty$ for $|x| < 1$.

In the language of Chapter 6, the series $\sum_{r=0}^{n} a(a-1)\cdots(a-r+1)\,x^r/r!$ is convergent for $|x| < 1$ with sum $(1+x)^a$, but we have not reached that stage yet. It is instructive to try (unsuccessfully!) to prove this result using the form of Taylor's theorem given in Section 3.4.

The expression $a(a-1)\cdots(a-r+1)/r!$ is the binomial coefficient, also written $\binom{a}{r}$.

Proof

Take $f(x) = (1+x)^a$. Then $f^{(r)}(x) = a(a-1)\cdots(a-r+1)(1+x)^{a-r}$ and so (4.8) gives

$$(1+x)^a = \sum_{r=0}^{n} \frac{a(a-1)\cdots(a-r+1)}{r!} x^r + s_n(x)$$

where

$$s_n(x) = \frac{a(a-1)\cdots(a-n)}{n!} \int_0^x (x-t)^n (1+t)^{a-n-1}\, dt, \tag{4.9}$$

and our aim is to show that $|s_n(x)| \to 0$ as $n \to \infty$ when $|x| < 1$. We want an estimate for the integral in (4.9):

$$\int_0^x (x-t)^n (1+t)^{a-n-1}\, dt \leq \text{const.}|x|^n \qquad \text{for } |x| < 1. \tag{4.10}$$

If $0 \leq x < 1$ then $0 \leq t \leq x$ and for large enough n, $a-n-1$ will be negative, so for these n,

$$\begin{aligned} \int_0^x (x-t)^n (1+t)^{a-n-1}\, dt &\leq \int_0^x (x-t)^n\, dt \\ &\leq \int_0^x x^n dt \leq x^n. \end{aligned}$$

If $-1 < x < 0$ then $x \leq t \leq 0$ and so $|(x-t)/(1+t)| \leq |x|$ (the value when $t=0$). Thus

$$\begin{aligned} \int_0^x (x-t)^n (1+t)^{a-n-1}\, dt &\leq |(x-t)/(1+t)|^n \int_0^x (1+t)^{a-1}\, dt \\ &\leq |x|^n \int_0^x (1+t)^{a-1}\, dt \end{aligned}$$

where the integral is finite since $x > -1$ and so the range of integration does not get close to the discontinuity at $t = -1$. Thus (4.10) is proved so we have

$$s_n(x) \leq \text{const.} \left| \frac{a(a-1)\cdots(a-n)}{n!} x^n \right| = a_n, \text{ say.}$$

But it is easy to see that $a_n \to 0$ as $n \to \infty$ since $|a_{n+1}/a_n| = |(a-n-1)x/(n+1)| \to |x|$ as $n \to \infty$. It follows that for $|x| < 1$, $s_n(x) \to 0$ as $n \to \infty$ as required. ■

EXERCISES

1. Find the values of the following integrals.

$$\int_0^1 \sqrt{1-x^2}dx, \int_1^2 \frac{\ln t}{t}dt, \int_0^{\pi/4} \tan u\, du,$$
$$\int_0^{\pi} x^2 \cos bx\, dx, \int_0^{2\pi} \frac{dx}{a+\cos x} \qquad (a>1).$$

2. Show, using Exercise 2, Chapter 3, that all regulated derivatives are continuous.

3. Show that if f is an even function ($f(x) = f(-x)$ for all x) then $\int_{-a}^{a} f(x)\, dx = 2\int_0^a f(x)\, dx$. If f is odd then $\int_{-a}^{a} f(x)\, dx = 0$.

4.* The length of a curve in $\mathbb{R}^2$ given parametrically by $x = f(t)$, $y = g(t)$, $a \le t \le b$ is given by (or is defined by, if one prefers that point of view) $l = \int_a^b \sqrt{f'^2 + g'^2}dt$. Check that this is reasonable by verifying that it gives (i) $\sqrt{(x_1-x_2)^2 + (y_1-y_2)^2}$ for the distance between the points (x_1, y_1) and (x_2, y_2), and (ii) $2\pi r$ for the circumference of the circle of radius r. Find the length of the cycloid given by $(x,y) = r(t - \sin t, 1-\cos t)$ for $0 \le t \le 2\pi$.

5. (i) Let $I_n = \int_0^{\pi/2} \sin^n x\, dx = \int_0^{\pi/2} \cos^n x\, dx$ for $n = 0, 1, 2, \ldots$ Integrate by parts to show that for $n \ge 2$, $nI_n = (n-1) I_{n-2}$ and hence that

$$I_{2n} = \frac{(2n-1)(2n-3)\cdots 3.1}{2n(2n-2)\cdots 4.2}\frac{\pi}{2},$$
$$I_{2n+1} = \frac{2n(2n-2)\cdots 4.2}{(2n+1)(2n-1)\cdots 3.1}. \tag{4.11}$$

(ii) Show that

$$\sqrt{n}I_n = \frac{n-1}{\sqrt{n(n-2)}}\sqrt{n-2}I_{n-2}$$

and hence that both sequences $(\sqrt{n}I_n)_{n \text{ even}}$ and $(\sqrt{n}I_n)_{n \text{ odd}}$ are increasing with limits a, b say, which may be finite or infinite.

(iii) Use the fact that $I_n \ge I_{n+1}$ to show that the limits a, b are equal.

(iv) Deduce from (4.11) that $I_{2n}I_{2n+1} = \pi/(2(2n+1))$ and hence $a = b = \sqrt{\pi/2}$.

6. Show that if f is continuous on $[a, b]$ and has a right-hand derivative on $[a, b)$ which is regulated then $\int_a^b f'_+ = f(b) - f(a)$. (This result is developed further in the solutions section.)

7. Show that for a real-valued regulated function on $[a, b]$, $\left|\int_a^b f\right| \leq \int_a^b |f|$. (Consider $f^+ = \max(f, 0)$ and $f^- = -\min(f, 0)$ separately). The corresponding result for complex valued functions is Exercise 20, Chapter 6.

5
Improper Integrals

The need to extend the definition of the integral beyond the limits imposed in Chapter 4 becomes apparent quite quickly, sometimes even without noticing if we are not too careful. For instance, if we want to integrate $f(x) = x^p$ over $[0,1]$ then we can write uncritically

$$\int_0^1 x^p dx = \left[\frac{x^{p+1}}{p+1}\right]_0^1 = \frac{1}{p+1}\left(1^{p+1} - 0^{p+1}\right) = \frac{1}{p+1}.$$

This is fine if $p \geq 0$, since then x^p is continuous on $[0,1]$. However if $p < 0$ then x^p is not bounded on $(0,1)$ and the theory of the last chapter does not apply. The calculation can be rescued in a way we shall develop further in this chapter by considering, for $0 < t < 1$, the integral

$$\int_t^1 x^p dx = \left[\frac{x^{p+1}}{p+1}\right]_t^1 = \frac{1}{p+1}\left(1^{p+1} - t^{p+1}\right).$$

If $-1 < p < 0$, then $t^{p+1} \to 0$ as $t \to 0+$, and we can write

$$\int_{0+}^1 x^p dx = \frac{1}{p+1}, \qquad -1 < p < 0$$

with this new interpretation. If $p \leq -1$ then this approach fails, though even here we are able to say *something*, as we shall see in Chapter 7.

5.1 Improper Integrals on an Interval

The example considered in the introduction illustrates the following definition.

Definition 5.1

(i) Let f be defined on $(a, b]$ and regulated on $[t, b]$ for each $t \in (a, b]$. Suppose that the integral $\int_t^b f$ has a limit l, say, as $t \to a+$. Then we say that f is improperly integrable over $[a, b]$ and write

$$\int_{a+}^{b} f(x)\,dx = l = \lim_{t \to a+} \int_t^b f(x)\,dx.$$

Similarly

$$\begin{aligned}\int_a^{b-} f(x)\,dx &= \lim_{u \to b-} \int_a^u f(x)\,dx, \text{ and} \\ \int_{a+}^{b-} f(x)\,dx &= \lim_{t \to a+} \int_t^c f(x)\,dx + \lim_{u \to b-} \int_c^u f(x)\,dx\end{aligned}$$

for any $c \in (a, b)$, if the relevant limits exist. We omit the $+/-$ signs in the limits if no confusion is possible.

(ii) More generally we say that f is improperly integrable over $[a, b]$ if $[a, b]$ is a finite union $[a_0, a_1] \cup [a_1, a_2] \cup \cdots \cup [a_{m-1}, a_m]$ of intervals on each of which f is improperly integrable in the sense of (i). In this case

$$\int_a^b f = \sum_{n=1}^{m} \int_{a_{n-1}}^{a_n} f.$$

Example 5.2

(i) x^p is improperly integrable over $[0, 1]$ if $p > -1$ with $\int_0^1 x^p dx = 1/(p+1)$.

(ii) $x^p (1-x)^q$ is improperly integrable over $[0, 1]$ if both $p, q > -1$.

(iii) $\ln x$ is improperly integrable over $[0, 1]$ (or any interval $[0, a]$, $a > 0$) with $\int_{0+}^1 \ln x\,dx = -1$.

(iv) $x^p \sin(1/x)$ is improperly integrable over $[0, 1]$ (or any interval $[0, a]$, $a > 0$) for $p > -2$.

Part (i) was already done in the introduction to this chapter. Part (ii) follows from (i) by comparing the integrands at both $x = 0$ and $x = 1$, using the result of Theorem 5.3(ii) below. (The value of this integral will be found in terms of Gamma functions in Section 5.3.) For (iii), consider $\int_t^1 \ln x \, dx = [x \ln x - x]_t^1 = -1 - (t \ln t - t) \to -1$ as $t \to 0_+$, where we used the result of Example 3.28 to show that $t \ln t \to 0$ as $t \to 0_+$. In (iv) put $x = 1/y$ to get

$$\lim_{t \to 0_+} \int_t^1 x^p \sin(1/x) \, dx = \lim_{a \to \infty} \int_1^a y^{-p-2} \sin y \, dy \tag{5.1}$$

which we will see in the next section exists for $p + 2 > 0$. ♦

Existence of improper integrals can often be recognised from one or other part of the next theorem.

Theorem 5.3

(i) If f is bounded on $[a, b]$ and regulated on $[a + \delta, b - \delta]$ for all $\delta > 0$, then f is improperly integrable over $[a, b]$.

(ii) If f is regulated on $[a + \delta, b - \delta]$ for all $\delta > 0$, and $|f(x)| \leq g(x)$ on $[a, b]$ where g is improperly integrable over $[a, b]$, then so is f.

Notice that in part (iv) of Example 5.2, x^p is not bounded if $p < 0$, and not improperly integrable over $[0, 1]$ if $p < -1$, so these conditions are sufficient but not necessary.

Proof

(i) is a 'Cauchy sequence' argument – compare Exercise 18, Chapter 2. Since f is bounded on $[a, b]$, $|f(x)| \leq M$ say, then for $a < p < q < b$, $\left|\int_p^q f(x) \, dx\right| \leq M|q - p|$ which $\to 0$ as both $p, q \to a$ or b. This implies the existence of both the limits $\lim_{t \to a} \int_t^c f(x) \, dx$ and $\lim_{u \to b} \int_c^u f(x) \, dx$, c being some arbitrarily chosen element of (a, b).

A similar argument gives (ii), where now $\left|\int_p^q f(x) \, dx\right| \leq \int_p^q g(x) \, dx \to 0$ as both $p, q \to a$ or b. ■

Example 5.4

(i) $\int_0^1 (\sin x) / x^a dx$ exists for all $a < 2$.

(ii) $\int_0^{\pi/2} \ln(\sin x) \, dx = -(\pi \ln 2)/2$.

Part (i) follows since $0 \leq \sin x \leq x$ on $[0, \pi/2]$ so the integral can be compared with $\int_0^1 x^{1-a} dx$ which exists for $1 - a > -1$. For (ii), let $I =$

$\int_0^{\pi/2} \ln(\sin x)\,dx$. The integral exists by Theorem 5.3(ii), since $\sin x \geq 2x/\pi$ on $[0, \pi/2]$ so $|\ln(\sin x)| \leq |\ln(2x/\pi)|$ and $\ln x$ is improperly integrable over $[0, a]$ for any $a > 0$ by (iii) of Example 5.2. Then also $I = \int_0^{\pi/2} \ln(\cos x)\,dx = \int_{\pi/2}^{\pi} \ln(\sin x)\,dx$ and

$$\begin{aligned} 2I &= \int_0^{\pi} \ln(\sin x)\,dx = 2\int_0^{\pi/2} \ln(\sin 2t)\,dt \qquad (\text{put } x = 2t), \\ I &= \int_0^{\pi/2} \ln(2\sin t\cos t)\,dt = \int_0^{\pi/2} (\ln 2 + \ln\sin t + \ln\cos t)\,dt \\ &= (\pi\ln 2)/2 + I + I, \end{aligned}$$

so $I = -(\pi\ln 2)/2$. ♦

5.2 Improper Integrals at Infinity

For integrals over an unbounded interval, for instance $(0, \infty)$, the definition is a variation on that of the previous section; we take a limit of a known integral $\int_0^b$, and let b go to infinity. When we come to integrals over $(-\infty, \infty)$ we take $\int_a^b$ and let both $a \to -\infty$ and $b \to \infty$ independently, as in the next definition. It is tempting, but a mistake, to take the limit of $\int_{-a}^{a}$ as $a \to \infty$; this leads to absurdities such as $\int_{-\infty}^{\infty} x\,dx = 0$.

Definition 5.5

We say that f is improperly integrable over (a, ∞), and we write

$$\int_a^{\infty} f(x)\,dx = l_1,$$

if for each $b > a$, f is improperly integrable on $[a, b]$ (in the sense of Definition 5.1 (i) and (ii)), and the limit

$$\lim_{b\to\infty} \int_a^b f(x)\,dx$$

exists and is equal to l_1.

Similarly f is improperly integrable over $(-\infty, b)$ or $(-\infty, \infty)$ if

$$\begin{aligned} \int_{-\infty}^{b} f(x)\,dx &= l_2 = \lim_{c\to-\infty} \int_c^b f(x)\,dx, \text{ or} \\ \int_{-\infty}^{\infty} f(x)\,dx &= \lim_{c\to-\infty} \int_c^a f(x)\,dx + \lim_{b\to\infty} \int_a^b f(x)\,dx \end{aligned} \tag{5.2}$$

and the relevant limits exist.

Note that it makes no difference what value is chosen for a in (5.2) – this is Exercise 1. The condition $f(x) \to 0$ as $x \to \infty$ is not necessary for convergence of the integral, as shown for instance by Example 5.7(ii) below. The result which corresponds to the vanishing condition for series (Proposition 6.4 below) is that $\int_a^{a+1} f(x)\,dx \to 0$ as $a \to \infty$.

Example 5.6

(i) $\int_1^\infty x^a dx$ exists and is equal to $-1/(a+1)$ for all $a < -1$. It does not exist for $a \geq -1$.

(ii) $\int_0^\infty x^a dx$ does not exist for any real a.

(iii) $\int_1^\infty (\sin x)/x^a dx$ exists for all $a > 0$.

(iv) $\int_0^\infty (\sin x)/x^a dx$ exists for $0 < a < 2$.

(v) $\int_0^\infty (\cos x)/x^a dx$ exists for $0 < a < 1$.

For $a \neq -1$, part (i) follows from $\int_1^b x^a dx = (b^{a+1} - 1)/(a+1)$ since $b^{a+1} \to 0$ as $b \to \infty$ only when $a < -1$. If $a = -1$ we can use the properties of logarithms, or notice that $\int_b^{2b} dx/x = \int_1^2 dt/t$ (put $x = bt$) does not tend to zero as $b \to \infty$. Part (ii) follows by combining this with Example 5.2(i). For part (iii), integrate by parts to get

$$\int_1^b \frac{\sin x}{x^a} dx = \left[-\frac{\cos x}{x^a}\right]_1^b - (a+1)\int_1^b \frac{\cos x}{x^{a+1}}.$$

Since $a > 0$, the integrated term on the right $\to \cos 1$ as $b \to \infty$, and the integral exists by comparison with $\int_1^b x^{-a-1} dx$. (We are using the analogue of Theorem 5.3 (ii) above – the reader is encouraged to state and prove this explicitly in Exercise 4.) Part (iv) follows by combining this with part (i) of Example 5.4, and part (v) follows similarly. ♦

The values of some of these integrals can be found explicitly – we state them now for interest, though complete justification will have to wait until later sections.

Example 5.7

(i) $\int_0^\infty (\sin x)/x\,dx = \pi/2$.

(ii) $\int_0^\infty \cos x^2 dx = \int_0^\infty \sin x^2 dx = \sqrt{\pi/8}$.

(iii) $\int_{-\infty}^{\infty} e^{-\pi x^2} dx = 1$.

For (i) consider

$$\int_0^b \frac{\sin x}{x} dx = \left[\frac{1-\cos x}{x}\right]_0^b + \int_0^b \frac{1-\cos x}{x^2} dx$$

where the integrated term is zero (use L'Hôpital's rule at $x = 0$) and the integral on the right converges as $b \to \infty$ by comparison with $\int_1^{\infty} x^{-2} dx$. Hence

$$\begin{aligned}
\int_0^{\infty} \frac{\sin x}{x} dx &= \int_0^{\infty} \frac{1-\cos x}{x^2} dx = \int_0^{\infty} \frac{2\sin^2(x/2)}{x^2} dx \\
&= \int_0^{\infty} \frac{\sin^2 t}{t^2} dt = \frac{1}{2}\int_{-\infty}^{\infty} \frac{\sin^2 t}{t^2} dt \\
&= \frac{1}{2}\sum_{n=-\infty}^{\infty} \int_{n\pi}^{(n+1)\pi} \frac{\sin^2 t}{t^2} dt = \frac{1}{2}\sum_{n=-\infty}^{\infty} \int_0^{\pi} \frac{\sin^2 t}{(t-n\pi)^2} dt \\
&= \frac{1}{2}\int_0^{\pi} \sum_{n=-\infty}^{\infty} \frac{\sin^2 t}{(t-n\pi)^2} dt = \frac{1}{2}\int_0^{\pi} dt = \frac{\pi}{2}.
\end{aligned}$$

The steps which require justification here are the interchange of integration and summation (for which see Theorem 6.65 below) and the identity $\sum_{-\infty}^{\infty} \sin^2 t\,(t-n\pi)^{-2} = 1$ which is part of Example 7.9(ii), the series being uniformly convergent by the M-test.

Part (ii) is a Fresnel integral, used in physical optics, and (iii) is a probability integral. Their values will be found in Corollary 7.53 and Example 5.13 respectively. ♦

Definition 5.8

We say that f is integrable over an interval (finite or infinite) to mean either (i) that the interval I is $[a, b]$ and f is regulated on I so that the integral exists in the sense of Definition 4.10, or (ii) that f is improperly integrable over the interval (finite or infinite) in the sense of Definitions 5.1 or 5.5.

The last result in this section is a technicality which will be important when we come to the study of Fourier transforms in Section 7.2.

Theorem 5.9

Let f be regulated on $\mathbb{R}$ and let $|f|$ be integrable over $\mathbb{R}$. Then $\int_{-\infty}^{\infty} |f(x+h) - f(x)|\,dx \to 0$ as $h \to 0$.

The hypothesis means that f (and so $|f|$ also) has right- and left-hand limits at all points, and $\int_{-\infty}^{\infty}|f|$ exists. Some unobvious examples of such functions are in Exercise 5. The result is true more widely – it is one of the most important results of the Lebesgue theory – but this version is enough for our needs in Chapter 7.

Proof

We begin by showing that the result holds for step functions, i.e. that for a given step function g and $\varepsilon > 0$ there is some $\delta > 0$ such that $\int_{-\infty}^{\infty}|g(x+h) - g(x)|dx < \varepsilon$ when $|h| < \delta$. Suppose that in the notation of Definition 4.1,

$$g = \sum_{j=1}^{n} c_j \chi_{(t_{j-1},t_j)},$$

and let $d = \min_j (t_j - t_{j-1})$. Let m be the least upper bound of $|g|$ on $\mathbb{R}$. Then for $|h| < d$, $g(x+h)$ is equal to $g(x)$ except on disjoint intervals T_j of length $|h|$ at each t_j (to be precise, $T_j = (t_j - h, t_j)$ if $h > 0$, or $T_j = (t_j, t_j - h)$ if $h < 0$). Then we have

$$\begin{aligned}\int_{-\infty}^{\infty}|g(x+h)-g(x)|dx &= \sum_{j=1}^{n}\int_{T_j}|g(x+h)-g(x)|dx \\ &\leq \sum_{j=1}^{n}\int_{T_j} 2m\,dx = 2mnh.\end{aligned}$$

Hence $\int_{-\infty}^{\infty}|g(x+h)-g(x)|dx < \varepsilon$ if $|h| < \min(d, \varepsilon/2mn)$. This gives the result for step functions.

Now suppose that f satisfies the conditions of the theorem. Since $\int_{-\infty}^{\infty}|f|$ exists then given $\varepsilon > 0$ there is some $A > 0$ such that $\int_{|x|>A}|f(x)|\,dx = \int_{-\infty}^{-A}|f| + \int_{A}^{\infty}|f| < \varepsilon$. Use Theorem 4.6 to find a step function g on $[-A, A]$ such that $|f(x) - g(x)| < \varepsilon/2A$ for all $x \in [-A, A]$, and hence $\int_{-\infty}^{\infty}|f-g| = \int_{-A}^{A}|f-g| + \int_{|x|>A}|f(x)|\,dx < 2\varepsilon$. For this g, choose $\delta > 0$ as above such that $\int_{-\infty}^{\infty}|g(x+h)-g(x)|dx < \varepsilon$ when $|h| < \delta$. Then for $|h| < \delta$ we have

$$\begin{aligned}\int_{-\infty}^{\infty}|f(x+h)-f(x)|dx &\leq \int_{-\infty}^{\infty}|f(x+h)-g(x+h)|dx \\ &\quad + \int_{-\infty}^{\infty}|g(x+h)-g(x)|dx \\ &\quad + \int_{-\infty}^{\infty}|g(x)-f(x)|dx \\ &\leq 2\varepsilon + \varepsilon + 2\varepsilon = 5\varepsilon\end{aligned}$$

and the result is proved. (Notice that since the choice of A depends on ε, it was necessary to have $|f(x) - g(x)| < \varepsilon/2A$ instead of just $< \varepsilon$.) ■

5.3 The Gamma Function

The factorial function $n! = 1.2.\cdots.n$ occurs so frequently that it is natural to want to extend it to non-integer values (we shall consider both real and complex values – look ahead to Section 6.4, or consult [14] for a quick refresher). A possible route might be to look at the product $x(x-1)(x-2)\cdots$, but there is an obvious problem of where to stop taking further terms. Instead we consider $x(x+1)(x+2)\cdots$ with the product continued to infinity; to ensure convergence we have to include other factors to balance the growth of the product.

The Gamma function is considered here since it uses improper integrals in an essential way. Some of the proofs however refer ahead to Chapter 6, so one possibility might be to read the results now and return for the proofs later.

Lemma 5.10

Let

$$f_n(x) = \frac{x(x+1)(x+2)\cdots(x+n-1)}{n!n^{x-1}}.$$

Then the limit $F(x) = \lim_{n\to\infty} f_n(x)$ exists for all $x \in \mathbb{C}$ and satisfies $xF(x+1) = F(x)$ and $F(1) = 1$. In particular for integer n,

$$F(n) = \begin{cases} 1/(n-1)! & \text{for } n \geq 1, \\ 0 & \text{for } n \leq 0. \end{cases}$$

Proof

Consider the ratio

$$\frac{f_{n+1}(x)}{f_n(x)} = \frac{x+n}{n+1}\left(\frac{n}{n+1}\right)^{x-1} = \frac{1+x/n}{(1+1/n)^x}$$

where we regard x as fixed and concentrate attention on the behaviour with respect to n. We replace n on the right side by a real variable where $y > 2|x|$ and $y > 2$; in particular this makes $|x/y| < 1$ and allows us to take (complex) logarithms, as in Definition 6.52, to get $g(y)$ where

$$\begin{aligned} g(y) &= \log\left(1+\frac{x}{y}\right) - x\log\left(1+\frac{1}{y}\right) \\ &= \sum_1^\infty \frac{(-1)^{n-1}}{n}\left\{\left(\frac{x}{y}\right)^n - x\left(\frac{1}{y}\right)^n\right\} \\ &= \sum_2^\infty \frac{(-1)^{n-1}}{n}\left\{\left(\frac{x}{y}\right)^n - x\left(\frac{1}{y}\right)^n\right\} \\ &\leq \frac{1}{2}\left\{\frac{|x|^2}{y^2(1-|x|/y)} - \frac{|x|}{y^2(1-1/y)}\right\} \leq \frac{|x|^2-|x|}{y^2} \end{aligned}$$

since $y > |x|$ and 1, and we used the expansion (6.17) for log and $\sum_0^\infty t^n$ for $(1-t)^{-1}$. Thus the series $\sum g(n) = \sum(\log f_{n+1}(x) - \log f_n(x))$ is convergent by comparison with $\sum n^{-2}$ (see Theorem 6.61) which shows that $\log f_n(x)$, and hence $f_n(x)$ also, have finite limits as $n \to \infty$.

To show that $xF(x+1) = F(x)$, consider

$$\begin{aligned} F(x+1) &= \lim_{n\to\infty} \frac{(x+1)(x+2)\cdots(x+n)}{n!n^x} \\ &= \lim_{n\to\infty} \frac{x+n}{xn}\frac{x(x+1)(x+2)\cdots(x+n-1)}{n!n^{x-1}} \\ &= \left(\lim_{n\to\infty}\frac{x+n}{xn}\right)F(x) = \frac{1}{x}F(x) \end{aligned}$$

as required. The value $F(n) = 1/(n-1)!$ follows from this and we are done. ■

This gives Euler's Gamma function $\Gamma(x)$ as the reciprocal of $F(x)$ and hence also the required extension of the factorial function.

Definition 5.11

The Gamma function Γ is defined for $x \in \mathbb{C}$ not equal to a negative integer by

$$\Gamma(x) = \frac{1}{F(x)} = \lim_{n\to\infty} \frac{n!n^{x-1}}{x(x+1)(x+2)\cdots(x+n-1)}.$$

It satisfies $\Gamma(x+1) = x\Gamma(x)$ for x not a negative integer, and $\Gamma(n) = (n-1)!$ for integers $n \geq 1$.

Two important types of integrals can be evaluated using the Gamma function, namely $\int_0^1 t^{x-1}(1-t)^{y-1}\,dt$ and $\int_0^\infty t^{x-1}e^{-t}dt$ which are called the Eulerian integrals of the first and second kinds respectively. To begin with

Theorem 5.12

The integral $\int_0^\infty t^{x-1}e^{-t}dt$ converges for $\Re x > 0$ with value $\Gamma(x)$.

Proof

The integral is the limit of $\int_0^n t^{x-1}(1-t/n)^n\,dt = n^x \int_0^1 u^{x-1}(1-u)^n\,du$ using dominated convergence (the argument is given in detail as Example 6.68). Repeated integration by parts gives

$$\begin{aligned} n^x \int_0^1 u^{x-1}(1-u)^n\,du &= n^x \frac{n}{x}\int_0^1 u^x (1-u)^{n-1}\,du \\ &= n^x \frac{n(n-1)}{x(x+1)}\int_0^1 u^{x+1}(1-u)^{n-2}\,du \\ &= n^x \frac{n(n-1)\cdots 1}{x(x+1)\cdots(x+n-1)}\int_0^1 u^{x+n-1}(1-u)^0\,du \\ &= n^{x-1}\frac{n!}{x(x+1)\cdots(x+n-1)}\frac{n}{x+n} \to \Gamma(x) \end{aligned}$$

from Definition 5.11, as required. ∎

Example 5.13

For $a, \Re x > 0$,

$$\int_0^\infty t^{x-1}e^{-at^2}dt = \frac{1}{2a^{x/2}}\Gamma(x/2). \tag{5.3}$$

In particular

$$\int_0^\infty e^{-at^2}dt = \frac{\Gamma(1/2)}{2\sqrt{a}} = \frac{1}{2}\sqrt{\frac{\pi}{a}} \tag{5.4}$$

and putting $a = \pi$ gives the result of Example 5.7(iii) above.

In the first integral, put $at^2 = v$ to give

$$\int_0^\infty t^{x-1}e^{-at^2}dt = \frac{1}{2a^{x/2}}\int_0^\infty v^{x/2-1}e^{-v}dv = \frac{1}{2a^{x/2}}\Gamma(x/2).$$

The second follows on putting $x = 1$, except for the value of $\Gamma(1/2)$ for which we have to do a little more work. The proof of Theorem 5.12 shows that

$$\begin{aligned} \Gamma(1/2) &= \lim_{n\to\infty} n^{1/2}\int_0^1 u^{-1/2}(1-u)^n\,du \\ &= \lim_{n\to\infty} 2n^{1/2}\int_0^{\pi/2}\cos^{2n+1}\theta\,d\theta, \qquad \text{putting } u = \sin^2\theta \\ &= \sqrt{2}\lim_{n\to\infty}\sqrt{2n}I_{2n+1} = \sqrt{\pi} \end{aligned}$$

using the notation and the result of Exercise 5, Chapter 4. ♦

The result of the last part is sufficiently striking to be worth stating separately.

Theorem 5.14

(i) $\Gamma(1/2) = \sqrt{\pi}$.

(ii) For integers $n \geq 1$,

$$\begin{aligned} \Gamma(n+1/2) &= (n-1/2)(n-3/2)\cdots(1/2)\sqrt{\pi}, \\ \Gamma(-n+1/2) &= \frac{\sqrt{\pi}}{(-n+1/2)(-n+3/2)\cdots(-1/2)}. \end{aligned}$$

Proof

Part (i) was proved in Example 5.13 above, and the others follow from $\Gamma(x+1) = x\Gamma(x)$. ■

The value of the derivative, $\Gamma'(1) = -\gamma$ is also important – this will be found, along with a general expression for Γ', in Example 6.70.

The Eulerian integral of the first kind defines the Beta function:

Definition 5.15

For $\Re x, \Re y > 0$,

$$B(x,y) = \int_0^1 t^{x-1}(1-t)^{y-1}\,dt.$$

Its value can be expressed in terms of the Gamma function.

Theorem 5.16

For $\Re x, \Re y > 0$,

(i) $xB(x,y) = (x+y)B(x+1,y)$.

(ii) $\lim_{t\to\infty} t^y B(t,y) = \Gamma(y)$.

(iii)

$$B(x,y) = \frac{\Gamma(x)\Gamma(y)}{\Gamma(x+y)}.$$

Proof

(i) Integration by parts gives for $\Re x, \Re y > 0$

$$\begin{aligned}
B(x+1,y) &= \int_0^1 t^x (1-t)^{y-1}\,dt \\
&= \frac{1}{y}\left[-t^x (1-t)^y\right]_0^1 + \frac{x}{y}\int_0^1 t^{x-1}(1-t)^y\,dy \\
&= \frac{x}{y} B(x, y+1),
\end{aligned}$$

and from direct addition of the integrands,

$$\begin{aligned}
B(x+1,y) + B(x,y+1) &= \int_0^1 t^{x-1}(1-t)^{y-1}\left[t + (1-t)\right]dt \\
&= B(x,y).
\end{aligned}$$

Combining these gives

$$\begin{aligned}
B(x,y) &= B(x+1,y) + \frac{y}{x} B(x+1,y) \\
&= \frac{x+y}{x} B(x+1,y),
\end{aligned}$$

which is (i). Applying this repeatedly gives

$$\begin{aligned}
B(x,y) &= \frac{x+y}{x} B(x+1,y) \\
&= \frac{(x+y)(x+y+1)}{x(x+1)} B(x+2,y) \\
&= \frac{(x+y)(x+y+1)\cdots(x+n-1)}{x(x+1)\cdots(x+n-1)} B(x+n,y) \\
&= \frac{(x+y)(x+y+1)\cdots(x+n-1)}{n!n^{x+y-1}} \times \\
&\quad \frac{n!n^{x-1}}{x(x+1)\cdots(x+n-1)} n^y B(x+n,y). \qquad (5.5)
\end{aligned}$$

The first two factors in (5.5) tend to $1/\Gamma(x+y)$ and $\Gamma(x)$, and the third tends to $\lim_{n\to\infty} (n/(n+x))^y (n+x)^y B(x+n,y)$ which is $\Gamma(y)$, assuming (ii). Thus (iii) will follow when we have proved (ii). For (ii), consider first $\lim_{n\to\infty} n^y B(n,y)$ so that the limit is taken through integer values only. Then applying (i) repeatedly,

$$\begin{aligned}
n^y B(n,y) &= n^y \frac{n-1}{n+y-1} B(n-1,y) \\
&= n^y \frac{(n-1)\cdots 1}{(n+y-1)\cdots(y+1)} B(1,y) \\
&= n^{y-1} \frac{n!}{(n+y-1)\cdots(y+1)\,y}
\end{aligned}$$

which has the limit $\Gamma(y)$ by the definition of the Gamma function. In particular, for $\Re y > 0$, $|B(n,y)| \to \infty$ as $n \to \infty$. Then for $n \le t \le n+1$ we have

$$\begin{aligned}\left|\frac{B(n,y)}{B(t,y)} - 1\right| &\le \frac{1}{|B(t,y)|}\int_0^1 \left|u^{n-1} - u^{t-1}\right|\left|(1-u)^{y-1}\right| du \\ &\le \frac{2}{|B(t,y)|}\int_0^1 (1-u)^{\Re y - 1}\, du = \frac{2}{\Re y\, |B(t,y)|}\end{aligned}$$

and so $B(n,y)/B(t,y) \to 1$ as $n \to \infty$ which gives the general result, $\lim_{t\to\infty} t^y B(t,y) = \Gamma(y)$. ■

Example 5.17

(i) For $a, b > -1$,

$$\int_0^{\pi/2} \sin^a x \cos^b x\, dx = \frac{\Gamma((a+1)/2)\,\Gamma((b+1)/2)}{2\Gamma((a+b)/2+1)}.$$

(ii)

$$\int_0^1 \frac{dt}{\sqrt{1-t^4}} = \frac{(\Gamma(1/4))^2}{4\sqrt{2\pi}}.$$

Both of these come almost directly from part (iii) of the theorem. For instance putting $t = \sin^2 x$ in the first gives

$$\begin{aligned}\int_0^{\pi/2} \sin^a x \cos^b x\, dx &= \int_0^1 t^{a/2}(1-t)^{b/2} \frac{dt}{2\sqrt{t(1-t)}} \\ &= \int_0^1 t^{(a+1)/2-1}(1-t)^{(b+1)/2-1}\, dt \\ &= \frac{\Gamma((a+1)/2)\,\Gamma((b+1)/2)}{2\Gamma((a+b)/2+1)}.\end{aligned}$$

The second has a sting in the tail. Put $t^4 = u$ to get

$$\begin{aligned}\int_0^1 \frac{dt}{\sqrt{1-t^4}} &= \frac{1}{4}\int_0^1 u^{-3/4}(1-u)^{-1/2}\, du \\ &= \frac{\Gamma(1/4)\,\Gamma(1/2)}{4\Gamma(3/4)} = \frac{(\Gamma(1/4))^2}{4\sqrt{2\pi}}\end{aligned}$$

where in the last step we have used both $\Gamma(1/2) = \sqrt{\pi}$ which is Theorem 5.14(i) above, and $\Gamma(1/4)\,\Gamma(3/4) = \pi/\sin(\pi/4) = \pi\sqrt{2}$ which will be proved in Section 6.7. This integral is important in the theory of elliptic functions, and is as near as we are able to get to the statement in the introduction that

$$\sum_{n=-\infty}^{\infty} e^{-n^2\pi} = \frac{\Gamma(1/4)}{\sqrt{2}\pi^{3/4}}.$$

The proof of this requires a special case of the identity

$$\int_0^1 \frac{dt}{\sqrt{(1-t^2)(1-k^2t^2)}} = \frac{\pi}{2}\theta_3^2(q), \qquad \text{where } k = \theta_2^2(q)/\theta_3^2(q) \text{ and}$$

$$\theta_2(q) = 2\sum_{n=0}^{\infty} q^{(n+1/2)^2}, \ \theta_3(q) = \sum_{n=-\infty}^{\infty} q^{n^2}$$

for a complete elliptic integral in terms of theta functions, and would unfortunately require a whole chapter to itself – the interested reader is referred to [16] or [17]. ♦

EXERCISES

1. Show that the value chosen for a in Definition 5.5 has no effect on the integrability of f or the value of the integral.
2. Find the values of the following (improper) integrals.
$$\int_0^1 \left(1-t^2\right)^{-1/2} dt, \ \int_0^1 t\left(1-t^2\right)^{-1/2} dt, \ \int_0^\infty e^{-ax}\cos(bx)\,dx,$$
$$\int_0^\infty e^{-ax}\sin(bx)\,dx.$$
3. Show that for $4b > a^2$,
$$\int_{-\infty}^{\infty} \frac{dx}{x^2+ax+b} = \frac{\pi}{\sqrt{b-a^2/4}}$$
and deduce, using partial fractions and $x^4+1 = (x^2+\sqrt{2}x+1)(x^2-\sqrt{2}x+1)$ that
$$\int_{-\infty}^{\infty} \frac{dx}{x^4+1} = \frac{\pi}{\sqrt{2}}.$$
Find similarly $\int_{-\infty}^{\infty} x^2/\left(x^4+1\right) dx$.
4. Show that if f, g are regulated on (b,∞), $|f(x)| \le g(x)$ for all $x \ge b$, and $\int_b^\infty g$ is finite, then $\int_b^\infty f$ exists. (This is the comparison test for improper integrals over (b,∞).)

5. Show that the function given by $f(x) = n$ on $[n, n + n^{-3})$, $n = 1, 2, 3, \ldots$, $f(x) = 0$ otherwise, gives an unbounded function which satisfies the conditions of Theorem 5.9, but that the function given by $f(x) = n^2$ on $[n, n + n^{-3}/2)$, $= -n^2$ on $[n + n^{-3}/2, n + n^{-3})$ is integrable (improperly) but does not satisfy the conditions, since $\int_{-\infty}^{\infty} |f|$ diverges.

6. Find $\int_0^1 t^2/\sqrt{1-t^4}dt$ (and more generally $\int_0^1 t^a \left(1 - t^b\right)^c dt$) in terms of Gamma functions. Show similarly that if $b, c > 0$ and $0 < \Re a < bc$, then

$$\int_0^\infty \frac{t^{a-1}}{\left(1+t^b\right)^c} dt = \frac{\Gamma(a/b)\,\Gamma(c-a/b)}{b\Gamma(c)}.$$

(This gives a different way of finding the result of Exercise 3.)

7. Let $x_n = (2n + 1/2)^2 \pi^2$. Show that $\int_0^{x_n} \sin(\sqrt{x})\, dx$ tends to a limit (and find its value) though $\int_0^\infty \sin(\sqrt{x})\, dx$ is undefined.

8.* g-summability. Let g be a decreasing function on $[0, \infty)$ with $g(x) \to g(0) = 1$ as $x \to 0+$ and $g(x) \to 0$ as $x \to \infty$. We say that an integral $\int_0^\infty f(x)\, dx$ is g-summable with value I if $I(t) = \int_0^\infty f(x)\, g(tx)\, dx$ exists for all $t > 0$ and $I(t) \to I$ as $t \to 0+$.

Show that this is compatible with ordinary convergence of integrals, in the sense that if $\int_0^\infty f(x)\, dx$ exists according to Definition 5.5 then it is g–summable (for any g) to the same value. However, many integrals which do not exist in the sense of Definition 5.5 are nonetheless g–summable for a suitable choice of g. For instance, show that $\int_0^\infty \cos(ax)\, dx$ and $\int_0^\infty \sin(ax)\, dx$ are g–summable with $g(x) = e^{-x}$, and find their values.

(The subject of summability of series and integrals is a very large one; the classic reference is [4].)

9. Let f be (improperly) integrable over $(0, \infty)$ and suppose that $f'(x) \to 0$ as $x \to \infty$. Show that also $f(x) \to 0$ as $x \to \infty$.

6
Series

A series is an expression of the form $a_1 + a_2 + \cdots + a_n + \cdots$, where we begin with a given sequence (a_n) and form a new one by adding the terms together in order – then we say the series is convergent if the new sequence of sums is convergent in the sense of Chapter 1. For most of us this is the way in which we first meet the definitions and processes of analysis; for instance in what sense does the infinite sum $1 - \frac{1}{2} + \frac{1}{4} - \frac{1}{8} + \cdots + \left(\frac{-1}{2}\right)^n + \cdots$, which we saw in the Preface, converge and why is its sum 2/3? And why does the sum $1 - \frac{1}{2} + \frac{1}{3} - \frac{1}{4} + \cdots + \frac{(-1)^{n-1}}{n} + \cdots$ converge while the corresponding sum $1 + \frac{1}{2} + \frac{1}{3} + \frac{1}{4} + \cdots + \frac{1}{n} + \cdots$ with positive terms does not? Our aim in this chapter is to put these ideas into a general framework and to derive the most commonly useful tests for convergence. For greater detail see for instance [1].

6.1 Convergence

The introduction shows that the following definition is a natural one.

Definition 6.1

Given a sequence $(a_n)_1^\infty$, we form a further sequence $s = (s_n)_1^\infty$ by adding the terms together in order:

$$s_n = a_1 + a_2 + \cdots + a_n, \qquad n \geq 1. \tag{6.1}$$

This determines the series s whose terms are the elements of the sequence (a_n). The sums s_n are called the partial sums of the series. We may write informally

$$s = \sum_{1}^{\infty} a_n = a_1 + a_2 + \cdots + a_n + \cdots \tag{6.2}$$

but nothing is to be read into this expression regarding convergence; it is simply another way of expressing the relation (6.1).

Notice that we have already used the convention (6.2) in the introduction to the chapter. We could equally well begin the sequence with a_0 (or any other term), when the partial sums would be given by $s_n = a_0 + a_1 + a_2 + \cdots + a_n$, $n \geq 0$. We shall often write just $\sum a_n$ when the starting value is clear from the context (or irrelevant).

For convergence of the series we require that the partial sums should have a limit.

Definition 6.2

The series $s = \sum_1^\infty a_n$ is said to be convergent with sum S if the sequence (s_n) of partial sums is convergent with limit S; equivalently $a_1 + a_2 + \cdots + a_n \to S$ as $n \to \infty$. A series which is not convergent is said to be divergent.

Example 6.3

(i) The series $1 - \frac{1}{2} + \frac{1}{4} - \frac{1}{8} + \cdots + \left(\frac{-1}{2}\right)^n + \cdots$ is convergent with sum $2/3$. More generally for any $r \in (-1, 1)$ the geometric series $\sum_0^\infty r^n$ is convergent with sum $1/(1-r)$.

(ii) The series $1 + \frac{1}{2} + \frac{1}{3} + \frac{1}{4} + \cdots + \frac{1}{n} + \cdots$ is not convergent; its partial sums tend to infinity.

For (i), we consider the partial sum $1 + r + r^2 + \cdots + r^n$ whose value is $\left(1 - r^{n+1}\right)/(1-r)$ as we see by induction, or by multiplying both sides by $1 - r$. As $n \to \infty$, $r^{n+1} \to 0$ (Example 1.14(ii)) so that the sum of the series is $1/(1-r)$. The first series is the special case in which $r = -1/2$.

For (ii), we show that the sums $s_n = 1 + \frac{1}{2} + \frac{1}{3} + \frac{1}{4} + \cdots + \frac{1}{n}$ do not form a Cauchy sequence. Consider the difference $s_{2n} - s_n = 1/(n+1) + 1/(n+2) + \cdots + 1/(2n)$. Each of these terms is at least equal to $1/(2n)$, and there are n of them, so the whole sum is at least n times $1/(2n)$, i.e. $s_{2n} - s_n \geq 1/2$ and the difference does not go to zero. Thus (s_n) is not a Cauchy sequence, and so is not convergent. Since (s_n) is increasing but not convergent it must tend to infinity. ♦

The series $1 + \frac{1}{2} + \frac{1}{3} + \frac{1}{4} + \cdots + \frac{1}{n} + \cdots$ is called the harmonic series, and its divergence is somehow non-intuitive for a beginner – 'after all the terms tend to zero, don't they?' is the common complaint. So it should be constantly borne in mind that the next result is strictly in one direction only.

Proposition 6.4 (Vanishing Condition)

The condition $a_n \to 0$ is necessary, but not sufficient, for the convergence of the series $\sum a_n$.

Proof

If the series converges, then $s_n \to S$ and $a_n = s_n - s_{n-1} \to S - S = 0$, so the terms must tend to zero. The harmonic series shows that the condition is not sufficient. ■

Example 6.5

The series $\sum (-1)^n$ is not convergent since the terms do not tend to zero. ♦

We shall find, as the chapter continues, that the series for which we can find an exact sum from the definition are quite few in number, so the next example is a pleasant discovery, and illustrates a useful trick.

Example 6.6

(i) The series $\sum_1^\infty \frac{1}{n(n+1)} = \frac{1}{1.2} + \frac{1}{2.3} + \cdots$ is convergent with sum 1.

(ii) The series $\sum_1^\infty 1/\left(\sqrt{n} + \sqrt{n+1}\right)$ is divergent.

The trick in (i) is to write each term as a difference, $\frac{1}{n(n+1)} = \frac{1}{n} - \frac{1}{n+1}$ so that the partial sum simplifies dramatically,

$$\begin{aligned}\frac{1}{1.2} + \frac{1}{2.3} + \cdots + \frac{1}{n\,(n+1)} &= \left(\frac{1}{1} - \frac{1}{2}\right) + \left(\frac{1}{2} - \frac{1}{3}\right) + \cdots + \left(\frac{1}{n} - \frac{1}{n+1}\right) \\ &= 1 - \frac{1}{n+1}\end{aligned}$$

since all other terms cancel (we say the series telescopes). Hence the partial sums tend to 1 as $n \to \infty$.

In (ii) we write $1/\left(\sqrt{n} + \sqrt{n+1}\right) = \sqrt{n+1} - \sqrt{n}$ so that $\sum_1^m 1/(\sqrt{n} + \sqrt{n+1}) = \sqrt{m+1} - 1$ which tends to infinity. ♦

6.2 Series with Positive Terms

When we come to consider general classes of series, the easiest to deal with are those whose terms are positive. This is because if each $a_n \geq 0$, then the sequence of partial sums in increasing, and Theorem 1.16 says that such a sequence is convergent if and only if its terms are bounded above. This observation is enough to prove the following result.

Theorem 6.7

A series $\sum a_n$ with positive terms is convergent if and only if its partial sums are bounded above.

Example 6.8

The series $\sum_1^\infty 1/n^2$ is convergent.

Notice that if $n \geq 2$, then $1/n^2 < 1/(n(n-1))$ and so

$$\sum_1^m 1/n^2 < 1 + \sum_2^m 1/(n(n-1)) < 1 + 1 - 1/m < 2,$$

using the result of Example 6.6(i) above. Hence the series is convergent with sum at most 2. (Its exact sum is $\pi^2/6 = 1.6449\ldots$ as claimed in the Preface. We will prove this in Section 7.1.) ♦

This example indicates a general method. Given a new series to consider, we look for a related series whose behaviour is known and try to use it as a standard of comparison. This is particularly suitable when the terms are positive, since the one sequence of partial sums will be a bound for the other, and we can use Theorem 6.7.

Theorem 6.9 (Comparison Test)

(i) Suppose that $0 \leq a_n \leq b_n$ for all $n \geq 1$. If the series $\sum b_n$ is convergent then the series $\sum a_n$ is also convergent; equivalently if $\sum a_n$ diverges then also $\sum b_n$ diverges.

(ii) If for some $K > 0$ and integer $N \geq 1$ we have $0 \leq a_n \leq Kb_n$ for $n \geq N$, then the same conclusions as in (i) apply.

This test is capable of the same sort of misuse as we saw earlier for the Vanishing Condition. It is *not* true that if $0 \leq a_n \leq b_n$ and $\sum a_n$ converges, then $\sum b_n$ converges also – consider our examples above in which $0 < 1/n^2 < 1/n$ and $\sum 1/n^2$ converges, $\sum 1/n$ diverges. But provided we get the result the right way round, it is easy to prove and to use.

Proof

Suppose $0 \leq a_n \leq b_n$ for all $n \geq 1$. Then the respective partial sums $s_n = a_1 + \cdots + a_n$ and $t_n = b_1 + \cdots + b_n$ are both increasing and satisfy $0 \leq s_n \leq t_n$ for all n. But we are supposing in the first part of (i) that (t_n) is convergent, so its limit is an upper bound for the partial sums (s_n) and the convergence of $\sum a_n$ follows from Theorem 6.7. The modifications required to prove (ii) are left to the reader. ■

Example 6.10

(i) The series $\sum 1/n^a$ is convergent for $a \geq 2$, divergent for $a \leq 1$.

(ii) For any positive function f, which is bounded on $(0, \infty)$, and $0 \leq r < 1$, the series $\sum f(n) r^n$ is convergent.

(i) If $a \geq 2$, then $1/n^a \leq 1/n^2$ and $\sum 1/n^2$ is convergent, so $\sum 1/n^a$ is convergent by comparison. Similarly if $a \leq 1$ then $1/n^a \geq 1/n$, and $\sum 1/n$ is divergent.

(ii) If $f(n) \leq A$ say then $f(n)r^n \leq Ar^n$ for all n, and (ii) of the theorem applies since $\sum r^n$ is convergent for $0 \leq r < 1$. ♦

The gap in (i) between $a = 1$ and $a = 2$ will be filled in Example 6.14 below.

Part (ii) of this theorem illustrates a general principle, that (just as for sequences) the convergence of a series is unaffected by the behaviour of a finite number of terms at the beginning, although the sum may be altered. For if $a_n = b_n$ for all $n \geq N$ say, then $s_n = a_1 + \cdots + a_n$ and $t_n = b_1 + \cdots + b_n$ have a constant difference for $n \geq N$, and hence the sequences (s_n) and (t_n) converge or diverge together.

A common form of series has the form $\sum g(n) r^n$ where g is some 'nuisance factor', for instance a power of n. If we recall Example 1.24(iii) which says that $n^a x^n \to 0$ when $|x| < 1$, irrespective of the value of a, then we might hope that the factor r^n will force convergence of $\sum g(n) r^n$ when $0 < r < 1$. This is true, and is the essential idea in the following simple test for convergence.

Theorem 6.11 (Ratio Test)

(i) Suppose that for some $r < 1$ and for all n (or for all $n \geq N$) $a_n > 0$ and

$a_{n+1}/a_n \le r$. Then the series $\sum a_n$ is convergent.

(ii) Suppose that $a_n > 0$ for all n (or for all $n \ge N$) and that the limit $\lim_{n\to\infty}(a_{n+1}/a_n)$ exists and is equal to r say. Then the series $\sum a_n$ is convergent if $r < 1$, divergent if $r > 1$. There is no conclusion if $r = 1$.

Proof

(i) We have $a_{N+1} \le ra_N$, $a_{N+2} \le ra_{N+1} \le r^2 a_N$, and similarly $a_{N+k} \le r^k a_N$. Thus

$$\sum_N^{N+k} a_n \le a_N\left(1 + r + \cdots + r^k\right) < \frac{a_N}{1-r}$$

and the partial sums of $\sum a_n$ are bounded above by $\sum_1^{N-1} a_n + a_N/(1-r)$. This proves convergence by Theorem 6.7. (We could have given the argument more briefly by saying simply 'compare $\sum a_n$ with the geometric series $\sum r^n$'.)

(ii) Given that $\lim_{n\to\infty} a_{n+1}/a_n$ exists and is equal to $r < 1$, choose r_1 with $r < r_1 < 1$ and N' so that $a_{n+1}/a_n < r_1$ for $n \ge N'$ (from the definition of a limit). Part (i) now applies with r_1 in place of r.

If $r > 1$ then (a_n) is increasing for sufficiently large n, and so cannot tend to zero and the series is divergent by the vanishing condition.

The two series $\sum 1/n$, which diverges, and $\sum 1/n^2$ which converges, show that there can be no conclusion when $r = 1$. ■

Notice that the ratio a_{n+1}/a_n is not required to have a limit in part (i) of the theorem.

Example 6.12

(i) The series

$$\frac{1}{3} + \frac{1}{6} + \frac{1}{9} + \frac{1}{18} + \frac{1}{27} + \frac{1}{54} + \cdots$$

where each term a_n is a_{n-1} multiplied by either $1/2$ if n is even, or by $2/3$ if n is odd, is convergent, by part (i) of the theorem with $r = 2/3$. The ratio a_{n+1}/a_n has no limit here.

(ii) $\sum_1^\infty n^a r^n$ is convergent for any real a if $0 < r < 1$, since

$$\frac{(n+1)^a r^{n+1}}{n^a r^n} = \left(1 + \frac{1}{n}\right)^a r \to r \qquad \text{as } n \to \infty. \blacklozenge$$

Neither part of the theorem says that if $a_{n+1}/a_n < 1$ for all n then $\sum a_n$ is convergent (another common misunderstanding). This condition implies only that the terms a_n are strictly decreasing – it does not even force $a_n \to 0$. Again, if the supposed result were true then since $n/(n+1) < 1$ it would follow that the series $\sum 1/n$ would be convergent which we know to be false.

The next result is useful for series of the form $\sum f(n)$ where f is a function defined on $[1, \infty)$. It says that the series $\sum f(n)$ and the (improper) integral $\int^{\infty} f(x)\,dx$, as defined in Section 5.2, have the same behaviour (both convergent or both divergent), provided only that f is positive and decreasing (f need not even tend to zero, but in this case the result is trivial: both series and integral diverge).

Theorem 6.13 (Integral Test)

Let f be positive and decreasing on $[1, \infty)$. Then the series $\sum_1^{\infty} f(n)$ and $\int_1^{\infty} f(x)\,dx$ are either both convergent or both divergent. In addition we have the inequalities

$$\sum_1^{m} f(n) \geq \int_1^{m+1} f(x)\,dx \geq \sum_2^{m+1} f(n), \tag{6.3}$$

and, if both sequence and series converge,

$$\sum_1^{\infty} f(n) \geq \int_1^{\infty} f(x)\,dx \geq \sum_2^{\infty} f(n) = \sum_1^{\infty} f(n) - f(1). \tag{6.4}$$

More generally, if f is positive and decreasing on $[m, \infty)$ then the corresponding results hold with the lower limits $1, 2$ replaced by $m, m+1$ respectively. If f is strictly decreasing then we have strict inequality throughout.

Proof

Since f is decreasing on $(n, n+1)$ we have $f(n) \geq f(x) \geq f(n+1)$ for $n \leq x \leq n+1$ and so

$$f(n) \geq \int_n^{n+1} f(x)\,dx \geq f(n+1). \tag{6.5}$$

(We can have equality in one or other place here if f is constant on $(n, n+1)$; note that f is not required to be continuous.) Add these inequalities for $n = 1, 2, \ldots, m$ and (6.3) follows.

If the series converges, then from the first inequality in (6.3), its sum (to infinity) is an upper bound for the integral $\int_1^{m+1} f$ which must therefore converge. Similarly if the integral converges, then from the second inequality in (6.3), its integral (to infinity) is an upper bound for the sum

$\sum_2^{m+1} f(n)$ which must therefore converge. When both converge, (6.4) follows by letting $m \to \infty$. Notice that if f is strictly decreasing then we have $\int_n^{n+1} f(x)\,dx \geq (f(n+1) + f(n))/2 > f(n)$ and so strict inequality in (6.5); moreover this positive difference in each term remains when we add to obtain the limit of the sums. ■

Example 6.14

(i) $\sum_1^\infty 1/n^a$ is convergent for $a > 1$ and

$$\frac{a}{a-1} > \sum_1^\infty 1/n^a > \frac{1}{a-1}.$$

(ii) $\sum_2^\infty 1/(n \ln^a n)$ is convergent for $a > 1$, divergent for $a \leq 1$.

The convergence in (i) follows by putting $f(x) = 1/x^a$ and using Example 5.6(i). The inequality follows from (6.4) since for $a > 1$, $\int_1^\infty x^{-a}dx = 1/(a-1)$. (For $a > 2$, a better inequality on the right is just $\sum_1^\infty 1/n^a > 1$; the sum of the series is greater than its first term. Thus the real interest is either the upper bound on the left, or for values near 1 when both sides are near $1/(a-1)$.)

For part (ii) consider $f(x) = 1/(x \ln^a x)$ on $[2, \infty)$ (we have to omit the term with $n = 1$ since $\ln 1 = 0$). Then

$$\int_2^b \frac{dx}{x \ln^a x} = \int_{\ln 2}^{\ln b} \frac{dy}{y^a}$$

(put $y = \ln x$), and the integral with respect to y is convergent if and only if $a > 1$. ♦

Example 6.15

Let $h_n = 1 + 1/2 + 1/3 + \cdots + 1/n$, the so-called harmonic numbers. Then

(i) $h_n \geq \ln n \geq h_n - 1$,

(ii) the sequence given by $c_n = h_n - \ln n$ is decreasing, with limit $\gamma \geq 1/2$.

Part (i) is (6.4) with $f(x) = 1/x$. For (ii) we have $c_n - c_{n+1} = -1/(n+1) + \ln((n+1)/n) = \int_n^{n+1} dt/t - 1/(n+1) > 0$ since $1/t > 1/(n+1)$ on $(n, n+1)$, so (c_n) is decreasing. To show that the limit is $\geq 1/2$, notice that on $(n, n+1)$, $1/t$ is concave upwards (i.e. its second derivative is positive – see Exercise 9, Chapter 3) and hence $\int_n^{n+1} dt/t < (1/n + 1/(n+1))/2$. Adding these inequalities from $n = 1$ to $m-1$ gives $\int_1^m dt/t < 1/2 + (1/2 + 1/3 + \cdots + 1/(m-1)) +$

$1/(2m)$ which is equivalent to

$$h_m - \ln m > \frac{1}{2} + \frac{1}{2m}$$

and the result follows. ♦

The limit γ in (ii) is called Euler's constant. Its value is $0.577\ldots$, and although it has been calculated to many thousands of decimal places, it is still a major unsolved problem to determine whether it is rational or not.

6.3 Series with Arbitrary Terms

In the introduction to this chapter we mentioned the series $1 - \frac{1}{2} + \frac{1}{3} - \frac{1}{4} + \cdots + \frac{(-1)^{n-1}}{n} + \cdots$ whose terms alternate in sign. We will write series of this type as $\sum_1^\infty a_n = \sum_1^\infty (-1)^{n-1} b_n$ where $b_n > 0$ for all n. It is a pleasant discovery that provided the terms tend *monotonically* to zero, the series with alternating signs converges (almost the converse of the vanishing condition, Proposition 6.4). This will be generalised later, but the argument in this special case is sufficiently attractive that it is worth giving separately.

Theorem 6.16 (Alternating Series Test)

Let (b_n) be a sequence which decreases to zero. Then the series $\sum_1^\infty (-1)^{n-1} b_n$ is convergent with sum S, and for all $n \geq 1$, $0 \leq (-1)^n (S - s_n) = |S - s_n| \leq b_{n+1}$, where s_m is the partial sum $s_m = \sum_1^m (-1)^{n-1} b_n$.

Proof

We can write the partial sum with an even number of terms in the form

$$s_{2n} = (b_1 - b_2) + (b_3 - b_4) + \cdots + (b_{2n-1} - b_{2n})$$

in which each bracketed term is positive. Hence (s_{2n}) forms an increasing subsequence of (s_n). Similarly for an odd number of terms,

$$s_{2n+1} = b_1 - (b_2 - b_3) - (b_4 - b_5) - \cdots - (b_{2n} - b_{2n+1})$$

and again each bracketed term is positive so (s_{2n+1}) forms a decreasing subsequence of (s_n). Then $s_1 \geq s_{2n+1} = s_{2n} + b_{2n+1} > s_{2n}$ so s_1 is an upper bound for (s_{2n}) which is therefore convergent with limit S say, and (s_{2n+1}) is convergent with the same limit since $s_{2n+1} - s_{2n} = b_{2n+1} \to 0$. Hence $s_{2n+1} \geq$

$S \geq s_{2n}$, and $s_{2n+1} = s_{2n+2} + b_{2n+2} \leq S + b_{2n+2}$ which is the required result for odd suffix. The case when the suffix is even is similar and is left to the reader (Exercise 4). ■

Example 6.17

(i) The series $\sum(-1)^{n-1}/n$, $\sum(-1)^{n-1}/(2n+1)$, and more generally $\sum(-1)^{n-1}/(an+b)$ where a, b are any positive real numbers, are all convergent. (We shall find the sums of the first two in Section 6.5.)

(ii) The series $\sum(-1)^{n-1}/n^a$, $\sum(-1)^{n-1}/(\ln n)^a$ are convergent for any $a > 0$.

(iii) The series $1 - 1 + 1/2 - 1/2^2 + 1/3 - 1/3^2 + \cdots$ whose terms are alternately $1/n$ and $-1/n^2$ is not convergent.

Parts (i) and (ii) are immediate from the theorem, since the terms tend monotonically to zero and alternate in sign. For (iii), notice that we can write the sum of $2m$ terms in the form $\sum_1^m 1/n - \sum_1^m 1/n^2$ and we know from Examples 6.3 and 6.8 that $\sum_1^m 1/n \to \infty$, while $\sum_1^m 1/n^2 < 2$. Thus the sum of $2m$ terms tends to infinity and the series diverges. ♦

The last example illustrates that we cannot deduce convergence when (b_n) does not tend to zero monotonically.

Theorem 6.16 is a special case of the next result which gives a much more general condition on a sequence (c_n) of multipliers such that $\sum b_n c_n$ is convergent whenever b_n decreases to zero.

Theorem 6.18 (Dirichlet's Test)

Let (c_n) be a sequence whose sums $C_m = \sum_1^m c_n$ are bounded, and let (b_n) be a decreasing sequence with limit zero. Then $\sum b_n c_n$ is convergent.

The alternating series test, Theorem 6.16, is the special case in which $(c_n) = \left(1, -1, 1, -1, \ldots, (-1)^{n-1}, \ldots\right)$, when $(C_n) = (1, 0, 1, 0, \cdots)$. Dirichlet's test can be used when the multipliers are formed by repetition of any set of numbers such as $\{1, 0, -3, 2\}$ whose sum is zero. For a less artificial application, see the example following the theorem.

Proof

Consider the partial sums $S_m = \sum_1^m b_n c_n$, which we shall show form a Cauchy sequence. For $r < s$ we have

$$\begin{aligned} S_r - S_s &= b_{r+1}c_{r+1} + \cdots + b_s c_s \\ &= b_{r+1}\left(C_{r+1} - C_r\right) + b_{r+2}\left(C_{r+2} - C_{r+1}\right) + \cdots \\ &\quad + b_{s-1}\left(C_{s-1} - C_{s-2}\right) + b_s\left(C_s - C_{s-1}\right). \end{aligned}$$

In this expression we expand each bracket, and then regroup, collecting the terms with the same suffix for C. This results in

$$\begin{aligned} S_r - S_s &= -b_{r+1}C_r + \left(b_{r+1} - b_{r+2}\right)C_{r+1} + \cdots \\ &\quad + \left(b_{s-1} - b_s\right)C_{s-1} + b_s C_s, \\ |S_r - S_s| &\le b_{r+1}M + \left(b_{r+1} - b_{r+2}\right)M + \cdots + \left(b_{s-1} - b_s\right)M + b_s M \\ &= 2b_{r+1}M \end{aligned}$$

where M is an upper bound for $|C_n|$. Hence since $b_r \to 0$, (S_n) is a Cauchy sequence as required. ■

Example 6.19

If (b_n) decreases to zero, then all of the series $\sum_1^\infty b_n \cos n\theta$, $\sum_1^\infty b_n \sin n\theta$, $\sum_1^\infty b_n e^{in\theta}$ are convergent for $0 < \theta < 2\pi$.

It is enough to show that $\sum_1^\infty b_n e^{in\theta}$ is convergent, since the other two series are its real and imaginary parts. (Look ahead to the next two sections for more on complex numbers.) From Dirichlet's test, it is enough to show that the partial sums $C_m = \sum_1^m e^{in\theta}$ are bounded for $0 < \theta < 2\pi$. But C_m is a geometric sum equal to $e^{i\theta}\left(1 - e^{im\theta}\right) / \left(1 - e^{i\theta}\right)$ and so

$$|C_m| \le \frac{2}{|1 - e^{i\theta}|} = \frac{4}{\sin(\theta/2)}$$

which is finite since $0 < \theta/2 < \pi$. ♦

There is an important difference between series such as $\sum (-1)^{n-1}/n$ whose convergence requires the alternating sign, and $\sum \pm/2^n$ which would converge whatever the choice of sign may be, as we see below. This difference is made precise in the next definition.

Definition 6.20

A series $\sum a_n$ such that $\sum |a_n|$ is convergent is said to be absolutely convergent. A series $\sum a_n$ which is convergent but for which $\sum |a_n|$ diverges is said to be conditionally convergent.

Example 6.21

$\sum (-1)^{n-1}/2^n$ is absolutely convergent, $\sum (-1)^{n-1}/n$ is conditionally convergent.

This follows since $\sum 1/2^n$ is convergent but $\sum 1/n$ is not. ♦

There is an unobvious point about Definition 6.20; the definition of absolute convergence does not require the series to be (ordinarily) convergent, though it would be a misuse of language if this were not so. We take care of this small point in the next proposition.

Proposition 6.22

An absolutely convergent series is convergent.

Proof

The general principle of convergence (Theorem 1.33) allows us to do this very quickly. For if $\sum |a_n|$ converges then for any $\varepsilon > 0$ there is N such that $\sum_{N+1}^{\infty} |a_n| < \varepsilon$. Then also $\left|\sum_{n+1}^{m} a_n\right| \leq \sum_{n+1}^{m} |a_n| < \varepsilon$ for $m, n \geq N$ so the sequence (s_n) of partial sums satisfies $|s_m - s_n| < \varepsilon$ for $m, n \geq N$ and is thus a Cauchy sequence as required. ■

In particular, a series which is absolutely convergent will remain absolutely convergent (and thus convergent) however the signs of the terms are altered; this establishes the remark above that $\sum \pm/2^n$ is convergent for an arbitrary choice of signs.

Combining this result with the ratio test gives the following useful sufficient condition for absolute convergence (whose proof is left to the reader).

Proposition 6.23

The series $\sum a_n$ is absolutely convergent if for some $N \geq 1$ and $r < 1$, $|a_{n+1}/a_n| < r$ for $n \geq N$.

An easy way to tell the difference between absolutely convergent series and conditionally convergent ones is to look at the positive and negative terms separately, as in the next result.

Proposition 6.24

(i) A series is absolutely convergent if and only if the two series formed from its positive and negative terms are both convergent.

(ii) If a series is conditionally convergent, then the two series formed from its positive and negative terms are both divergent.

(iii) A series for which one of the two series formed from its positive and negative terms is convergent, while the other is divergent, must diverge.

Notice that the result in part (i) is both necessary and sufficient, while the result in (ii) is one way only. The result of (i) is sometimes taken as the definition of absolute convergence.

Proof

(i) Given the series $\sum a_n$, let

$$p_n = \begin{cases} a_n & \text{if } a_n > 0, \\ 0 & \text{otherwise,} \end{cases}$$

$$q_n = \begin{cases} -a_n & \text{if } a_n < 0, \\ 0 & \text{otherwise,} \end{cases}$$

so that $\sum p_n$ and $\sum q_n$ are the series formed from the positive and negative terms of $\sum a_n$.

Then $|a_n| = p_n + q_n$ and so if $\sum |a_n|$ is convergent then both $\sum p_n$, $\sum q_n$ are convergent since $0 \leq p_n, q_n \leq |a_n|$. Conversely if both $\sum p_n$, $\sum q_n$ are convergent then $\sum |a_n| = \sum p_n + \sum q_n$ is convergent also. This proves (i).

Since both $p_n, q_n \geq 0$ the sums $\sum p_n$, $\sum q_n$ must either converge or tend to infinity. If one converges and the other does not then since $a_n = p_n - q_n$ it follows that $\sum a_n$ must diverge. This proves (iii), and hence (ii) also, since if the series is not absolutely convergent then by (i) at least one of $\sum p_n$, $\sum q_n$ must diverge, and if only one diverges, then (iii) shows that the series is divergent. ∎

Notice that the series $\sum (-1)^n$ (or less trivially, $1 - 1/\sqrt{2} + 1/3 - 1/\sqrt{4} + \cdots + 1/(2n-1) - 1/\sqrt{2n} + \cdots$) is not convergent and the series of positive and negative terms both diverge. Hence the converse of (ii) is false.

It is tempting (and useful) to consider rearranging the terms of a series, that is to add together the same terms but in a different order. (More formally, a series $\sum b_j$ is a rearrangement of $\sum a_n$ if there is some bijection ϕ of $\mathbb{N}$ such that $b_j = a_{\phi(j)}$ for all j.) This has no effect on the sum of an absolutely convergent series, as we shall show in Theorem 6.27, but the effect on a conditionally convergent series may be dramatic.

Theorem 6.25

Let $\sum a_n$ be a conditionally convergent series. Then its terms may be rearranged to converge to any real number or to $\pm\infty$.

Proof

Let a real number x, say be given. We know from the proof of Proposition 6.24 that the series $\sum p_n, \sum q_n$ are both divergent so that starting at any value m, the partial sums $\sum_m^k p_n, \sum_m^k q_n$ both tend to infinity with k. To form the required rearrangement with sum x, first take terms of $\sum p_n$, say $\sum_1^{k_1} p_n$, taking the smallest value of $k_1 \geq 1$ for which $\sum_1^{k_1} p_n > x$. Then take terms of $\sum q_n$, say $\sum_1^{j_1} q_n$, taking the smallest value of $j_1 \geq 1$ for which $\sum_1^{k_1} p_n - \sum_1^{j_1} q_n < x$. Continue in this way, taking the least integer $k_2 > k_1$ for which $\sum_1^{k_1} p_n - \sum_1^{j_1} q_n + \sum_{k_1+1}^{k_2} p_n > x$, and then the least $j_2 > j_1$ for which $\sum_1^{k_1} p_n - \sum_1^{j_1} q_n + \sum_{k_1+1}^{k_2} p_n - \sum_{j_1+1}^{j_2} q_n < x$, and so on. Since at least one extra term is used at each stage this gives a genuine rearrangement of the series. Also, since the terms tend to zero (the series is conditionally convergent), the rearranged series tends to x. The modification required to show that the sums of the rearranged series may tend to $\pm\infty$ is left as an exercise. ■

Example 6.26

Let the sum of the series $\sum_1^{\infty} (-1)^{n-1}/n$ be S. Then $S = \ln 2$, and the sum of

$$1 + \frac{1}{3} - \frac{1}{2} + \frac{1}{5} + \frac{1}{7} - \frac{1}{4} + \cdots + \frac{1}{4n+1} + \frac{1}{4n+3} - \frac{1}{2n+2} + \cdots$$

is $3\ln 2/2$.

In Example6.15(ii) we showed that $h_n = 1 + 1/2 + \cdots + 1/n = \ln n + c_n$ where $c_n \to \gamma$ as $n \to \infty$. It follows that

$$\begin{aligned} 1 - 1/2 + 1/3 - 1/4 + \cdots + 1/(2n-1) - 1/(2n) &= \\ 1 + 1/2 + 1/3 + 1/4 + \cdots + 1/(2n-1) + 1/(2n) - & \\ 2(1/2 + 1/4 + \cdots + 1/(2n)) &= h_{2n} - h_n = \\ \ln(2n) - \ln n + c_{2n} - c_n &\to \ln 2 \text{ as } n \to \infty. \end{aligned}$$

Similarly

$$\begin{aligned}1+\frac{1}{3}-\frac{1}{2}+\frac{1}{5}+\frac{1}{7}-\frac{1}{4}+\cdots+\frac{1}{4n+1}+\frac{1}{4n+3}-\frac{1}{2n+2} &= \\ 1+\frac{1}{3}+\frac{1}{5}+\frac{1}{7}+\cdots+\frac{1}{4n+1}+\frac{1}{4n+3}- & \\ (1/2+1/4+\cdots+1/(2n+2)) &= \\ h_{4n+4}-h_{2n+2}/2-h_{n+1}/2 &= \\ \ln(4n+4)-(\ln(2n+2)+\ln(n+1))/2+ & \\ c_{4n+4}-c_{2n+2}/2-c_{n+1}/2 &\to \\ \ln\left(4/\sqrt{2}\right) &= 3\ln 2/2\end{aligned}$$

as $n\to\infty$. ♦

Theorem 6.27

Any rearrangement of an absolutely convergent series is absolutely convergent with the same sum.

Proof

Let $\sum a_n$ be the given series with sum s and let $(p_n),(q_n)$ be as in the proof of Proposition 6.24, so that $s=\sum p_n-\sum q_n$. Let $\sum b_j$ be any rearrangement of $\sum a_n$, and suppose that $\varepsilon>0$ is given. Then there is some N such that $\sum_{N+1}^{\infty}p_n$ and $\sum_{N+1}^{\infty}q_n$ are both $<\varepsilon$ and hence both $\sum_S p_n, \sum_S q_n<\varepsilon$ where the sums are over any finite set S of integers, all of whose elements are $>N$; in particular $\left|s-\sum_1^N a_n\right|<\varepsilon$. Now take N_1 such that $\{b_1,b_2,\ldots,b_{N_1}\}$ contains all of $a_1,a_2,\ldots,a_N$. Then for $k\geq N_1$, $\sum_1^k b_j-\sum_1^N a_n$ contains only terms b_j which correspond to terms a_n with $n>N$, so $-\varepsilon<\sum_1^k b_j-\sum_1^N a_n<\varepsilon$, and it follows that $\left|s-\sum_1^k b_j\right|<2\varepsilon$ as required. ■

Related to the rearrangement of series is the question of double series. These are of the form $\sum\sum a_{m,n}$ where m,n are integers and the sums may be taken over the positive integers, or over all integers. (Now is a good time to take a quick look back at the second half of Section 1.5 which considers double sequences – the definitions and results here are direct analogues.) We shall concentrate on series of the form $\sum_{n\geq 0}\sum_{m\geq 0}a_{m,n}$ leaving the others as exercises.

By analogy with ordinary series we consider $\sum_{n=0}^{N}\sum_{m=0}^{M}a_{m,n}$ and investigate what happens when $M,N\to\infty$.

Definition 6.28

(i) The double sum

$$\sum_{n=0}^{\infty}\sum_{m=0}^{\infty} a_{m,n}$$

(no brackets! – compare (ii)) is defined as the double limit (if it exists) of $\sum_{n=0}^{N}\sum_{m=0}^{M} a_{m,n}$ as $M, N \to \infty$ in the sense of Definition 1.35.

(ii) Suppose that for each $n \geq 0$ the series $\sum_{m=0}^{\infty} a_{m,n}$ is convergent with sum s_n, and that the series $\sum_{n=0}^{\infty} s_n$ is convergent with sum S. Then we write

$$S = \sum_{n=0}^{\infty} s_n = \sum_{n=0}^{\infty}\left(\sum_{m=0}^{\infty} a_{m,n}\right)$$

and call S the iterated sum over m, then n. The iterated sum

$$T = \sum_{m=0}^{\infty}\left(\sum_{n=0}^{\infty} a_{m,n}\right)$$

is defined similarly.

After Section 1.5 we expect that in general the values of these sums (if they exist) may be different.

Example 6.29

For $m, n \geq 0$, let $a_{m,n} = 1$ if $m = n$, $a_{m,n} = -1$ if $n = m+1$, otherwise $a_{m,n} = 0$.

Then for all m, $\sum_{n=0}^{\infty} a_{m,n} = 0$ so $\sum_{m=0}^{\infty}(\sum_{n=0}^{\infty} a_{m,n})$ exists and is zero. For $n = 0$, $\sum_{m=0}^{\infty} a_{m,n} = 1$, otherwise $\sum_{m=0}^{\infty} a_{m,n} = 0$, so $\sum_{n=0}^{\infty}(\sum_{m=0}^{\infty} a_{m,n})$ exists and equals 1. The double sum does not exist. ♦

We concentrate on the cases in which the sums are equal. One such result follows at once from Theorem 1.39(ii).

Proposition 6.30

If $a_{m,n} \geq 0$ for all m, n then the three sums in Definition 6.28 are either all finite and equal, or all infinite.

Proof

Since $a_{m,n} \geq 0$, the sums $s_{M,N} = \sum_{n=0}^{N}\sum_{m=0}^{M} a_{m,n}$ are increasing in both M and N and Theorem 1.39(ii) applies. ■

The other important case in which all the double sums are equal is when one (and hence all) of the series are absolutely convergent.

Theorem 6.31

Suppose $(a_{m,n})$ is such that any one of the sums $\sum\sum |a_{m,n}|$ in Definition 6.28 is finite. Then all of the sums $\sum\sum a_{m,n}$ exist and are equal.

Proof

Let $p_{m,n}$ and $q_{m,n}$ be as in the proof of Proposition 6.24. Then since both are positive and $\leq |a_{m,n}|$ it follows by the proposition that all the sums $\sum\sum p_{m,n}$ are equal and finite, and the same is true for $\sum\sum q_{m,n}$ (and the real and imaginary parts if the terms are complex valued). Hence all the sums $\sum\sum a_{m,n} = \sum\sum p_{m,n} - \sum\sum q_{m,n}$ are finite and equal. ■

The next two sections contain several striking examples of this result in action. Notice that the rearrangement of terms in the double sums, which is implicit in Definition 6.28, is different from the rearrangements which were considered in Theorems 6.25 and 6.27, since now if we fix attention on one term, say $a_{r,s}$, then infinitely many terms may be added before $a_{r,s}$ is reached (but see Exercise 12 for a result which includes both of these cases). See Exercise 6 for the effect of grouping of terms of a series.

6.4 Power Series

In this section and the next, we shall need complex numbers for a proper understanding of what is going on. We summarise the properties which we shall need; for a more detailed introduction the entertaining discussion in [14] is recommended.

To begin with we need only the Cartesian form of a complex number; the polar form is considered later. We recall the following properties.

(i) A complex number z is an ordered pair of real numbers (x, y), where x, y are (illogically) called the real and imaginary parts of $z : x = \Re z,\ y = \Im z$. The set of complex numbers is denoted by $\mathbb{C}$.

(ii) Addition and multiplication in $\mathbb{C}$ are defined by

$$\begin{aligned}(x,y)+(u,v) &= (x+u, y+v)\\ (x,y)(u,v) &= (xu-yv, xv+yu).\end{aligned}$$

The real number x is identified with the complex number $(x, 0)$, so that the definitions of addition and multiplication are compatible with those for real and complex numbers, and we may regard $\mathbb{R}$ as being embedded in $\mathbb{C}$. The operations are commutative and associative, and multiplication is distributive over addition.

The complex number $(0,1)$ is denoted by i, so that $i^2 = (0,1)(0,1) = (-1,0) = -1$. Each complex number $z = (x, y)$ can thus be written $z = (x, 0) + (0, y) = x + iy$ (and it will always be implied when we write $z = x + iy$ that x, y are real: $x = \Re z$, $y = \Im z$).

(iii) Each $(x, y) \neq (0,0)$ has a multiplicative inverse given by

$$(x,y)^{-1} = \left(\frac{x}{x^2+y^2}, \frac{-y}{x^2+y^2}\right).$$

(iv) There is no ordering on $\mathbb{C}$ which extends the usual ordering on $\mathbb{R}$ and at the same time has the properties required of an ordering (in such an ordering we would have either $i > 0$ or $i < 0$ and both lead to a contradiction since $i^2 = -1$). Thus inequalities between complex numbers are meaningless.

(v) The absolute value (modulus) $|z|$ of a complex number $z = x + iy$ is the positive root $|z| = \sqrt{x^2 + y^2}$. In the Argand diagram (the plane $\mathbb{R}^2$ with the complex number $x + iy$ identified with the point of $\mathbb{R}^2$ whose coordinates are x, y), $|z|$ represents the distance of z from the origin, and $|z - w|$ represents the distance from z to w. For $a \in \mathbb{C}$, $r > 0$ (notice that it is not necessary to say that r is real – the inequality $r > 0$ implies this) the disc $\{z : |z - a| < r\}$ will be denoted by $S(a, r)$.

(vi) The absolute value satisfies the inequalities (between real numbers!)

$$\text{for } z = x + iy, \qquad \max(|x|, |y|) \le |z| \le |x| + |y|, \tag{6.6}$$

$$\text{for } z, w \in \mathbb{C}, \qquad ||z| - |w|| \le |z + w| \le |z| + |w|. \tag{6.7}$$

Convergence of complex sequences and series is defined in the natural way:

Definition 6.32

The sequence (z_n) of complex numbers is convergent with limit l if $|z_n - l| \to 0$ as $n \to \infty$ (for each $\varepsilon > 0$ there is an integer N such that $|z_n - l| < \varepsilon$ if $n \ge N$).

The series $\sum_1^\infty c_n$ (with complex terms) is convergent if the partial sums $s_n = \sum_1^n c_j$ form a convergent sequence.

The series $\sum_1^\infty c_n$ is absolutely convergent if the series $\sum_1^\infty |c_j|$ is convergent.

Fortunately these definitions relate to the convergence of the real and imaginary part of the terms in the expected way.

Proposition 6.33

(i) Let (z_n) be a sequence with $z_n = x_n + iy_n$. Then (z_n) is convergent with limit $l = l_1 + il_2$ if and only if $(x_n), (y_n)$ are convergent with limits l_1, l_2 respectively. Similarly a series $\sum_1^\infty c_n$ where $c_n = a_n + ib_n$ is convergent if and only if $\sum_1^\infty a_n$ and $\sum_1^\infty b_n$ are convergent and

$$\sum_1^\infty c_n = \sum_1^\infty a_n + i\sum_1^\infty b_n.$$

(ii) A series $\sum_1^\infty c_n$ where $c_n = a_n + ib_n$ is absolutely convergent if and only if $\sum_1^\infty a_n$ and $\sum_1^\infty b_n$ are absolutely convergent.

Proof

This all follows as an exercise in the use of (6.6). For instance in (ii),

$$\sum_1^\infty \max\left(|a_n|, |b_n|\right) \le \sum_1^\infty |c_n| = \sum_1^\infty |a_n| + \sum_1^\infty |b_n|.$$

The details (if necessary!) are left to the reader. ∎

A series of the form $\sum a_n z^n$, or more generally $\sum a_n (z - z_0)^n$, in which (a_n) is a sequence of real or complex numbers, and z, z_0 are real or complex, is called a power series. We shall show that as a function of z, its region of convergence is always a disc with a subset (possibly empty) of its boundary, as in the following examples.

Example 6.34

For any real a, the series $\sum_{n=1}^\infty n^a z^n$ is convergent for all (complex) $|z| < 1$, divergent for $|z| > 1$. If $a \ge 0$ the series is divergent when $|z| = 1$. If $0 > a \ge -1$ it is divergent at $z = 1$, convergent for other z with $|z| = 1$. If $a < -1$ it is absolutely convergent for all z with $|z| = 1$.

For $c_n = n^a z^n$ we have $|c_{n+1}/c_n| = (1 + 1/n)^a |z| \to |z|$ as $n \to \infty$. Hence Proposition 6.23 shows that $\sum c_n$ is absolutely convergent (and hence convergent also) for $|z| < 1$ and divergent for $|z| > 1$.

If $a \ge 0$ the terms do not tend to zero if $|z| = 1$ and so the series diverges there. If $0 > a \ge -1$ the series is divergent (Example 6.10(i)) if $z = 1$, but convergent for other z with $|z| = 1$ from Example 6.19. If $a < -1$ the series is absolutely convergent when $|z| = 1$ since $\sum n^a$ is convergent. ♦

Proposition 6.35

Suppose that for some $z \neq 0$, the series $\sum a_n z^n$ is convergent. Then the series is absolutely convergent for all w with $|w| < |z|$.

Proof

If $\sum a_n z^n$ is convergent, then in particular the terms are bounded, say $|a_n z^n| \leq M$ for all n. Then for w with $|w| < |z|$, $|a_n w^n| \leq M\,|w/z|^n$ and so $\sum a_n w^n$ is absolutely convergent by comparison with the geometric series $\sum |w/z|^n$. Thus as soon as we have convergence for some z, we have absolute convergence for all $w \in S(0, |z|)$. ■

It follows that the region of convergence is always a circular disc whose radius R is the least upper bound of the values of r for which $(a_n r^n)$ is bounded. Outside this disc (i.e. for $|z - z_0| > R$), the series is divergent.

Definition 6.36

The radius of convergence of the series $\sum a_n (z - z_0)^n$ is the radius R, of the largest disc $S(z_0, R)$ on which it is absolutely convergent. $S(z_0, R)$ is called the disc (or sometimes the circle) of convergence.

The proof of Proposition 6.35 gives a little more, which is stated as part (i) of the next result.

Proposition 6.37 (Abel's Lemma)

(i) Let $r > 0$ be such that the sequence $(a_n r^n)$ is bounded. Then the series $\sum a_n z^n$ is absolutely convergent for $|z| < r$.

(ii) Suppose $|a_{n+1}/a_n| \to l$ as $n \to \infty$. Then if $0 < l < \infty$, the series $\sum a_n z^n$ is absolutely convergent for $|z| < 1/l$, divergent for $|z| > 1/l$. If $l = 0$, the series is absolutely convergent for all z; if $l = \infty$, the series is convergent at $z = 0$ only.

Proof

Part (i) is already proved, and (ii) is immediate from the ratio test. ■

Example 6.38

The series $\sum z^n/n!$ has $l = 0$ and is absolutely convergent for all z; the series $\sum n!z^n$ has $l = \infty$ and is convergent only for $z = 0$. All the series in Example 6.34 have $l = 1$. ♦

Neither part of the theorem gives any information about what happens when $|z| = R$; this must be investigated for each series individually as we did in Example 6.34.

The striking fact about a function which is given by the sum of a power series is that it is infinitely differentiable, and moreover its Taylor series is absolutely convergent to f at each point in the disc of convergence. We prove this now by a somewhat strenuous elementary argument based on rearrangement of absolutely convergent series. A reader who does not want to get their hands dirty can instead look ahead to Section 6.6 where there is an alternative proof using uniformity of convergence.

Theorem 6.39

Let $f(z) = \sum_{n=0}^{\infty} a_n (z - z_0)^n$ where the series has radius of convergence $R > 0$. Then

(i) f has derivatives of all orders on $S(z_0, R)$ given by

$$f^{(k)}(z) = \sum_{n=k}^{\infty} n(n-1)\cdots(n-k+1)\, a_n (z - z_0)^{n-k}, \quad k = 1, 2, \ldots, \text{ and} \tag{6.8}$$

(ii) for each point $w \in S(z_0, R)$, f is equal to the sum of the Taylor series about w,

$$f(z) = \sum_{n=0}^{\infty} \frac{f^{(n)}(w)}{n!} (z - w)^n$$

whose radius of convergence is at least $R - |w - z_0|$.

Proof

Most of the effort goes into proving (6.8) with $k = 1$. To simplify the writing we take $z_0 = 0$ (this involves no loss of generality, since we can otherwise take a new variable $w = z - z_0$) so that $f(z) = \sum_0^{\infty} a_n z^n$. Note to begin with that all the differentiated series (6.8) have the same radius of convergence, since if $(a_n r^n)$

is bounded, then so is $\left(n^k a_n t^n\right)$ for any $t < r$ and $k \geq 1$. For $z, w \in S(z_0, R)$,

$$f(w) - f(z) = \sum_1^\infty a_n (w^n - z^n),$$

$$\frac{f(w) - f(z)}{w - z} - \sum_1^\infty n a_n z^{n-1} = \sum_1^\infty a_n \left\{ \frac{w^n - z^n}{w - z} - n z^{n-1} \right\}$$

$$= \sum_1^\infty a_n \left\{ \left(w^{n-1} + \cdots + z^{n-1}\right) - n z^{n-1} \right\}$$

$$= \sum_2^\infty a_n \{ z^{n-2} (w - z) + z^{n-3} \left(w^2 - z^2\right) + \cdots + z^0 \left(w^{n-1} - z^{n-1}\right) \}$$

$$= (w - z) \sum_2^\infty a_n \{ z^{n-2} + z^{n-3} (w + z) + \cdots + z^0 \left(w^{n-2} + \cdots + z^{n-2}\right) \}.$$

Hence with $\rho = \max(|z|, |w|)$,

$$\left| \frac{f(w) - f(z)}{w - z} - \sum_1^\infty n a_n z^{n-1} \right| \leq |w - z| \sum_2^\infty |a_n| \rho^{n-2} (1 + 2 + \cdots (n-1))$$

$$\leq \frac{|w - z|}{2} \sum_2^\infty n^2 |a_n| \rho^{n-2}$$

where the series on the right is convergent as noted above. Hence

$$f'(z) = \lim_{w \to z} \frac{f(w) - f(z)}{w - z} = \sum_1^\infty n a_n z^{n-1}$$

as required for the case $k = 1$.

Applying this result repeatedly gives

$$f''(z) = \sum_2^\infty n(n-1) a_n z^{n-2}$$

$$\cdots$$

$$f^{(k)}(z) = \sum_{n=k}^\infty n(n-1) \cdots (n-k+1) a_n z^{n-k}$$

and the rest of (i) is proved.

For (ii) let $|w| < R$ and $|h| < R - |w|$. Then the series $\sum |a_n| (|w| + |h|)^n$ is convergent and so the following rearrangements (using Theorem 6.31) are valid:

$$\begin{aligned} f(w+h) &= \sum_{n=0}^{\infty} a_n (w+h)^n \\ &= \sum_{n=0}^{\infty} a_n \sum_{k=0}^{n} \binom{n}{k} w^{n-k} h^k \qquad \text{(by the binomial theorem)} \\ &= \sum_{k=0}^{\infty} \frac{h^k}{k!} \sum_{n=k}^{\infty} a_n n (n-1) \cdots (n-k+1) w^{n-k} \\ &= \sum_{k=0}^{\infty} \frac{f^{(k)}(w)}{k!} h^k \end{aligned}$$

which gives (ii). ■

An example with $>$ in (ii) is given in Section 6.5.

A particularly observant reader may have noticed that Theorem 6.39 involves a new kind of derivative, namely

$$f'(z) = \lim_{h \to 0} \frac{f(z+h) - f(z)}{h}$$

where the limit is taken as $h \to 0$ through complex values – i.e. for all h in some disc $S(0, \delta)$. If you noticed this then you are ready for a first course in complex analysis – if on the other hand it makes you feel nervous, then you can close your eyes to the complex numbers, and read the theorem as if only real numbers were involved (when for instance the disc $S(a, r)$ would be replaced by the interval $(a - r, a + r) \subset \mathbb{R}$); the statement and proof are unaltered with this understanding.

6.5 Exponential and Trigonometric Functions

In many of the examples so far we have used the elementary exponential, logarithmic and trigonometric functions in an informal way, relying on the reader's previous knowledge and experience. At the same time it is important to know how they are defined precisely, and we shall do that in this section. We begin with the exponential function and its most immediate properties – others will appear as we go along.

Definition 6.40

The exponential function $\exp z$ is the sum of the series $\sum_{n=0}^{\infty} z^n/n!$ which is absolutely convergent for all $z \in \mathbb{C}$.

Notice that we are avoiding the notation e^x until we have defined general powers below.

Theorem 6.41

(i) exp is its own derivative: $\exp' z = \exp z$ for all $z \in \mathbb{C}$.

(ii) For all $z, w \in \mathbb{C}$, $\exp(z + w) = \exp z \exp w$; for integer n, $(\exp z)^n = \exp(nz)$.

(iii) For real x, $\exp x$ is positive and strictly increasing on $\mathbb{R}$ with $\exp 0 = 1$, $\exp x \to 0$ as $x \to -\infty$, $\exp x \to \infty$ as $x \to \infty$.

Proof

Theorem 6.39(i) allows us to differentiate the series to get

$$\exp' z = \sum_{n=1}^{\infty} z^{n-1}/(n-1)! = \exp z$$

which proves (i). Then Theorem 6.39(ii) gives

$$\exp(z + w) = \sum_{n=0}^{\infty} \exp z \, w^n/n! = \exp z \exp w$$

which is (ii), and the second statement follows by induction. (Alternatively we could simply multiply the series for $\exp z, \exp w$ and rearrange using Theorem 6.31 and the binomial theorem to get $\exp(z + w)$.)

If x is real then $\exp x$ is real and $\exp x = (\exp(x/2))^2 \geq 0$. But $\exp x$ cannot be zero for any x (real or complex) since $\exp x \exp(-x) = \exp 0 = 1$, so $\exp x$ is strictly positive for real x. Then exp is strictly increasing on $\mathbb{R}$ since $\exp' = \exp > 0$. It tends to infinity as $x \to \infty$ since for instance if $x > 0$ then $\exp x = \sum_{n=0}^{\infty} x^n/n! > x$, and finally $\exp x \to 0$ as $x \to -\infty$ since $\exp x \exp(-x) = 1$. ■

The logarithm may be defined either as the inverse of exp on $(0, \infty)$, or as an integral, or a power series. Slightly inconsistently, we use ln (after Euler) for the real logarithm on $(0, \infty)$, and log for its extension to complex values.

Definition 6.42

The real, or natural logarithm is defined on $(0,\infty)$ as the inverse of the real exponential:

$$\text{for } x>0,\ y=\ln x \text{ if and only if } x=\exp y.$$

The definition makes sense since we have just shown that exp is strictly increasing from $\mathbb{R}$ to $(0,\infty)$.

Theorem 6.43

For $x>0$,

(i) $\ln x$ is strictly increasing with $\ln 1=0$, $\ln x\to-\infty$ as $x\to 0+$, $\ln x\to\infty$ as $x\to\infty$, and

(ii) $\ln' x=1/x$ for all $x>0$ and $\ln x=\int_1^x dt/t$.

(iii) For all $u,v>0$, $\ln(uv)=\ln u+\ln v$.

(iv) For $-1<x\le 1$, $\ln(1+x)=\sum_{n=1}^{\infty}(-1)^{n-1}x^n/n$; more generally for $a>0$ and $-a<x\le a$, $\ln(a+x)=\ln a+\sum_{n=1}^{\infty}(-1)^{n-1}(x/a)^n/n$.

(The sum $\ln 2$ for $\sum_{n=1}^{\infty}(-1)^{n-1}/n$ was found in Example 6.26.)

Proof

Part (i) is immediate from part (iii) of Theorem 6.41 and (ii) comes from the rule for differentiating inverse functions, Theorem 3.18: if $f(x)=\ln x=y$ then $f^{-1}(y)=\exp y$ and so

$$f'(x)=\frac{1}{(f^{-1})'(y)}=\frac{1}{\exp y}=\frac{1}{x}.$$

For (iii), put $u=\exp x$, $v=\exp y$ so that $uv=\exp x\exp y=\exp(x+y)$, and $\ln(uv)=x+y=\ln u+\ln v$.

To get the power series for $g(x)=\ln(1+x)$ we calculate the derivatives, $g^{(n)}(x)=(-1)^{n-1}(n-1)!/(1+x)^n$, and use Taylor's theorem with integral remainder, Theorem 4.23. This gives

$$\ln(1+x)=x-\frac{x^2}{2}+\cdots+\frac{(-1)^{n-1}}{n}x^n+r_n(x)$$

where

$$r_n(x)=\frac{1}{n!}\int_0^x(x-t)^n g^{(n+1)}(t)\,dt=\frac{(-1)^n}{n!}\int_0^x\frac{(x-t)^n n!}{(1+t)^{n+1}}dt.$$

For $0 \leq x \leq 1$ this gives $|r_n(x)| \leq \int_0^x (1+t)^{-n-1}\,dt \leq 1/n \to 0$ as $n \to \infty$. For $x < 0$, put $x = -a$, with $0 < a < 1$. Then $|r_n(x)| = \int_0^a (a+t)^n / (1+t)^{n+1}\,dt \leq (2a/(1+a))^n \to 0$ as $n \to \infty$ since $(a+t)/(1+t)$ is increasing and $0 < 2a < a+1$. The second series follows from $\ln(a+x) = \ln a + \ln(1+x/a)$. ■

Suppose that we are given some $a > 0$ and rational $r = m/n$ where m, n are integers with $n > 0$. Then $a = \exp b$ for some real b and so $a^m = (\exp b)^m = \exp(mb) = \exp(nrb) = (\exp(rb))^n$ from Theorem 6.41(ii). Thus $\exp(rb)$ is the real n^{th} root of a^m, and we have shown that for rational r and $a > 0$, $a^r = a^{m/n} = \exp(rb) = \exp(r \ln a)$. This motivates the following definition of general real powers.

Definition 6.44

For real x, and $a > 0$, define $a^x = \exp(x \ln a)$. In particular $\exp x = e^x$ where $e = \exp 1$.

Powers have the expected properties (whose proofs are Exercise 16). See Definition 6.52 for the corresponding definition of complex powers.

Proposition 6.45

(i) For $a > 0$ and real x, y, $a^{x+y} = a^x a^y$. In particular $a^0 = 1$ and $a^{-x} = 1/a^x$.

(ii) For $a, b > 0$ and real x, $(ab)^x = a^x b^x$ and $\left(a^b\right)^x = a^{(bx)}$.

The exponential function increases sufficiently fast on the real axis as $x \to \infty$ that it outpaces any power of x – correspondingly ln increases so slowly that it is outpaced by any positive power of x as we see in the next result.

Proposition 6.46

For any $a > 0$, $x^{-a} \exp x \to \infty$ and $x^{-a} \ln x \to 0$ as $x \to \infty$.

Proof

Choose an integer $n > a$. For $x > 0$, $\exp x > x^n/n!$ (from the power series) so $x^{-a} \exp x > x^{n-a}/n! \to \infty$ as $x \to \infty$. Now put $y = \ln x$ for $x > 1$ so that $y > 0$. Then $x^{-a} \ln x = y/(\exp y)^a = \left(y^{1/a}/\exp y\right)^a$ which tends to zero as we have just proved. ■

At this stage, we have defined the exponential function on $\mathbb{C}$, but the natural logarithm only on $\mathbb{R}$. When we want to investigate the behaviour of the logarithm for complex values we find that we need to consider the trigonometric functions too. We shall use a geometric definition – for a treatment which is wholly analytic see for instance [11].

Definition 6.47

On the unit circle $x^2+y^2=1$ measure a distance t along the perimeter, starting with $t=0$ at the point $(1,0)$, where $t>0$ corresponds to the anticlockwise, $t<0$ to the clockwise direction (Figure 6.1). The coordinates $P=(x,y)$ of the point reached in this way define the (real) trigonometric functions $x=\cos t$, $y=\sin t$; for all t we have $\cos^2 t+\sin^2 t = x^2+y^2=1$. The value of t defines the radian measure of the angle POA between OP and the x-axis. For $x\neq 0$, $\tan t$ is defined as $y/x=\sin t/\cos t$.

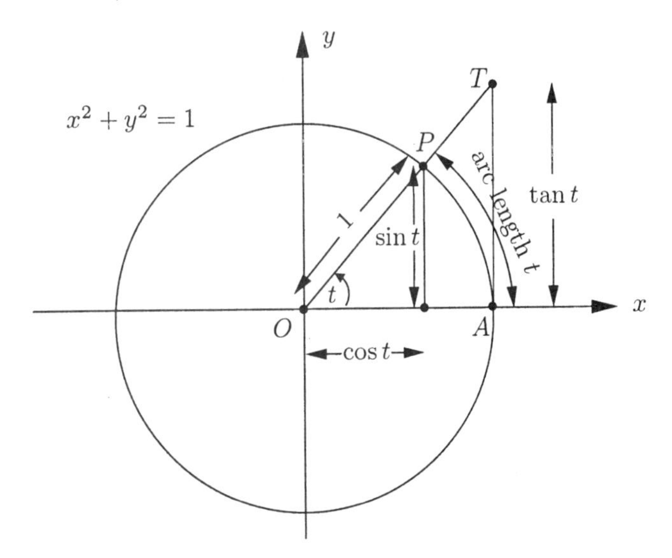

Figure 6.1 Definition 6.47. Adapted from *Handbook of Applicable Mathematics*, Volume 4: Analysis, W. Ledermann and S. Vajda (Eds), 1982. ©John Wiley & Sons Limited. Reproduced with permission.

It is important to notice that this definition defines the trigonometric functions for all real values of t at once – we do not have to start in the first quadrant and then extend to other values of t. The facts (i) that the perimeter of each quadrant of the unit circle is of length $\pi/2$ and (ii) that each of the coordinates varies continuously and monotonically in each quadrant, give us immediately the following properties of cos, sin.

Proposition 6.48

Each of the functions cos, sin is continuous on $\mathbb{R}$ and has period 2π. Also

(i) cos is even and strictly decreasing from 1 to -1 on $[0, \pi]$, strictly increasing from -1 to 1 on $[\pi, 2\pi]$; $\cos t = 0$ if and only if t is an odd multiple of $\pi/2$.

(ii) sin is odd and strictly increasing from -1 to 1 on $[-\pi/2, \pi/2]$, strictly decreasing from 1 to -1 on $[\pi/2, 3\pi/2]$; $\sin t = 0$ if and only if t is an integer multiple of π.

(iii) tan is continuous on $\mathbb{R}$ except at odd multiples of $\pi/2$ where cos is zero. It has period π, and is strictly increasing from $-\infty$ to ∞ on $(-\pi/2, \pi/2)$.

The next step is to establish the addition theorems for cos and sin and the familiar rules for their derivatives.

Proposition 6.49

(i) For all real t, u,

$$\cos(t+u) = \cos t \cos u - \sin t \sin u, \tag{6.9}$$

$$\sin(t+u) = \sin t \cos u + \cos t \sin u. \tag{6.10}$$

(ii) For $0 < t < \pi/2$, $\sin t < t < \tan t$. For $0 < |t| < \pi/2$, $\cos t < (\sin t)/t < 1$. Also

$$\lim_{t \to 0} \frac{\sin t}{t} = 1. \tag{6.11}$$

(iii) Each of cos, sin has derivatives of all orders, with $\cos' t = -\sin t$, $\sin' t = \cos t$, and $\cos^{(2k)} t = (-1)^k \cos t$, $\sin^{(2k)} t = (-1)^k \sin t$ for all integers k.

Proof

(i) Consider the points $P_1 = (x_1, y_1) = (\cos t, \sin t)$ and $P_2 = (x_2, y_2) = (\cos u, \sin u)$. The distance P_1P_2 is given by

$$\begin{aligned}(P_1P_2)^2 &= (\cos t - \cos u)^2 + (\sin t - \sin u)^2 \\ &= 2 - 2\cos t \cos u - 2 \sin t \sin u.\end{aligned}$$

If we subtract u from each argument we get the points $P_3 = (\cos(t-u), \sin(t-u))$ and $A = (1, 0)$. This does not affect the distance between the points:

$P_1P_2 = P_3A$. But

$$\begin{aligned}(P_3A)^2 &= (\cos(t-u)-1)^2 + (\sin(t-u)-0)^2 \\ &= 2 - 2\cos(t-u),\end{aligned}$$

and comparing these expressions gives $\cos(t-u) = \cos t \cos u + \sin t \sin u$. Now replace u by $-u$ to get the result for $\cos(t+u)$. Putting $u = \pm\pi/2$ gives $\cos(t \pm \pi/2) = \mp \sin t$, and then $u - \pi/2$ for u in (6.9) gives (6.10).

(ii) In Figure 6.1, the area of the triangle AOP is $(\sin t)/2$ (half base times height), the area of the sector of the circle bounded by OA, OP is $t/2$ (the area of a sector with angle t in a circle of radius r is $r^2t/2$ since the area of the whole circle is $\pi r^2 = r^2(2\pi)/2$), and finally the area of the triangle OAT is $(\tan t)/2$. Comparing these gives the first inequality. Rearranging gives $\cos t < (\sin t)/t < 1$ for $0 < t < \pi/2$ and since both sides are now even, the same is true for $0 > t > -\pi/2$. The limit follows at once since $\cos t \to 1$ as $t \to 0$.

(iii) To show that cos, sin are differentiable, consider

$$\begin{aligned}\frac{\cos(t+h)-\cos t}{h} &= \frac{\cos t\cos h - \sin t \sin h - \cos t}{h} \\ &= \cos t\frac{\cos h - 1}{h} - \sin t\frac{\sin h}{h} \\ &= -\cos t\frac{\sin^2(h/2)}{h/2} - \sin t\frac{\sin h}{h}.\end{aligned}$$

We just proved that $(\sin h)/h \to 1$ as $h \to 0$ and so $\left(\sin^2(h/2)\right)/(h/2) = \sin(h/2)(\sin(h/2)/(h/2)) \to 0$ which gives $\cos' = -\sin$. The result for sin is proved similarly, and the rest follows by repeated application of the first derivatives. ■

Since we are not writing a textbook of trigonometry we shall assume all the familiar consequences of (i) – for instance the double angle formulae $\cos 2t = \cos^2 t - \sin^2 t = 1 - 2\sin^2 t$ and $\sin 2t = 2\sin t\cos t$ without further explanation or comment – indeed we already did so in the proof of (iii) above.

Example 6.50

Show that

$$\begin{aligned}\lim_{t\to 0}\frac{1-\cos t}{t^2} &= \frac{1}{2}, \\ \lim_{t\to 0}\frac{t\cos t - \tan t}{t^3} &= -\frac{5}{6}.\end{aligned}$$

These both come from L'Hôpital's rule. For instance

$$\lim_{t\to 0}\frac{1-\cos t}{t^2}=\lim_{t\to 0}\frac{\sin t}{2t}=\frac{1}{2},$$

where the reader is expected to check that the conditions for L'Hôpital's rule are satisfied, and to find the second limit similarly. ♦

Theorem 6.51

(i) The functions cos, sin have the expansions in Taylor series

$$\begin{aligned}\cos z &= \sum_{n=0}^{\infty}(-1)^n\frac{z^{2n}}{(2n)!},\\ \sin z &= \sum_{n=0}^{\infty}(-1)^n\frac{z^{2n+1}}{(2n+1)!}\end{aligned}\qquad(6.12)$$

which converge for all complex z, and hence define the extensions of cos, sin to $\mathbb{C}$.

(ii) For all $z\in\mathbb{C}$, $\exp iz=\cos z+i\sin z$ and

$$\begin{aligned}\cos z &= \frac{\exp(iz)+\exp(-iz)}{2},\\ \sin z &= \frac{\exp(iz)-\exp(-iz)}{2i}.\end{aligned}\qquad(6.13)$$

The addition formulae for cos, sin with complex arguments follow from this and the addition formulae for exp.

(iii) $\exp z=1$ if and only if $z=2k\pi i$ for some integer k. Hence exp is periodic with period $2\pi i$ in $\mathbb{C}$. The real zeros of cos, sin noted in Proposition 6.48 are their only zeros in $\mathbb{C}$.

Proof

All derivatives of cos, sin are equal to $\pm\cos$, $\pm\sin$ and so are bounded by 1 on $\mathbb{R}$. Hence the remainder in Taylor's theorem 3.23 tends to zero for all real x and so the series converge to cos, sin on $\mathbb{R}$. The ratio test shows that in fact the series converge for all complex z, giving the extensions to $\mathbb{C}$.

For (ii), we combine the series to get

$$\begin{aligned}\cos z+i\sin z &= \sum_{n=0}^{\infty}(-1)^n\frac{z^{2n}}{(2n)!}+i\sum_{n=0}^{\infty}(-1)^n\frac{z^{2n+1}}{(2n+1)!}\\ &= \sum_{n=0}^{\infty}\frac{(iz)^{2n}}{(2n)!}+\sum_{n=0}^{\infty}\frac{(iz)^{2n+1}}{(2n+1)!}=\sum_{n=0}^{\infty}\frac{z^n}{n!}=\exp z\end{aligned}$$

and (6.13) follows by combining the results for z and $-z$. It follows that for real y, $|\exp iy|^2 = \cos^2 y + \sin^2 y = 1$, and hence that $|\exp(x+iy)| = \exp x$ is > 1 if $x > 0$, < 1 if $x < 0$. Thus $\exp(x+iy) = \exp 0 = 1$ if and only if both $x = 0$ and $\exp iy = \cos y + i \sin y = 1$, which implies $y = 2k\pi i$. Thus generally, $\exp z = \exp w$ if and only if $z = w + 2k\pi i$ as required for the first part of (iii). The results for cos, sin are left to the reader (Exercise 17). ■

Now that we have the trigonometric functions, we can put a general complex number into polar form. For $z = x + iy \neq 0$, let $r = |z| = \sqrt{x^2 + y^2} \neq 0$, so $u = (x + iy)/r$ has $|u| = 1$ and so u is a point of the unit circle. It follows from Definition 6.47 that there are infinitely many values of t with $u = \cos t + i \sin t$, and if we fix the unique $t_0 \in (-\pi, \pi]$ then all other values of t are of the form $t = t_0 + 2k\pi i$ for some integer k. (The requirement that $t = \tan^{-1}(y/x)$ has to be used with care since it determines t only to within a multiple of π.) Such values of t are called arguments of z; t_0 is called the principal argument and denoted by $\arg z$. This leads to the next definition.

Definition 6.52

(i) **(Polar Form)** Any $z \in \mathbb{C}$, $z \neq 0$ can be written as $z = r(\cos t + i \sin t)$ where $r = |z| > 0$ and $t = t_0 + 2k\pi i$ for uniquely defined $t_0 = \arg z \in (-\pi, \pi]$ and integer k. If $z = 0$, then $\arg z$ is undefined.

(ii) **(Complex Logarithms)** Any $z \in \mathbb{C}$, $z \neq 0$ can be written as $z = \exp w$ where $w = \ln|z| + it$, t being any argument of z. The value $\ln|z| + it_0$ where $t_0 = \arg z \in (-\pi, \pi]$ is called the principal logarithm, and denoted $\log z$.

(iii) **(Complex Powers)** For any $z, w \in C$, $z \neq 0$, the (principal) complex power z^w is defined as $\exp(w \log z)$. Other values are given by $\exp(w(\log z + 2k\pi i))$ for integer k.

Parts (ii) and (iii) are the extension to complex numbers of the earlier Definitions 6.42 and 6.44. The following examples are for the reader to play with.

Example 6.53

(i) $i = \exp(i\pi/2)$, $\log i = i\pi/2$, $i^i = e^{-\pi/2}$.

(ii) $-1 = \exp(i\pi)$, $\log(-1) = i\pi$, $(-1)^{-1/2} = \exp(-i\pi/2) = -i$. ♦

The notation $\exp z = e^z$ is now available to us, but shall stay with exp for the rest of the section for the sake of continuity.

Proposition 6.54

(i) If both $|\arg z|, |\arg w| < \pi/2$, then $\log(zw) = \log z + \log w$.

(ii) If both $|\arg z|, |\arg w| < \pi/2$, then $(zw)^s = z^s w^s$ for all complex s.

(iii) For all complex $s \neq 0$, $s^z s^w = s^{(z+w)}$.

(iv) For all complex z and integer n, $(\cos z + i \sin z)^n = \cos(nz) + i \sin(nz)$.

Proof

The restriction on the arguments is to ensure that $|\arg(z+w)| < \pi$. Then both (i) and (ii) are immediate from the definition. No such restriction is needed for (iii). Part (iv) is the complex form of de Moivre's theorem and follows at once from Theorems 6.41(ii) and 6.51(ii). ■

Without the restriction on the arguments these results may be false – for instance if $z = w = \exp(3i\pi/4)$ then $zw = \exp(3i\pi/2) = -i$ and $\arg(zw) = -\pi/2 \neq \arg z + \arg w$.

Although everything seems to have been covered, there is still one point to take care of. The power series $\ln x = \sum_{n=1}^{\infty} (-1)^{n-1} x^n/n$ in Theorem 6.43(iv) is absolutely convergent for (real) $|x| < 1$ and hence is also convergent for complex $|x| < 1$. We need to be sure that this is consistent with the definition of complex logarithms given above. A method, due to Cauchy, which could be used at this point is to consider the binomial expansion, Theorem 4.24, for $(1+z)^a$ and rearrange this in powers of a (really!), when the limit

$$\lim_{a \to 0} \frac{(1+z)^a - 1}{a}$$

turns out to be both $\log(1+z)$ and $\sum (-1)^{n-1} z^n/n$. But this is seriously messy to justify, and the method which we use in the next section (Example 6.66) is much simpler.

6.6 Sequences and Series of Functions

It often happens in mathematics that an idea which is useful for one class of objects can then be applied more widely. Consider for instance convergence, which was defined for sequences and series of numbers and will now be considered for sequences of functions. To keep things simple we shall concentrate on the case in which the functions are defined on an interval $[a, b]$ of $\mathbb{R}$.

The most obvious form of convergence is to say that the sequence (f_n) of functions is convergent with limit f if $f_n(x) \to f(x)$ at each point of $[a, b]$. This gives our first definition.

Definition 6.55

We say that the sequence (f_n) of functions is convergent pointwise on $[a, b]$ with limit f if for each $x \in [a, b]$, $f_n(x) \to f(x)$ as $n \to \infty$.

If we analyse this definition we find that for each $x \in [a, b]$ and $\varepsilon > 0$ we require an integer N depending on ε and x such that $|f_n(x) - f(x)| < \varepsilon$ when $n \geq N$. This dependence on x turns out to be a weakness, and pointwise convergence fails to have many of the properties we should like as the following examples illustrate.

Example 6.56

(i) Let $f_n(x) = x^n$ on $[0, 1]$. Each f_n is continuous and $f_n(x) \to 0$ for $0 \leq x < 1$, $f_n(1) \to 1$. The limit function is not continuous on $[0, 1]$

(ii) Let $f_n(x) = (\sin nx)/n$ on $[0, \pi]$. Each f_n is differentiable and $f_n(x) \to 0$ for all $x \in [0, \pi]$. The limit function is 0 and is differentiable, but the sequence of derivatives $(\cos nx)$ does not converge at any point of $(0, \pi)$ (Exercise 15, Chapter 1).

(iii) Let $f_n(x) = \sqrt{x^2 + 1/n^2}$ on $\mathbb{R}$. Each f_n is differentiable and $f_n(x) \to |x|$ for $x \in \mathbb{R}$. The limit function is not differentiable at $x = 0$.

(iv) Let $f_n(x) = n^2 x^n (1 - x)$ on $[0, 1]$. Each f_n is continuous on $[0, 1]$ and $f_n(x) \to 0$ for $0 \leq x \leq 1$, but $\int_0^1 f_n$ does not $\to 0$.

All these examples are immediate, except possibly for (iv) in which we used Example 1.24(iii) to show that $f_n \to 0$, and where

$$\int_0^1 f_n = n^2 \int_0^1 \left(x^n - x^{n+1}\right) dx = n^2 \left(\frac{1}{n+1} - \frac{1}{n+2}\right) \to 1 \text{ as } n \to \infty. \blacklozenge$$

To rescue the situation we have to replace pointwise by uniform convergence. This was Definition 4.8 and is restated here for convenience.

A sequence (f_n) of functions such that for each $\varepsilon > 0$ there is some integer N with $|f_n(x) - f(x)| < \varepsilon$ for all $x \in [a, b]$ and $n \geq N$ is said to be uniformly *convergent with limit f on $[a,b]$.*

By contrast with the definition of pointwise convergence, the same value of N works for all $x \in [a,b]$. It gives the result we want on continuity.

Theorem 6.57

Let (f_n) be a sequence of functions on $[a,b]$ which converges uniformly to f on $[a,b]$, and suppose that each f_n has a right-hand limit, l_n say, at a. Then (i) f has a right-hand limit at a, (ii) the sequence (l_n) has a limit as $n \to \infty$, and (iii) the limits are equal,

$$f(a+) = \lim_{n\to\infty} l_n.$$

The corresponding result holds for left-hand limits at b.

Proof

The definition of uniform convergence gives that for each $\varepsilon > 0$ there is some N such that

$$|f_n(x) - f(x)| < \varepsilon \qquad \text{for all } n \geq N \text{ and } x \in [a,b].$$

It follows that for $m, n \geq N$,

$$|f_n(x) - f_m(x)| \leq |f_n(x) - f(x)| + |f(x) - f_m(x)| < 2\varepsilon$$

and letting $x \to a+$ shows that $|l_n - l_m| \leq 2\varepsilon$. Hence (l_n) is a Cauchy sequence, $l_n \to L$ say, as $n \to \infty$, and letting $m \to \infty$ gives $|l_n - L| \leq 2\varepsilon$ if $n \geq N$.

Since $f_N(x) \to l_N$ as $x \to a+$, there is some $\delta > 0$ with $|f_N(x) - l_N| < \varepsilon$ for $a < x < a + \delta$. For such x,

$$\begin{aligned} |f(x) - L| &\leq |f(x) - f_N(x)| + |f_N(x) - l_N| + |l_N - L| \\ &\leq \varepsilon + \varepsilon + 2\varepsilon \end{aligned}$$

so $f(x) \to L$ as $x \to a+$ as required. ■

If we apply this result at each point of an interval we get the following corollary, which is the usual form in which the result is needed.

Corollary 6.58

(i) The limit of a uniformly convergent sequence of continuous functions is continuous.

(ii) The limit of a uniformly convergent sequence of regulated functions is regulated.

Theorem 6.57 can be stated in the form

$$\lim_{n\to\infty}\lim_{x\to a+} f_n(x) = \lim_{x\to a+}\lim_{n\to\infty} f_n(x)$$

and so belongs to the class of results which give conditions under which two limiting processes may be interchanged. Many examples have already shown that this is not generally true, but now at least we have a rule of thumb (which we shall not attempt to formulate precisely) – to interchange two limiting processes, look for some uniformity in one of the variables.

Going back to Example 6.56(i), it can be proved that the limit of a pointwise convergent sequence of continuous functions must have *some* points of continuity, but this proof is not at all easy – see [5, Example (6.92)].

The next example illustrates the usual way of showing that a sequence is uniformly convergent: find the maximum of $|f_n - f|$ and show that it tends to zero.

Example 6.59

The sequence given by $f_n(x) = x^n(1-x)$ is uniformly convergent with limit zero on $[0,1]$.

The maximum of $x^n(1-x)$ is at $x = n/(n+1)$ where $0 \le f_n(x) = (n/(n+1))^n/(n+1) < 1/(n+1) \to 0$ as $n \to \infty$. ♦

This and similar examples suggest that a sequence of continuous functions which converges monotonically to a continuous limit must converge uniformly (continuity of the limit is required to avoid the situation shown in Example 6.56(i)).

Theorem 6.60 (Dini's Theorem)

Let (f_n) be a sequence of continuous functions on $[a,b]$ which converges monotonically to a continuous function f on $[a,b]$. Then f_n converges uniformly to f on $[a,b]$.

Proof

Assume that $f = 0$ and that $f_n(x)$ decreases to 0 for each $x \in [a,b]$ (otherwise consider $\pm(f_n - f)$). Let m_n be the least upper bound of f_n; m_n is a decreasing sequence and our aim is to prove that $m_n \to 0$. If not then for some $\delta > 0$, $m_n > \delta$ for all n, and hence for each n, there is some point x_n with $f_n(x_n) > \delta$. We know from Corollary 1.29 that some subsequence of (x_n) is convergent; let z be its limit. The hypothesis tells us that $f_n(z) \to 0$ so there is some N with $f_N(z) < \delta$, and since f_N is continuous at z, there is some $\varepsilon > 0$ such that $f_N(x) < \delta$ for all x with $|x - z| < \varepsilon$. Then the monotonicity shows that $f_n(x) < \delta$ for all x with $|x - z| < \varepsilon$ and $n \geq N$. Since z is the limit of some subsequence of (x_n) we can choose x_n with $n \geq N$ and $|x_n - z| < \varepsilon$, when we would have both $f_n(x_n) < \delta$, and $f_n(x_n) > \delta$ from the definition of (x_n), which is impossible, and the result is proved. ■

A simple way of showing that a series is uniformly convergent is to estimate each term separately, hoping to get a convergent series of real numbers. This motivates the next result, the Weierstrass M-test.

Theorem 6.61 (M-test)

The series $\sum_{n=1}^{\infty} f_n(x)$ is uniformly convergent on $[a, b]$ if there is some convergent series $\sum_{n=1}^{\infty} m_n$ such that $|f_n(x)| \leq m_n$ for all $x \in [a, b]$.

Proof

Another Cauchy sequence argument: given $\varepsilon > 0$ there is some N such that $\sum_{j=n+1}^{m} m_j < \varepsilon$ for $m, n \geq N$. Then with $s_n(x) = \sum_{j=1}^{n} f_j(x)$, $|s_m(x) - s_n(x)| = \left|\sum_{n+1}^{m} f_j(x)\right| \leq \sum_{n+1}^{m} |f_j(x)| \leq \sum_{j=n+1}^{m} m_j < \varepsilon$. Thus for each x, $(s_n(x))$ is a Cauchy sequence which must therefore have a limit, $s(x)$ say. Then letting $m \to \infty$ we get $|s(x) - s_n(x)| \leq \varepsilon$ for $n \geq N$ and so $s_n(x) \to s(x)$ uniformly as required. ■

In the course of the proof we established another small technicality which is worth stating separately:

Proposition 6.62

A sequence $(f_n(x))$ is uniformly convergent on $[a, b]$ if and only if it is 'uniformly Cauchy': for each $\varepsilon > 0$ there is some N such that $|f_n(x) - f_m(x)| < \varepsilon$ for all $m, n \geq N$ and $x \in [a, b]$.

Example 6.63

(i) The sum of the series $\sum_{n=0}^{\infty} (\cos(2^n x)) / 2^n$ is continuous on $\mathbb{R}$.

(ii) More generally, if $\sum a_n$ is absolutely convergent and (b_n) is any sequence of real numbers, then the series $\sum a_n \cos(b_n x)$ and $\sum a_n \sin(b_n x)$ are uniformly convergent, and their sums are continuous on $\mathbb{R}$.

(iii) For each $n \geq 1$, define a function f_n on $[0,1]$ by $f_n = 0$ on $[0, 1/(n+1)]$ and $[1/n, 1]$, $f_n(p_n) = 1/n$ where $p_n = 1/(n+1/2)$, and the graph of f_n is given by straight lines joining $(1/(n+1), 0)$ to $(p_n, 1/n)$, and $(p_n, 1/n)$ to $(1/n, 0)$ (Figure 6.2). Then $\sum f_n(x)$ is uniformly convergent on $[0,1]$ but the M-test is not satisfied.

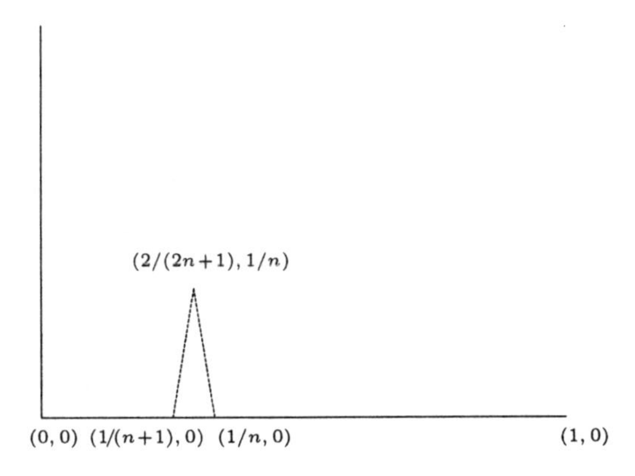

Figure 6.2 Example 6.63(iii).

Part (ii) follows from the M-test since for real b_n and x, $|\cos(b_n x)| \leq 1$, etc. and (i) is a special case. The series in (iii) is pointwise convergent with sum s, say, since for any x, at most one term $f_n(x)$ is non-zero. It is uniformly convergent since $\left|s(x) - \sum_{j=1}^{n} f_j(x)\right| \leq 1/(n+1)$, this being the maximum of the largest term remaining. ♦

We promised in the introduction to Chapter 3 to prove the existence of continuous nowhere differentiable functions. Such examples were at one time regarded as pathological monstrosities and an affront to intuition, but with time they have become accepted and find applications in modelling such natural phenomena as Brownian motion of particles subject to repeated collisions. Many different constructions are known; the following is one of the simplest.

Example 6.64

Let $f_1(x) = |x|$ for $|x| \leq 1/2$, and be extended to $\mathbb{R}$ with period 1 so that f_1 gives the distance from x to the nearest integer. For $n \geq 2$, let $f_n(x) =$

$2^{-n+1}f_1(2^{n-1}x)$, and let $f(x) = \sum_1^\infty f_n(x)$. Then f is continuous on $\mathbb{R}$ with period 1 by the M-test, but is not differentiable at any point.

The property of f is most easily established by considering the so-called dyadic rational numbers, i.e. those of the form $k/2^n$ for integers k, n with $n \geq 0$. Let $E_n = \{k/2^j : k \in Z, 0 \leq j \leq n\}$ and observe that the complement of E_n consists of disjoint intervals of length 2^{-n}. Each f_n is continuous on $\mathbb{R}$, and has slope alternately ± 1 on the intervals complementary to E_n, starting with $+1$ on $[0, 2^{-n}]$. Let $s_n(x) = \sum_1^n f_j(x)$ be the partial sum of the series defining f; note that if $x \in E_n$ then $f_j(x) = 0$ for $j > n$ and so $f(x) = s_n(x)$ for such x. Each s_n is continuous on R, and its slope at a point not in E_n is given by taking the slope of s_{n-1} and adding ± 1 to it on the two halves resulting from the division of the intervals of E_{n-1}. For instance s_1 has slopes $\{+1, -1\}$ on $[0, 1/2]$, $[1/2, 1]$ and s_2 has slopes given by $\{+2, 0, 0, -2\}$ on $[0, 1/4]$, $[1/4, 1/2]$, $[1/2, 3/4]$, $[3/4, 1]$. Similarly s_3 and s_4 have slopes given by $\{3, 1, 1, -1, 1, -1, -1, -3\}$ and $\{4, 2, 2, 0, 2, 0, 0, -2, 2, 0, 0, -2, 0, -2, -2, 4\}$ and so on. With this preparation we can show that f is nowhere differentiable.

The argument takes different forms depending on whether the point t under consideration is itself a dyadic rational or not. Suppose first that it is not. We construct sequences $(x_n), (y_n)$ with $x_n < t < y_n$ such that x_n, y_n both $\to t$ as $n \to \infty$ but $(f(y_n) - f(x_n)) / (y_n - x_n)$ has no limit as $n \to \infty$; if f were differentiable at t, this would contradict the result of Exercise 3, Chapter 3 which showed that the limit of such a sequence must be the value of the derivative at the point. The construction of the sequences $(x_n), (y_n)$ is easy: we just define (x_n, y_n) to be the interval of E_n which contains t. The discussion above shows that $(f(y_n) - f(x_n)) / (y_n - x_n) = (s_n(y_n) - s_n(x_n)) / (y_n - x_n)$ is an integer k_n say, with $k_n - k_{n-1} = \pm 1$ so (k_n) has no limit, as required.

In case t is a dyadic rational, we define x_n, y_n as the elements of E_n immediately to the left and right of t. We leave it to the reader to show that in this case the limits $(f(y_n) - f(t)) / (y_n - t)$ and $(f(t) - f(x_n)) / (t - x_n)$ are equal to $\pm\infty$ respectively, and hence that f is not differentiable at t. ♦

Many of the most interesting applications of uniform convergence involve integration or differentiation.

Theorem 6.65

If each f_n is regulated on $[a, b]$ and $f_n \to f$ uniformly as n$\to \infty$, then f is regulated and

$$\int_a^b f_n \to \int_a^b f.$$

Proof

This one is easy. We already know that f is regulated from Corollary 6.58. Since the convergence is uniform, given $\varepsilon > 0$ there is some N with $|f_n(x) - f(x)| < \varepsilon$ for $n \geq N$ and $x \in [a, b]$. It follows that for $n \geq N$,

$$\left| \int_a^b f_n - \int_a^b f \right| = \left| \int_a^b (f_n - f) \right| \leq \int_a^b |f_n - f| \leq (b - a)\,\varepsilon$$

and the result is proved. ∎

Easy as it is, this result has lots of nice consequences like the following.

Example 6.66

Let $r \in [0, 1)$. Then

(i) the series

$$\begin{aligned} \sum_{n=0}^{\infty} r^n \cos n\theta &= \frac{1 - r\cos\theta}{1 - 2r\cos\theta + r^2}, \\ \sum_{n=1}^{\infty} r^n \sin n\theta &= \frac{r\sin\theta}{1 - 2r\cos\theta + r^2} \end{aligned} \tag{6.14}$$

are uniformly convergent for $\theta \in \mathbb{R}$ to the given sums.

(ii) Similarly

$$\sum_{n=1}^{\infty} \frac{r^n}{n} \cos n\theta = -\frac{1}{2} \ln\left(1 - 2r\cos\theta + r^2\right), \tag{6.15}$$

$$\sum_{n=1}^{\infty} \frac{r^n}{n} \sin n\theta = \tan^{-1}\left(\frac{r\sin\theta}{1 - r\cos\theta}\right). \tag{6.16}$$

(iii) For $|z| < 1$,

$$\log(1 + z) = \sum_{n=1}^{\infty} \frac{(-1)^{n-1}}{n} z^n. \tag{6.17}$$

For $|z| < 1$ we know that $1/(1 - z) = \sum_{n=0}^{\infty} z^n$, and hence putting $z = re^{i\theta}$, $0 \leq r < 1$, $\theta \in R$,

$$\frac{1}{1 - r(\cos\theta + i\sin\theta)} = \frac{1 - r\cos\theta + i\sin\theta}{1 - 2r\cos\theta + r^2} = \sum_{n=0}^{\infty} r^n (\cos n\theta + i\sin n\theta)$$

and (6.14) follows on equating real and imaginary parts. The series are uniformly convergent by the M–test since $r < 1$.

It follows by subtracting 1 from both sides, that

$$\sum_{n=1}^{\infty} r^n \cos n\theta = \frac{r\cos\theta - r^2}{1 - 2r\cos\theta + r^2} = \frac{d}{d\theta}\left(\tan^{-1}\left(\frac{-r\sin\theta}{1 - r\cos\theta}\right)\right)$$

(which should be checked by the reader), and the second result in (6.15) follows on integrating termwise using Theorem 6.65. Similarly

$$\frac{d}{d\theta}\left(\ln\left(1 - 2r\cos\theta + r^2\right)\right) = \frac{2r\sin\theta}{1 - 2r\cos\theta + r^2}$$

and the other part follows. But from Definition 6.52(ii),

$$\begin{aligned}
\log(1-z) &= \log(1 - r\cos\theta - i\sin\theta) \\
&= \frac{1}{2}\ln\left((1 - r\cos\theta)^2 + (r\sin\theta)^2\right) + i\tan^{-1}\left(\frac{-r\sin\theta}{1 - r\cos\theta}\right) \\
&= -\sum_{n=1}^{\infty}\frac{r^n}{n}\cos n\theta - i\sum_{n=1}^{\infty}\frac{r^n}{n}\sin n\theta = -\sum_{n=1}^{\infty}\frac{z^n}{n}
\end{aligned}$$

using de Moivre's theorem 6.54(iv), and so putting $-z$ for z

$$\log(1+z) = \sum_{n=1}^{\infty}\frac{(-1)^{n-1}}{n}z^n. \ \blacklozenge$$

With a little care, we can extend Theorem 6.65 to infinite intervals.

Theorem 6.67 (Dominated Convergence)

Let f_n, f, g be regulated functions on $[a,\infty)$ such that (i) for every $b > a$, $f_n \to f$ uniformly on $[a,b]$, (ii) for all $x \geq a$, $|f_n(x)| \leq g(x)$, and (iii) $\int_a^\infty g$ is finite. Then $\int_a^\infty f$ exists and $\int_a^\infty f_n \to \int_a^\infty f$ as $n \to \infty$.

The restriction (iii) is needed to avoid the situation in Example 6.56(iv). The function g is said to *dominate* the sequence (f_n), which gives the theorem its name. The result remains true if the condition of uniform convergence in (i) is replaced by pointwise convergence; however it is then very deep and one of the most important theorems of the Lebesgue theory. Our version is quite enough to give many interesting applications.

Proof

Inevitably we begin with $\varepsilon > 0$. By (iii) we can choose $b > a$ with $\int_b^\infty g < \varepsilon$, and by (i) there is some N such that $|f_n(x) - f(x)| < \varepsilon/(b-a)$ for $n \geq N$ and $x \in [a, b]$. Then

$$\begin{aligned} \left| \int_a^\infty f_n - \int_a^\infty f \right| &\leq \int_a^b |f_n - f| + \int_b^\infty |f_n - f| \\ &\leq \int_a^b \varepsilon/(b-a) + 2\int_b^\infty g = 3\varepsilon \end{aligned}$$

as required. ■

The theorem holds also if $n \to \infty$ is replaced by a continuous parameter h, say with $f_h \to f$ as $h \to 0$. This version will be useful in the next chapter.

Example 6.68

For $\Re x > 0$, $\int_0^n t^{x-1}(1-t/n)^n\,dt \to \int_0^\infty t^{x-1}e^{-t}dt$ as $n \to \infty$.

We can show, as in the discussion of Example 1.17(ii), that on any interval $[0, b]$ the sequence $(1-t/n)^n$ is increasing with limit e^{-t}, and is hence uniformly convergent on $[0, b]$ by Dini's theorem 6.60. For $\Re x \geq 1$ this allows us to deduce the result directly from the theorem, since then t^{x-1} is bounded and $t^{x-1}(1-t/n)^n\,dt \to t^{x-1}e^{-t}$ uniformly. If $0 < \Re x < 1$ then, given $\varepsilon > 0$, choose $\delta > 0$ such that $\left|\int_0^\delta t^{x-1}dt\right| < \varepsilon$, and apply the above reasoning on $[\delta, \infty]$. ♦

The corresponding result for differentiation is as follows.

Theorem 6.69

Suppose (i) each f_n is differentiable and f_n' is continuous on (a, b), (ii) $f_n(c)$ is convergent for some $c \in (a, b)$, and (iii) f_n' is uniformly convergent to ϕ on (a, b). Then on (a, b), (f_n) is uniformly convergent to a differentiable function f and $f' = \phi$.

The result is true without supposing f_n' is continuous, but is harder to prove (there is a proof in [1]), and we can manage without it. Notice that in Example 6.56(iii) the sequence $\sqrt{x^2 + 1/n^2}$ is uniformly convergent but the derivatives fail to converge, so that the condition on the derivatives is necessary here. The condition on $f_n(c)$ is required to avoid such examples as $f_n(x) = n$ on $[0, 1]$.

The conclusion can be stated as

$$\frac{d}{dx}(\lim f_n(x)) = \lim\left(\frac{d}{dx}f_n(x)\right).$$

Proof

Since we assume continuity of f_n', ϕ must be continuous and we can write

$$f_n(x) = f_n(c) + \int_c^x f_n'(t)\,dt = f_n(c) + \int_c^x (f_n'(t) - \phi(t))\,dt + \int_c^x \phi(t)\,dt.$$

Given $\varepsilon > 0$, choose N such that $|f_n'(x) - \phi(x)| < \varepsilon$ for all $x \in (a,b)$ and $n \geq N$. This gives

$$\left|f_n(x) - f_n(c) - \int_c^x \phi(t)\,dt\right| < \varepsilon(b-a)$$

so (f_n) is uniformly convergent to $f(x) = \lim f_n(c) + \int_c^x \phi(t)\,dt$ and hence $f' = \phi$. ■

From this result, we can deduce from Example 6.66 the sums of $\sum nr^n \cos n\theta$, etc. (Exercise 21). Another important application is to the derivative of the Gamma function. We shall need the result of Example 6.15 in which $h_n = 1 + 1/2 + \cdots + 1/n$ and $\gamma = \lim_{n\to\infty}(h_n - \ln n)$.

Example 6.70

For $x \in \mathbb{R}$, not an integer ≤ 0,

$$\frac{\Gamma'(x)}{\Gamma(x)} = -\gamma + \sum_{n=1}^{\infty}\left(\frac{1}{n} - \frac{1}{x+n-1}\right). \tag{6.18}$$

In particular, $\Gamma'(1) = -\gamma$.

The formal calculation runs as follows. We have from Definition 5.11 that

$$\frac{1}{\Gamma(x)} = \lim_{n\to\infty}\frac{x(x+1)(x+2)\cdots(x+n-1)}{n!n^{x-1}} = \lim_{n\to\infty} f_n(x)$$

say, where f_n is differentiable and (see Exercise 21, Chapter 3)

$$\begin{aligned}
\frac{f_n'(x)}{f_n(x)} &= \frac{1}{x} + \frac{1}{x+1} + \cdots + \frac{1}{x+n-1} - \ln n \\
&= (h_n - \ln n) + \left(\frac{1}{x} - 1\right) + \left(\frac{1}{x+1} - \frac{1}{2}\right) + \cdots + \left(\frac{1}{x+n-1} - \frac{1}{n}\right) \\
&\to \gamma - \sum_{n=1}^{\infty}\left(\frac{1}{n} - \frac{1}{x+n-1}\right) \qquad \text{as } n \to \infty
\end{aligned}$$

and the result follows. To make this rigorous we need to show that the limits of (f_n) and (f_n') are attained uniformly. For (f_n), we re-examine the argument which proves Lemma 5.10. In that proof the estimate $|g(y)| \leq |x(1-x)|/y^2$ was established for all $y > 2|x|$ and so the M–test gives uniform convergence of the limit of (f_n) which defines Γ. For f_n' the series in (6.18) also converges uniformly by the M–test since $|1/n - 1/(x+n-1)| \leq 2|x-1|/n^2$ for $n \geq 2|x-1|$.

(The result is also true for complex x, but we kept to real values since Theorem 6.69 is proved in this case only. This should not worry us unduly, since the only use we shall make of it is the value of $\Gamma'(1)$.) ♦

Theorem 6.67 on integration has a corresponding result for series (and another for products, Theorem 6.76) which is quite straightforward, and makes a good way to round off this section.

Theorem 6.71

Let $(t_{m,n})_{m,n=1}^{\infty}$ be a double sequence of real or complex terms such that

(i) for each n, $\sum_{m=1}^{\infty} t_{m,n}$ is convergent with sum s_n,

(ii) for each m, $(t_{m,n})$ is convergent with $t_{m,n} \to u_m$ as $n \to \infty$, and

(iii) there is a convergent series $\sum_{m=1}^{\infty} a_m$ of positive terms with $|t_{m,n}| \leq a_m$ for all m, n.

Then $\sum_{m=1}^{\infty} u_m$ is convergent with sum s say, and $s_n \to s$ as $n \to \infty$.

The final conclusion can be written

$$\lim_{n\to\infty} \left(\sum_{m=1}^{\infty} t_{m,n} \right) = \sum_{m=1}^{\infty} \left(\lim_{n\to\infty} u_m \right)$$

and is thus another of the results which allows us to interchange the order of two limiting operations. It is surely false without the existence of the dominant series $\sum a_m$ as can be seen for instance if $t_{m,n} = 1/n$ for $1 \leq m \leq n$, $t_{m,n} = 0$ otherwise, when $u_m = 0$ for all m, but $s_n = 1$ for all n.

Proof

Since $t_{m,n} \to u_m$ and $|t_{m,n}| \leq a_m$ it follows that $|u_m| \leq a_m$ and hence $\sum_{m=1}^{\infty} u_m$ is (absolutely) convergent; let its sum be s. Given $\varepsilon > 0$, choose M such that $\sum_{M+1}^{\infty} |u_m| \leq \sum_{M+1}^{\infty} a_m < \varepsilon$ and hence $\sum_{M+1}^{\infty} |t_{m,n}| < \varepsilon$ for all n

also. For this M, choose N such that $|t_{m,n} - u_m| < \varepsilon/M$ when $m = 1, 2, \ldots, M$ and $n \geq N$. Then

$$\begin{aligned} s_n - s &= \sum_{m=1}^{\infty} t_{m,n} - \sum_{m=1}^{\infty} u_m, \\ |s_n - s| &\leq \sum_{m=1}^{M} |t_{m,n} - u_m| + \sum_{m=M+1}^{\infty} |t_{m,n} - u_m| \\ &\leq M(\varepsilon/M) + 2\varepsilon = 3\varepsilon \end{aligned}$$

as required. ■

Example 6.72

For $z \in \mathbb{C}$, $(1 + z/n)^n \to e^z$.

We already did this for real z, so this extension to complex values both illustrates the power of the dominated converge theorem, and gives a unified proof covering all cases.

Consider the binomial expansion

$$\begin{aligned} \left(1 + \frac{z}{n}\right)^n &= \sum_{k=0}^{n} \binom{n}{k} \left(\frac{z}{n}\right)^k = \sum_{k=0}^{n} \frac{n(n-1)(n-2)\cdots(n-k+1)}{k!} \left(\frac{z}{n}\right)^k \\ &= \sum_{k=0}^{n} \frac{z^k}{k!} \left(1 - \frac{1}{n}\right)\left(1 - \frac{2}{n}\right)\cdots\left(1 - \frac{k-1}{n}\right). \end{aligned}$$

In this series, each of the bracketed terms is < 1, and so the series is dominated by $\sum_0^{\infty} |z|^k / k!$ which is convergent. Hence Theorem 6.71 allows us to take the limit of each term on the right to get

$$\left(1 + \frac{z}{n}\right)^n \to \sum_{k=0}^{\infty} \frac{z^k}{k!} = e^z$$

as required. ♦

6.7 Infinite Products

When we defined series, we did so by taking a sequence (a_n) and forming a new sequence by adding the terms together in order. If we use multiplication in place of addition we get a product whose convergence can be investigated in the same way as for series. Before proceeding, notice that we already came very close to doing this in the case of the Gamma function, when in Definition 5.11 we wrote

$$\Gamma(x) = \frac{1}{F(x)} = \lim_{n\to\infty} \frac{n! n^{x-1}}{x(x+1)(x+2)\cdots(x+n-1)}.$$

In this expression, write $n^x = e^{x \ln n}$ and recall the result of Example 6.15 in which $h_n = 1 + 1/2 + \cdots + 1/n = \ln n + c_n$ where $c_n \to \gamma$ as $n \to \infty$. Then we have successively

$$\begin{aligned}
\frac{1}{\Gamma(x)} &= \lim_{n\to\infty} \frac{x(x+1)(x+2)\cdots(x+n)}{n! n^x} \\
&= x \lim_{n\to\infty} \left\{(1+x/1)(1+x/2)\cdots(1+x/n)\, e^{-x(h_n - c_n)}\right\} \\
&= x \lim_{n\to\infty} e^{xc_n} \lim_{n\to\infty} \left\{(1+x/1)\, e^{-x}(1+x/2)\, e^{-x/2}\cdots(1+x/n)\, e^{-x/n}\right\} \\
&= xe^{\gamma x} \prod_{n=1}^{\infty} \left\{(1+x/n)\, e^{-x/n}\right\}
\end{aligned}$$

in the (self-explanatory) notation shortly to be defined. But we are getting ahead too quickly.

Definition 6.73

Let (b_n) be a sequence of real or complex numbers, and let $p_n = b_1 b_2 \cdots b_n$. The we say that the product $\prod_1^\infty b_n$ is convergent with value P if either (i) all $b_n \neq 0$ and $p_n \to P \neq 0$ as $n \to \infty$, or (ii) for some N, $b_n \neq 0$ for $n \geq N$, and $q_n = b_N b_{N+1} \cdots b_n \to Q \neq 0$ as $n \to \infty$ when $P = b_1 b_2 \cdots b_{N-1} Q$.

The reason for the concern about the zero values in the definition is that we don't want to produce a kind of 'false convergence' because of the presence of one or more zero factors while the rest of the factors behave badly. For instance we do not want to say either that $0.1.2.\cdots n$ converges because of the one zero term, or that

$$\frac{1}{2}\frac{2}{3}\cdots\frac{n-1}{n}$$

converges when the limit is zero (this would spoil Theorem 6.74(i) below).

The analogue of the Vanishing Condition, Proposition 6.4, is that for convergence it is necessary (though not sufficient) that $b_n \to 1$ as $n \to \infty$, and for this reason it is conventional to write $b_n = 1 + a_n$ in infinite products. The following simple result will be sufficient for our purposes – the general theory which is quite delicate can be found in [1] or [10].

Theorem 6.74

(i) If all $a_n \geq 0$ then the products $\prod(1+a_n)$ and $\prod(1-a_n)$ are convergent if and only if the series $\sum a_n$ is convergent.

(ii) The product $\prod(1+a_n)$ is convergent if the series $\sum a_n$ is absolutely convergent.

The absolute convergence in (ii) is sufficient but not necessary as can be seen by considering the product

$$\begin{aligned}\prod_{n=1}^{\infty}\left(1+\frac{(-1)^{n-1}}{n}\right) &= (1+1)\left(1-\frac{1}{2}\right)\left(1+\frac{1}{3}\right)\left(1-\frac{1}{4}\right)\cdots \\ &= 2.\frac{1}{2}.\frac{4}{3}.\frac{3}{4}.\cdots\left(1+\frac{1}{2k-1}\right)\left(1-\frac{1}{2k}\right)\cdots\end{aligned}$$

which is obviously (trivially?) convergent, with value 1, though $\sum(-1)^{n-1}/n$ is not absolutely convergent.

Proof

(i) Since $a_n \geq 0$ for all n, $p_n = (1+a_1)(1+a_2)\cdots(1+a_n)$ is increasing, and so is convergent if and only if it has an upper bound. Multiplying out the terms (and discarding most of them) gives $a_1 + a_2 + \cdots + a_n \leq p_n$ and so if the product is convergent, so is $\sum a_n$. To prove the reverse implication we need that for $x \geq 0$, $e^x \geq 1 + x$ which is immediate from the power series for exp. Then $p_n \leq \exp(a_1 + a_2 + \cdots + a_n)$ which gives an upper bound for p_n when the series converges.

For $\prod(1-a_n)$, consider the partial product $q_n = (1-a_1)(1-a_2)\cdots(1-a_n)$. If the product is convergent then $a_n \to 0$ (since $q_n/q_{n-1} \to 1$) so there can be only finitely many terms with $a_n \geq 1$ and we can discard these without affecting the convergence. Then q_n is decreasing and tends to a non-zero limit Q say. Again for $0 \leq x < 1$, $e^x \leq 1/(1-x)$ by comparing the power series, so $0 < Q \leq q_n \leq \exp(-(a_1 + a_2 + \cdots + a_n))$ and the series must converge. In the other direction, we have the elementary inequality (Exercise 22) that if $c_1 + c_2 + \cdots + c_n \leq 1$ then

$$(1-c_1)(1-c_2)\cdots(1-c_n) \geq 1 - (c_1 + c_2 + \cdots + c_n).$$

Hence if the series is convergent, choose N such that $\sum_N^{\infty} a_n < 1$, when for $n \geq N$,

$$(1-a_N)(1-a_{N+1})\cdots(1-a_n) \geq 1 - \sum_N^{\infty} a_n$$

gives a positive lower bound for $(1-a_N)(1-a_{N+1})\cdots(1-a_n)$ and the product converges.

For (ii) we assume that $\sum a_n$ is absolutely convergent and so there is some N with $|a_n|<1$ for $n\geq N$, and to prove convergence we can disregard the terms with $n<N$. By (i) the product $\prod(1-|a_n|)$ is convergent, and since

$$|(1+a_N)(1+a_{N+1})\cdots(1+a_n)|\geq(1-|a_N|)(1-|a_{N+1}|)\cdots(1-|a_n|)$$

it follows that the partial products $r_n=(1+a_N)(1+a_{N+1})\cdots(1+a_n)$ are bounded away from zero. Hence to prove that (r_n) has a limit it is sufficient to show that $r_m/r_n\to1$ as $m,n\to\infty$. But for $m>n$, it follows as above that

$$\begin{aligned}\left|\frac{r_m}{r_n}-1\right| &= |(1+a_{n+1})\cdots(1+a_m)-1| \\ &\leq \exp\left(\sum_{n+1}^{m}|a_n|\right)-1\end{aligned}$$

which tends to zero as $m,n\to\infty$ as required. ■

Example 6.75

The product $\prod_{n=1}^{\infty}(1+q^nz)$ is convergent for all $z,q\in\mathbb{C}$ with $|q|<1$.

This is immediate from (ii) of the theorem since $\sum|q|^n$ is convergent for $|q|<1$. ♦

There is also a version of dominated convergence for products which we shall find useful.

Theorem 6.76

Let $(t_{m,n})_{m,n=1}^{\infty}$ satisfy

(i) for each n, the product $\prod_{m=1}^{\infty}(1+t_{m,n})$ is convergent with value p_n,

(ii) for each m, $t_{m,n}\to l_m$ as $n\to\infty$, and

(iii) there is a convergent series of positive terms $\sum_{m=1}^{\infty}a_m$ such that for all m,n, $|t_{m,n}|\leq a_m$.

Then the product $\prod_{m=1}^{\infty}(1+l_m)$ and the sequence (p_n) both converge, and

$$\lim_n p_n=\prod_{m=1}^{\infty}(1+l_m).$$

Proof

Since $|t_{m,n}| \le a_m$ and $t_{m,n} \to l_m$ as $n \to \infty$ it follows that both $\sum_m |t_{m,n}|$ and $\sum |l_m|$ are convergent, and so $\prod_m (1 + t_{m,n})$ and $\prod (1 + l_m)$ are convergent by Theorem 6.74(ii); let $\prod_{m=1}^{\infty} (1 + l_m) = p$. Similarly since $\sum a_m$ is convergent, then given $\varepsilon > 0$, we can choose M such that $\prod_{M+1}^{\infty} (1 \pm a_m) \in (1 - \varepsilon, 1 + \varepsilon)$. Since $t_{m,n} \to l_m$ as $n \to \infty$, we can choose N such that

$$\left| \prod_{m=1}^{M} (1 + t_{m,n}) - \prod_{m=1}^{M} (1 + l_m) \right| < \varepsilon \tag{6.19}$$

for $n \ge N$ (this uses Proposition 1.20 on the product of sequences). It follows that

$$\begin{aligned} p_n - p &= \prod_{m=1}^{M} (1 + t_{m,n}) \prod_{m=M+1}^{\infty} (1 + t_{m,n}) - \prod_{m=1}^{M} (1 + l_m) \prod_{m=M+1}^{\infty} (1 + l_m) \\ &= \left(\prod_{m=1}^{M} (1 + t_{m,n}) \prod_{m=M+1}^{\infty} (1 + t_{m,n}) \prod_{m=M+1}^{\infty} (1 + l_m)^{-1} \right. \\ &\quad \left. - \prod_{m=1}^{M} (1 + l_m) \right) \prod_{m=M+1}^{\infty} (1 + l_m) \end{aligned}$$

which is small from (6.19) since all the products $\prod_{M+1}^{\infty}$ are in $(1 - \varepsilon, 1 + \varepsilon)$. ■

Example 6.77

(i) For all $z \in \mathbb{C}$,

$$\begin{aligned} \sin z &= z \prod_{n=1}^{\infty} \left(1 - \frac{z^2}{n^2 \pi^2} \right), \\ \cos z &= \prod_{n=0}^{\infty} \left(1 - \frac{z^2}{(n + 1/2)^2 \pi^2} \right). \end{aligned}$$

(ii)

$$\Gamma(z) \Gamma(1 - z) = \frac{\pi}{\sin \pi z}.$$

We begin with the observation that for any integer $k \ge 1$, $z = e^{2\pi i j/k}$, $j = 0, 1, \ldots, k-1$ gives k distinct complex numbers with $z^k = 1$, and hence $z^k - 1 = \prod_{j=0}^{k-1} \left(z - e^{2\pi i j/k} \right)$. With $k = 2n$ this gives

$$w^{2n} - 1 = \prod_{m=0}^{2n-1} \left(w - \exp \frac{2\pi i m}{2n} \right)$$

$$= (w-1)(w+1)\prod_{m=1}^{n-1}\left(w-\exp\frac{\pi i m}{n}\right)\left(w+\exp\frac{\pi i m}{n}\right),$$

$$\frac{w^{2n}-1}{w^2-1} = \prod_{m=1}^{n-1}\left(w^2-2w\cos\frac{\pi m}{n}+1\right). \tag{6.20}$$

Taking the limit as $w \to 1$ gives

$$n = \prod_{m=1}^{n-1}\left(2-2\cos\frac{\pi m}{n}\right) = 4^{n-1}\prod_{m=1}^{n-1}\sin^2\frac{\pi m}{2n}. \tag{6.21}$$

Divide (6.20) by (6.21) and by w^n to get

$$\frac{w^n-w^{-n}}{n\left(w-w^{-1}\right)} = 4^{1-n}\prod_{m=1}^{n-1}\frac{w-2\cos(\pi m/n)+w^{-1}}{\sin^2(\pi m/(2n))}.$$

Since this is true for any complex w, we can put $w=\exp(iz/n)$ to get

$$\begin{aligned}\frac{\sin z}{n\sin(z/n)} &= 4^{1-n}\prod_{m=1}^{n-1}\frac{2\cos(z/n)-2\cos(\pi m/n)}{\sin^2(\pi m/(2n))}\\ &= \prod_{m=1}^{n-1}\frac{\sin^2(\pi m/(2n))-\sin^2(z/(2n))}{\sin^2(\pi m/(2n))}\\ &= \prod_{m=1}^{n-1}\left(1-\frac{\sin^2(z/(2n))}{\sin^2(\pi m/(2n))}\right).\end{aligned} \tag{6.22}$$

Now to finish the argument, we need to take a limit on both sides as $n \to \infty$. On the left this is easy; we just get $(\sin z)/z$. On the right we obtain formally the required product $\prod_{m=1}^{\infty}\left(1-z^2/(\pi m)^2\right)$, where the passage to the limit has to be justified by dominated convergence. We make use of the elementary inequalities (i) $|\sin z| \le 2|z|$ for (complex) $|z| \le 1$, and (ii) $\sin x > 2x/\pi$ for (real) $x \in (0,\pi/2)$ which the reader has to establish in Exercise 15. Then given $z \in \mathbb{C}$, choose $n > |z|$, when the terms on the right of (6.22) satisfy

$$\left|\frac{\sin^2(z/(2n))}{\sin^2(\pi m/(2n))}\right| \le 4\frac{|z/(2n)|^2}{(m/n)^2} = \frac{|z|^2}{m^2}.$$

But the series $\sum_m |z|^2/m^2$ is convergent, so the conditions of the Dominated Convergence Theorem 6.76 are satisfied and the factorisation of the sine is established.

The result for the cosine is immediate on putting $z+\pi/2$ for z and rearranging, and (ii) follows by multiplying the products for $\Gamma(x)$ and for $\Gamma(1-x)$, to get the product for the sine. ♦

EXERCISES

1. Find the sum of $\sum_{n=1}^{\infty} n^k r^n$ for $-1 < r < 1$ and $k = 1, 2, 3$.

2. Find the sum of $\sum_{n=1}^{\infty} 1/(n(n+k))$ for integer $k \geq 2$ ($k = 1$ was done in Example 6.6).

3. In Example 6.14, find the inequality for the sum in (ii) which corresponds to the given inequality in (i).

4. Complete the proof of Theorem 6.16.

5. Show by modifying the proof of Theorem 6.25 that a conditionally convergent series may be rearranged to give sum $\pm\infty$.

6.* Grouping of Terms. Given a series $\sum_{n=1}^{\infty} a_n$ with partial sums $s_n = \sum_{m=1}^{n} a_m$, and any subsequence (n_j) of the positive integers, define $t_k = s_{n_k} = \sum_{m=1}^{n_k} a_m$. Show that if $b_j = \sum_{m=n_{j-1}+1}^{n_j} a_m = s_{n_j} - s_{n_{j-1}}$ then $t_k = \sum_{j=1}^{k} b_j$. Thus b_j is the sum of the terms from $n_{j-1}+1$ to n_j, and (t_k) is the subsequence of (s_{n_k}) of (s_n). This ensures that if $\sum_{n=1}^{\infty} a_n$ is convergent then also $\sum_{j=1}^{\infty} b_j$ is convergent with the same sum. Show by examples that $\sum_{j=1}^{\infty} b_j$ may be convergent when $\sum_{n=1}^{\infty} a_n$ is not.

7.* The Cauchy product of two series $\sum a_n$, $\sum b_n$ is the series $\sum c_n$ where $c_n = \sum_{j+k=n} a_j b_k$. (For instance the series for $\exp(z+w)$ is the Cauchy product of the series for $\exp z$, $\exp w$.) Show that if both series are absolutely convergent, then the Cauchy product is also absolutely convergent, and that $\sum c_n = \left(\sum a_n\right)\left(\sum b_n\right)$. (The result is also true if only one series is absolutely convergent, but is then much harder to prove – see [1].) Show that if $a_n = (-1)^n/\sqrt{n}$ then the Cauchy product of $\sum a_n$ with itself is not convergent.

8.** Find a necessary and sufficient condition on $\sum a_n$ such that the Cauchy product with $\sum b_n$ is convergent (absolutely convergent) whenever $\sum b_n$ is convergent (absolutely convergent).

9.* Bernoulli Polynomials. Expand $te^{tx}/(e^t - 1)$ in powers of t to get

$$\frac{te^{tx}}{e^t - 1} = \sum_{n=0}^{\infty} \frac{B_n(x)}{n!} t^n. \tag{6.23}$$

(i) Multiply (6.23) by $e^t - 1$ to obtain $B_0(x) = 1$, $B_1(x) = x - 1/2$, $B_2(x) = x^2 - x + 1/6$, and show generally that B_n are polynomials (the Bernoulli polynomials) of degree n, which satisfy the recurrence relation

$$B_n'(x) = nB_{n-1}(x).$$

(ii) Show that for each $n \geq 1$, $\int_0^1 B_n(x)\,dx = 0$, and

$$\sum_{k=0}^{n} \binom{n+1}{k} B_k(x) = (n+1)x^n. \tag{6.24}$$

10.* Bernoulli Numbers.

(i) Put $x = 0$ in (6.23) to get

$$\frac{t}{e^t - 1} = \sum_{n=0}^{\infty} \frac{B_n(0)}{n!} t^n$$

where $B_0(0) = 1$, $B_1(0) = -1/2$, $B_2(0) = 1/6$, $B_4(0) = -1/30$, $B_6 = 1/42$, etc. Write B_n for $B_n(0)$ (the Bernoulli numbers). Since

$$\frac{t}{e^t - 1} + \frac{t}{2} = \frac{t(2 + e^t - 1)}{2(e^t - 1)} = \frac{t\left(e^{t/2} + e^{-t/2}\right)}{2\left(e^{t/2} - e^{-t/2}\right)}$$

is an even function of t, we find that all $B_n = 0$ for odd $n \geq 3$. They are calculated from the recurrence (6.24) with $x = 0$.

(ii) Put $t = 2iz$ to get the power series for the cotangent

$$\cot z = \sum_{n=0}^{\infty} (-4)^n \frac{B_{2n}}{(2n)!} z^{2n-1}.$$

11. Find the power series for the secant, using the same idea as in Exercise 10, starting from the expansion of $2e^{tx}/(e^t + 1)$. This generates the Euler numbers.

12.* Countable Partitions.

(i) Let A be a set, and f be any real or complex valued function on A. We say that f is summable over A if there is some number S such that for any $\varepsilon > 0$ there is a finite subset $B \subset A$ such that for any finite set $C \supset B$, $\left|\sum_{x \in C} f(x) - S\right| < \varepsilon$. Show that f is summable over A if and only if the set E on which $f \neq 0$ is countable, and $\sum_E |f(a)|$ is finite (i.e. the set of values of $\sum_B |f(a)|$ for all $B \subset A$ is bounded above). (Consider first the case in which f is positive.)

(ii) Now let A be countable, and $A = \cup_{n=1}^{\infty} A_n$ where the sets A_n are disjoint, and may be finite or infinite. Then f is summable over A if and only if it is summable over each A_n and $\sum_{n=1}^{\infty} \left(\sum_{a \in A_n} |f(a)|\right)$ is finite. In this case

$$\sum_A f(a) = \sum_{n=1}^{\infty} \left(\sum_{a \in A_n} f(a) \right).$$

(iii) Show that (ii) contains both Theorems 6.27 and 6.31 as special cases.

13. Show that if the series $\sum_0^{\infty} a_n z^n$ has radius of convergence r, then $1/r = \limsup_{n\to\infty} |a_n|^{1/n}$, including the cases in which $r = 0, \infty$ when the lim sup is respectively $\infty, 0$.

14. Find the maximum value of the nowhere differentiable function defined in Example 6.64. At what point is the maximum attained? Find the value of the function at some other points, for instance $f(2/5)$.

15. Prove the inequalities $|\sin z| \leq 2|z|$ for (complex) $|z| \leq 1$ and $\sin x > 2x/\pi$ for (real) $x \in (0, \pi/2)$ which are required in the proof of Example 6.77. Show similarly that $\log(1+z) = z + cz^2$ where $|c| \leq 1$ for $|z| \leq 1/2$.

16. Prove Theorem 6.45.

17. Show that the zeros of cos, sin on the real axis are the only zeros in $\mathbb{C}$.

18.* Hyperbolic Functions. Define cosh, sinh as the 90° rotations of cos, sin

$$\cosh z = \cos(iz) = \frac{e^z + e^{-z}}{2},$$
$$\sinh z = -i \sin(iz) = \frac{e^z - e^{-z}}{2}.$$

Show that $\cos(x+iy) = \cos x \cosh y - i \sin x \sinh y$, and express $\sin(x+iy)$, $\tan(x+iy)$ similarly.

19. Show that $\exp\log z = z$ for all complex z, but $\log\exp z = z$ if and only if $-\pi < \Im z \leq \pi$.

20. Show that $\int cf = c\int f$ for a complex valued regulated function f and a complex constant c. Deduce that $\left|\int f\right| \leq \int |f|$ for complex valued regulated functions.

21. Find the sums of $\sum_{n=1}^{\infty} nr^n \cos n\theta$, $\sum_{n=1}^{\infty} nr^n \sin n\theta$, by differentiation of the series (6.14).

22. Show that if all $c_j \in (0,1)$ then

$$(1-c_1)(1-c_2)\cdots(1-c_n) \geq 1-(c_1+c_2+\cdots+c_n).$$

23. Show that for any complex a_j, $\left|\prod_n^m (1+a_n) - 1\right| \leq \prod_n^m (1+|a_n|) - 1$.

24. Show that for all complex z, $|e^z - 1| \leq |z|\, e^{|z|}$.

25.* Inverse Trigonometric Functions.

(i) We have shown that the function given by $y = f(x) = \sin x$ is monotone increasing from $[-\pi/2, \pi/2]$ to $[-1,1]$, and hence has a well-defined inverse function given by $x = g(y) = \arcsin y = \sin^{-1} y$ which is increasing from $[-1,1]$ to $[-\pi/2, \pi/2]$. Show that $g'(y) = \left(1-y^2\right)^{-1/2}$ on $(-1,1)$ and deduce from the Binomial Theorem 4.24 that

$$\arcsin y = \sum_{n=0}^{\infty} \binom{-1/2}{n} \frac{y^{2n+1}}{2n+1} \qquad \text{for } -1 \leq y \leq 1,$$

and

$$\frac{\pi}{2} = \sum_{n=0}^{\infty} \binom{-1/2}{n} \frac{1}{2n+1}.$$

(The cases $y = \pm 1$ require special consideration.)

(ii) Show similarly that the inverse tangent $\arctan = \tan^{-1}$ is well defined from $(-\infty, \infty)$ to $(-\pi/2, \pi/2)$. Deduce that

$$\arctan y = \sum_{n=0}^{\infty} (-1)^n \frac{y^{2n+1}}{2n+1} \qquad \text{for } -1 \leq y \leq 1,$$

and

$$\frac{\pi}{4} = \sum_{n=0}^{\infty} \frac{(-1)^n}{2n+1}.$$

26. Integrate the inequality $\cos x \leq 1$ twice over $(0,x)$ to obtain $\sin x \leq x$ and $\cos x \geq 1 - x^2/2$ for $x \geq 0$. Continue this process to show that

for $x \geq 0$,

$$x - \frac{x^3}{3!} + \cdots - \frac{x^{4n-1}}{(4n-1)!} \leq \sin x \leq x - \frac{x^3}{3!} + \cdots - \frac{x^{4n-1}}{(4n-1)!} + \frac{x^{4n+1}}{(4n+1)!},$$

$$1 - \frac{x^2}{2!} + \cdots - \frac{x^{4n-2}}{(4n-2)!} \leq \cos x \leq 1 - \frac{x^2}{2!} + \cdots - \frac{x^{4n-2}}{(4n-2)!} + \frac{x^{4n}}{(4n)!}.$$

27. Let (f_n) be a sequence of continuous functions which converges uniformly to f on $[a, b]$ and let m_n be the maximum of f_n. Show that (m_n) converges to the maximum of f. An example of a sequence of continuous functions which converges pointwise, and for which the result fails was given in Exercise 15, Chapter 3.

28. g-summability. Let g satisfy the same conditions as in Exerecise 8, Chapter 5. A series $\sum a_n$ is said to be g-summable with value s if $\sum a_n g(tn) \to s$ as $t \to 0+$. Show that this is compatible with ordinary convergence of series. Find the g-values of $\sum_0^\infty (-1)^n$ and $\sum_1^\infty (-1)^{n-1} n$ for $g(x) = e^{-x}$. (See again [4] for the theory of summability.)

7
Applications

This final chapter is intended as a brief introduction to several topics which show analysis in action. As stated in the preface, we do not intend to be at all complete; we are simply illustrating the theory in use and will refer to the literature for further developments.

7.1 Fourier Series

A Fourier series is a series of one of the forms $\sum a_n \cos nx$, $\sum b_n \sin nx$, $\sum c_n e^{inx}$ which we met in Example 6.19. In each series the terms are periodic (with period 2π); part of the reason why Fourier series play a central role in so many applications of mathematics is that, in a sense which we shall try to make clear, they allow us to represent a general periodic phenomenon, for instance the amplitude of a radio signal or the vibration of a structure, in terms of simple components (fundamental frequencies, in engineering language).

In addition, Fourier series are ideally suited to the solution of partial differential equations by the separation of variables, and in complex analysis, they give the restriction of a power series to a circle when written in the form $f(z) = f\left(re^{it}\right) = \sum c_n z^n = \sum c_n r^n e^{int}$. With all of these uses (and many others), as well as deep and unexpected mathematical properties, it is not surprising that they have played such a central and important role in the development of mathematics.

We shall investigate only the simplest conditions which ensure convergence to the correct sum; these are already enough to derive some striking results (look ahead to Examples 7.8 and 7.9).

Definition 7.1

(i) Let f be a (real or complex valued) regulated 2π-periodic function on $\mathbb{R}$. The series

$$Sf(x) = \sum_{n=-\infty}^{\infty} c_n e^{inx} \qquad \text{where } c_n = \frac{1}{2\pi}\int_0^{2\pi} f(t)\, e^{-int} dt \qquad (7.1)$$

is called the Fourier series of f, and we write $S_m f(x)$ for the partial sum $\sum_{-m}^{m} c_n e^{inx}$ of Sf.

(ii) If we put $e^{inx} = \cos nx + i \sin nx$ then we have the alternative form

$$Sf(x) = \frac{1}{2}a_0 + \sum_{n=1}^{\infty} (a_n \cos nx + b_n \sin nx) \qquad (7.2)$$

where

$$a_n,\, b_n = \frac{1}{\pi}\int_0^{2\pi} f(t)\,(\cos nt,\, \sin nt)\, dt \qquad (7.3)$$

respectively. If f is real valued then $(a_n)_0^{\infty}, (b_n)_1^{\infty}$ are sequences of real numbers.

(iii) If f is even ($f(x) = f(-x) = f(2\pi - x)$ for all x) then all $b_n = 0$, and

$$a_n = \frac{2}{\pi}\int_0^{\pi} f(t) \cos ntdt, \qquad n \geq 0. \qquad (7.4)$$

If f is odd ($f(x) = -f(-x) = -f(2\pi - x)$ for all x) then all $a_n = 0$, and

$$b_n = \frac{2}{\pi}\int_0^{\pi} f(t) \sin ntdt, \qquad n \geq 1. \qquad (7.5)$$

(iv) The coefficients $(c_n)_{-\infty}^{\infty}$, or $(a_n)_0^{\infty}, (b_n)_1^{\infty}$ as appropriate, are called the Fourier coefficients of f.

The definition also makes sense if $|f|$ is only improperly integrable over $[0, 2\pi]$ since the functions $e^{inx}, \cos nx, \sin nx$ are all bounded and so the corresponding integrals exist. If the function has some other period, $2a$ say, for some $a > 0$, then a change of variable gives

$$Sf(x) = \sum_{n=-\infty}^{\infty} c_n e^{in\pi x/a} \qquad \text{where } c_n = \frac{1}{2a}\int_0^{2a} f(t)\, e^{-in\pi t/a} dt.$$

Two points about periodic functions are important here. If we say that a function f is continuous with period $2a$, then it is assumed that $f(0) = f(2a)$, in other words the continuity extends to the end points of the interval. In addition, if we integrate, as above, over a period, then it makes no difference whether the integral is over $[0, 2a]$ or $[-a, a]$; any interval of length $2a$ will give the same result.

Example 7.2

Let f be 2π-periodic, and let $f(x) = 1$ on $(0, \pi)$, $= -1$ on $(-\pi, 0)$ and $= 0$ at $0, \pi$. Then

$$Sf(x) = \frac{4}{\pi} \sum_{n \text{ odd}}^{\infty} \frac{1}{n} \sin nx = \frac{4}{\pi} \sum_{k=0}^{\infty} \frac{\sin(2k+1)x}{2k+1}.$$

The function is odd so we have only to find the sine-coefficients for which (7.5) gives

$$\begin{aligned} b_n &= \frac{2}{\pi} \int_0^{\pi} f(x) \sin nx \, dx = \frac{2}{\pi} \int_0^{\pi} \sin nx \, dx \\ &= \frac{-2}{n\pi} [\cos nx]_0^{\pi} = \frac{-2}{n\pi} \left((-1)^n - 1\right) \\ &= \begin{cases} \dfrac{4}{n\pi} & \text{if } n \text{ odd}, \\ 0 & \text{if } n \text{ even}, \end{cases} \end{aligned}$$

as required. ♦

It is important to realise that at this stage, although we have found the Fourier series Sf, we have no information about its sum (or even whether it converges, though Example 6.19 could be modified to give this). Thus it would be quite unjustified to write $f(x)$ for the sum of the series.

The choice of coefficients in (7.1) is based on the following result which expresses the *orthogonality* of the sequence $\left(e^{inx}\right)_{-\infty}^{\infty}$ on the interval $[0, 2\pi]$.

Proposition 7.3

(i) Let $e_n(x) = e^{inx}$. Then

$$\int_0^{2\pi} e_n(x) e_{-m}(x) \, dx = \int_0^{2\pi} e_n(x) \bar{e}_m(x) \, dx = \begin{cases} 2\pi & \text{if } m = n, \\ 0 & \text{otherwise}. \end{cases}$$

(ii) If the series $\sum_{-\infty}^{\infty} c_n e^{inx}$ is uniformly convergent with sum f then the coefficients (c_n) are given as in (7.1).

Proof

(i) If $m = n$, then $\int_0^{2\pi} e_n(x)\, e_{-m}(x)\, dx = \int_0^{2\pi} dx = 2\pi$. Otherwise $\int_0^{2\pi} e_n(x)$ $e_{-m}(x) dx = \int_0^{2\pi} e^{i(n-m)x} dx = \left[e^{i(n-m)x}\right]_0^{2\pi} / (i(n-m)) = 0$.

(ii) Since the series is uniformly convergent we can multiply $f(x)$ by e^{-imx} and integrate termwise by Theorem 6.65. This gives

$$\begin{aligned}\int_0^{2\pi} f(x)\, e^{-imx} dx &= \int_0^{2\pi} \left(\sum_{-\infty}^{\infty} c_n e^{inx}\right) e^{-imx} dx \\ &= \sum_{-\infty}^{\infty} c_n \int_0^{2\pi} e^{inx} e^{-imx} dx = 2\pi c_m\end{aligned}$$

by (i), as required. ■

To avoid misunderstanding, notice that the condition in (ii) is on the series, and is thus the reverse of the expected situation in which f is given, and we want to deduce properties of the series. As a first step in this direction we shall show that if all the coefficients are zero, the function itself must be zero, except possibly at a countable number of points where it may be discontinuous (recall Definition 2.30 of a regulated function). We need a preliminary definition.

Definition 7.4

A finite sum of the form $P(x) = \sum_{-N}^{N} p_n e^{inx}$, where the (p_n) are arbitrary real or complex numbers, is called a trigonometric polynomial. P may be written in the equivalent form $\sum_0^N (q_n \cos nx + r_n \sin nx)$ where $q_0 = p_0$, and $q_n = p_n + p_{-n}$, $r_n = i(p_n - p_{-n})$ for $n \geq 1$.

Lemma 7.5

Any sum or product of trigonometric polynomials is again a trigonometric polynomial. Any positive integer power of a trigonometric polynomial is again a trigonometric polynomial.

Proof

This is evident on writing P in the form $P(x) = \sum_{-N}^{N} p_n z^n$ where $z = e^{ix}$. ■

Proposition 7.6

Let f be a regulated function on $[0, 2\pi]$ such that all its Fourier coefficients are equal to zero. Then $f(x) = 0$ at all points x at which f is continuous (and so at all except a countable number of points).

Proof

The given condition implies that $\int_0^{2\pi} f(x) P(x)\, dx = 0$ for any trigonometric polynomial P. Suppose that there is some x_0 at which f is continuous and $f(x_0) \neq 0$. We can suppose that f is real valued (otherwise consider the real and imaginary parts separately) and that $f(x_0) > 0$ (else consider $-f$). Using the definition of continuity, choose $h, \delta > 0$ such that $f(x) > h$ on $I = (x_0 - \delta, x_0 + \delta)$.

Consider $Q(x) = \cos(x - x_0) - \cos\delta + 1$ which is a trigonometric polynomial with the property that $Q(x) > 1$ on I, $Q(x) \leq 1$ elsewhere. We shall show that $\int_I fQ^n \to \infty$ while $\int_{I'} fQ^n$ is bounded as $n \to \infty$, which contradicts the hypothesis that $\int_0^{2\pi} fQ^n = 0$. The boundedness of $\int_{I'} fQ^n$ is immediate from $|\int_{I'} fQ^n| \leq \int_0^{2\pi} |f|$, since $Q(x) \leq 1$ on I'. On I we have

$$\int_I fQ^n \geq \int_{x_0-\delta/2}^{x_0+\delta/2} hQ^n \geq h\delta(\cos(\delta/2) - \cos\delta + 1)^n \to \infty$$

as required. ■

From this we get our first genuine theorem on Fourier series.

Theorem 7.7

Let f be continuous and 2π-periodic (recall that this includes the requirement that $f(0) = f(2\pi)$), and suppose that $Sf(x) = \sum_{-\infty}^{\infty} c_n e^{inx}$ is uniformly convergent. Then the sum of the series is $f(x)$ for all $x \in [0, 2\pi]$.

Proof

Let g be the sum of the Fourier series. By Proposition 7.3(ii) $c_n = \int_0^{2\pi} g(x) e^{-inx} dx$, and g is continuous since it is the sum of a uniformly convergent series of continuous functions. But Sf is given as the Fourier series of f, so also $c_n = \int_0^{2\pi} f(x) e^{-inx} dx$, and thus $\int_0^{2\pi} (f(x) - g(x)) e^{-inx} dx = 0$ for all n. It follows from Proposition 7.6 that $f - g = 0$ which is what we want. ■

Following the earlier comment about whether the hypothesis should be on the function or the series, Theorem 7.7 is a kind of hybrid since it has conditions on both f and Sf. But it allows us to deduce some striking consequences.

Example 7.8

(i) Let $f(x) = |x|$ on $[-\pi, \pi]$ Then

$$Sf(x) = \frac{\pi}{2} - \frac{4}{\pi}\sum_{k=1}^{\infty} \frac{\cos((2k-1)nx)}{(2k-1)^2} \tag{7.6}$$

and the series is uniformly convergent with sum $|x|$ on $[-\pi, \pi]$.

(ii)

$$\sum_{k=1}^{\infty} \frac{1}{(2k-1)^2} = \frac{\pi^2}{8}, \qquad \sum_{n=1}^{\infty} \frac{1}{n^2} = \frac{\pi^2}{6}. \tag{7.7}$$

Since f is even, we can use the formulae (7.4) to give $a_n = \frac{2}{\pi}\int_0^{\pi} x \cos nx\, dx$. If $n = 0$, $a_0 = \frac{2}{\pi}\int_0^{\pi} x\, dx = \pi$. For $n \geq 1$,

$$\begin{aligned} a_n &= \frac{2}{\pi}\int_0^{\pi} x \cos nx\, dx = \frac{2}{n\pi}[x \sin nx]_0^{\pi} - \frac{2}{n\pi}\int_0^{\pi} \sin nx\, dx \\ &= \frac{2}{n^2\pi}[\cos nx]_0^{\pi} = \begin{cases} \frac{-4}{n^2\pi}, & \text{if } n \text{ is odd,} \\ 0 & \text{if } n \text{ is even.} \end{cases} \end{aligned}$$

The series is uniformly convergent by comparison with $\sum 1/n^2$, and (i) follows from Theorem 7.7. For (ii) put $x = 0$ when $f(x) = 0$ and (7.6) gives

$$0 = \frac{\pi}{2} - \frac{4}{\pi}\sum_{k=1}^{\infty} \frac{1}{(2k-1)^2}.$$

To find the sum $s = \sum_1^{\infty} 1/n^2$, we use a standard trick: note that $s - s/4 = \sum_1^{\infty} 1/n^2 - \sum_1^{\infty} 1/(2n)^2 = \sum_{k=1}^{\infty} 1/(2k-1)^2 = \pi^2/8$, so $s = (4/3)(\pi^2/8) = \pi^2/6$. ♦

Example 7.9

(i) Let $f(x) = \cos ax$ on $[-\pi, \pi]$, where a is any non-integer complex number. Then

$$Sf(x) = \frac{\sin a\pi}{\pi}\left[\frac{1}{a} + 2a\sum_{n=1}^{\infty} \frac{(-1)^n}{a^2 - n^2} \cos nx\right] \tag{7.8}$$

where the series is uniformly convergent with sum $\cos ax$ on $[-\pi, \pi]$.

(ii) For any complex a which is not an integer,

$$\frac{\pi}{\sin a\pi} = \frac{1}{a} + 2a\sum_{n=1}^{\infty}\frac{(-1)^n}{a^2-n^2} = \frac{1}{a} + \sum_{n=1}^{\infty}(-1)^n\left(\frac{1}{a-n}+\frac{1}{a+n}\right), \quad (7.9)$$

$$\frac{\pi}{\tan a\pi} = \frac{1}{a} + 2a\sum_{n=1}^{\infty}\frac{1}{a^2-n^2} = \frac{1}{a} + \sum_{n=1}^{\infty}\left(\frac{1}{a-n}+\frac{1}{a+n}\right), \quad (7.10)$$

$$\frac{\pi^2}{\sin^2 a\pi} = \sum_{-\infty}^{\infty}\frac{1}{(a-n)^2}. \quad (7.11)$$

The formulae (7.4) give for $n \geq 0$,

$$\begin{aligned} a_n &= \frac{2}{\pi}\int_0^{\pi}\cos ax\cos nx\,dx = \frac{1}{\pi}\int_0^{\pi}(\cos(a-n)x + \cos(a+n)x)\,dx \\ &= \frac{1}{\pi}\left[\frac{\sin(a-n)x}{a-n} + \frac{\sin(a+n)x}{a+n}\right]_0^{\pi} = \frac{(-1)^n\sin a\pi}{\pi}\frac{2a}{a^2-n^2} \end{aligned}$$

for all $n \geq 0$. Notice that all coefficients are well defined since a is not an integer. Hence as before

$$Sf(x) = \cos ax = \frac{\sin a\pi}{\pi}\left[\frac{1}{a} + 2a\sum_{n=1}^{\infty}\frac{(-1)^n}{a^2-n^2}\cos nx\right].$$

Putting $x = 0, \pi$ gives (7.9) and (7.10). It is tempting to differentiate with respect to a, using Theorem 6.69 to get (7.11), but the theorem applies only to derivatives on $\mathbb{R}$ and so would not give the full result on $\mathbb{C}$. Instead we could use the Fourier series for $x\sin ax$ (the formal derivative of $\cos ax$ with respect to a) to get (7.11).

Note finally that if in (7.11) we multiply up by $\sin^2 a\pi$ the resulting series $\sum \sin^2 a\pi/(a-n)^2$ is uniformly convergent on $\mathbb{R}$ by the M-test. ♦

At this point we have to warn the reader against too much optimism concerning the convergence of Fourier series for wider classes of functions. It was shown by Ch. de la Vallée Poussin in the 1870s that there is a continuous function whose Fourier series is divergent at a point, and by Kolmogorov in the 1930s that there is a function which is integrable in the sense of Lebesgue whose Fourier series is divergent at every point of $[0, 2\pi]$. It was an open question until relatively recently whether the Fourier series of a continuous function need have any points of convergence at all, but in 1966 L. Carlesen showed that if $|f|^2$ is integrable (and in particular if f is continuous) then the Fourier series of f must be convergent at all points except those of a set of measure zero. (A set E has measure zero if for every $\varepsilon > 0$, there is a countable set (A_n) of intervals $A_n = (b_n, c_n)$ such that $E \subset \cup A_n$ and $\sum(c_n - b_n) < \varepsilon$.) The exact nature of the sets on which the Fourier series of a continuous function may diverge is still not fully understood. Thus the theorem above, and the further results to follow, must impose fairly heavy restrictions on the function in order to deduce convergence. In fact, at this point in our development, any function with a point of discontinuity at which left-hand and right-hand limits are unequal is outside our scope. To find some results to cover this case, suppose that f is bounded and improperly integrable over $[0, 2\pi]$, for instance $f(x) = \sin(1/x)$ for $x \neq 0$. In particular it is enough if f is regulated on $[0, 2\pi]$. Then we have

Theorem 7.10

Let f be bounded and improperly integrable over $[0, \pi]$. Then

$$\frac{1}{2\pi}\int_0^{2\pi} |f|^2 = \frac{1}{2\pi}\int_0^{2\pi} |f - S_m f|^2 + \sum_{-m}^{m} |c_n|^2. \tag{7.12}$$

In particular, the series $\sum_{-\infty}^{\infty} |c_n|^2$ is convergent and

$$\sum_{-\infty}^{\infty} |c_n|^2 \leq \frac{1}{2\pi}\int_0^{2\pi} |f|^2. \tag{7.13}$$

The statement (7.13) is *Bessel's inequality*: a little more work would show that under the stated conditions it is actually an equality (Bessel's equation), but the above is sufficient for our needs.

Proof

This is just an elaborate repetition of the identity $|z|^2 = z\bar{z}$ for real or complex z. Thus the integral on the right of (7.12) exists and can be written

$$\int (f - S_m)(f - S_m)^- = \int ff^- - \int f(S_m)^- - \int f^- S_m + \int S_m (S_m)^- \tag{7.14}$$

where we have written S_m for $S_m f$, $\int$ for $\int_0^{2\pi}$ and f^- for $\bar{f}$. The definition of the Fourier coefficients shows that

$$\begin{aligned} \int f(S_m)^- &= \sum_{-m}^{m} \bar{c}_n \int f(x) e^{-inx} dx = 2\pi \sum_{-m}^{m} \bar{c}_n c_n \\ &= 2\pi \sum_{-m}^{m} |c_n|^2 = \int f^- S_m \end{aligned}$$

similarly, and the orthogonality of (e^{inx}) gives

$$\int S_m (S_m)^- = \int \sum_{-m}^{m} c_j e^{ijx} \sum_{-m}^{m} \bar{c}_n e^{-inx} dx = 2\pi \sum_{-m}^{m} |c_n|^2$$

also. Hence (7.14) can be written

$$\frac{1}{2\pi} \int_0^{2\pi} |f - S_m f|^2 = \frac{1}{2\pi} \int_0^{2\pi} |f|^2 - \sum_{-m}^{m} |c_n|^2$$

which gives (i). Since $\int |f - S_m f|^2 \geq 0$, it follows that

$$\sum_{-m}^{m} |c_n|^2 \leq \frac{1}{2\pi} \int_0^{2\pi} |f|^2$$

and (ii) follows if we let $m \to \infty$. ■

Corollary 7.11

If f is bounded and improperly integrable over $[0, 2\pi]$, then its Fourier coefficients $c_n \to 0$ as $|n| \to \infty$.

Proof

Since the series $\sum_{-\infty}^{\infty} |c_n|^2$ is convergent, its terms must tend to zero. ■

Since as stated above, continuity alone is not sufficient to ensure convergence, we have to introduce a stronger requirement.

Definition 7.12

We say that f is Lipschitz continuous at x if there are $M, \delta > 0$ such that

$$|f(x+h) - f(x)| \leq M|h| \qquad \text{for } 0 \leq |h| < \delta. \tag{7.15}$$

We say that f is Lipschitz continuous on an interval I if there is some $M, \delta > 0$ such that (7.15) holds for all $x, x+h \in I$ with $|h| < \delta$.

Example 7.13

(i) $f(x) = \sqrt{|x|}$ is continuous but not Lipschitz continuous at $x = 0$.

(ii) $f(x) = |x|^a \sin(x^b)$ is Lipschitz continuous at $x = 0$ for $a > 0, b > 0$, and $a + b \geq 1$.

(i) is immediate from the definition, and (ii) follows since $|\sin x| \leq |x|$ and so $|x|^a |\sin(x^b)| \leq |x|^{a+b}$. ♦

Proposition 7.14

Any function which has a bounded derivative on $[a, b]$ is Lipschitz continuous on $[a, b]$.

Proof

If $|f'(x)| \leq M$ on $[a, b]$ then the Mean Value Theorem shows that $f(x+h) - f(x) = hf'(c)$ for some c between x and $x + h$, and so $|f(x+h) - f(x)| \leq M|h|$ as required. ∎

We are going to show that a function which is Lipschitz continuous at a point has a Fourier series which converges at that point. The result is perhaps unexpected since it shows that convergence is a local property, in other words that convergence at a point is determined by what happens in an arbitrarily small neighbourhood of the point. We need an explicit expression for the partial sum $S_n f(x)$ of the Fourier series.

Proposition 7.15

(i) Let $D_n(x) = \sum_{j=-n}^{n} e^{ijx}$. Then $\int_0^{2\pi} D_n = 2\pi$, and

$$D_n(x) = \begin{cases} \sin((n+1/2)x) / \sin(x/2) & \text{for } 0 < x < 2\pi, \\ 2n+1 & \text{for } x = 0, 2\pi. \end{cases}$$

(ii)

$$S_n f(x) = \frac{1}{2\pi} \int_0^{2\pi} f(x-t) D_n(t)\, dt.$$

Proof

(i) The integral $\int_0^{2\pi} D_n(x)\, dx = \sum_{j=-n}^{n} \int_0^{2\pi} e^{ijx} dx = \int_0^{2\pi} 1 dx = 2\pi$, since all terms with $j \neq 0$ have integral zero.

For $x = 0$, $D_n(0) = \sum_{j=-n}^{n} 1 = 2n + 1$, while if $0 < x < 2\pi$,

$$\begin{aligned} D_n(x) &= \sum_{j=-n}^{n} e^{ijx} = e^{-inx}\left(1 + e^{it} + \cdots + e^{2nix}\right) \\ &= e^{-inx}\frac{e^{(2n+1)ix} - 1}{e^{ix} - 1} = \frac{\sin((n + 1/2)x)}{\sin(x/2)}. \end{aligned}$$

(ii)

$$\begin{aligned} S_n f(x) &= \sum_{j=-n}^{n} c_j e^{ijx} = \sum_{j=-n}^{n} \left(\frac{1}{2\pi}\int_0^{2\pi} f(t)e^{-ijt}dt\right) e^{ijx} \\ &= \frac{1}{2\pi}\int_0^{2\pi} f(t)\left(\sum_{j=-n}^{n} e^{ij(x-t)}\right) dt = \frac{1}{2\pi}\int_0^{2\pi} f(t)D_n(x-t)dt \\ &= \frac{1}{2\pi}\int_0^{2\pi} f(x-t)D_n(t)dt \end{aligned}$$

where in the last line we have put $x - t$ for t and used the periodicity of f and D_n. ■

The sequence (D_n) of functions is called the Dirichlet kernel. Our main result is as follows.

Theorem 7.16

(i) Let f be a bounded improperly integrable function on $[0, 2\pi]$, and let a be a point of Lipschitz continuity of f. Then the Fourier series of f is convergent at $x = a$ with sum $f(a)$.

(ii) If at a point a, f has right-hand and left-hand limits l_1, l_2, such that for some $M, \delta > 0$, both $|f(a+h) - l_1|$ and $|f(a-h) - l_2|$ are $< Mh$ for $0 < h < \delta$, then the Fourier series of f is convergent at a with sum $s = (l_1 + l_2)/2$.

Proof

The partial sums $S_n f$ can be written

$$\begin{aligned} S_n f(x) &= \sum_{j=-n}^{n} c_j e^{ijx} = \sum_{j=-n}^{n} e^{ijx}\frac{1}{2\pi}\int_0^{2\pi} f(t)\, e^{-ijt}dt \\ &= \frac{1}{2\pi}\int_0^{2\pi} f(t)\sum_{j=-n}^{n} e^{ij(x-t)}dt = \frac{1}{2\pi}\int_0^{2\pi} f(t)\, D_n(x-t)\, dt \end{aligned}$$

$$= \frac{1}{2\pi}\int_{x-2\pi}^{x} f(x-u)\,D_n(u)\,du = \frac{1}{2\pi}\int_{0}^{2\pi} f(x-t)\,D_n(t)\,dt$$

where $t = x - u$, and in the last step we used the periodicity of f and D_n. Hence from the proposition we can write

$$\begin{aligned} S_n f(a) - f(a) &= \frac{1}{2\pi}\int_{-\pi}^{\pi} \left(f(a-t) - f(a)\right) D_n(t)\,dt \\ &= \frac{1}{2\pi}\int_{-\pi}^{\pi} \frac{f(a-t)-f(a)}{t}\,\frac{t}{\sin(t/2)}\sin(n+1/2)\,t\,dt \\ &= \frac{1}{2\pi}\int_{-\pi}^{\pi} \phi(t)\sin(n+1/2)\,t\,dt, \qquad \text{say,} \\ &= \frac{1}{2\pi}\int_{-\pi}^{\pi} \phi(t)\cos(t/2)\sin nt\,dt + \frac{1}{2\pi}\int_{-\pi}^{\pi} \phi(t)\sin(t/2)\cos nt\,dt \end{aligned}$$

where $\phi(t) = (f(a-t) - f(a))/(\sin(t/2))$. This represents $S_n f - f$ as Fourier coefficients of $\phi(t)\cos(t/2)$ and $\phi(t)\sin(t/2)$. In this expression, $(f(a-t)-f(a))/t$ and $t/(\sin(t/2))$ are bounded near $t = 0$ since f is Lipschitz continuous, and $t/(\sin(t/2)) \to 2$. It follows that $\phi(t)\cos(t/2)$ and $\phi(t)\sin(t/2)$ satisfy the conditions of Corollary 7.11 and so their Fourier coefficients tend to zero, and $S_n f(a) - f(a) \to 0$ as $n \to \infty$ which proves (i). In part (ii) we have similarly

$$\begin{aligned} S_n f(a) - s &= \frac{1}{2\pi}\int_{-\pi}^{\pi} \left(f(a-t) - s\right) D_n(t)\,dt \\ &= \frac{1}{2\pi}\int_{0}^{\pi} \left(f(a-t) + f(a+t) - 2s\right) D_n(t)\,dt \\ &= \frac{1}{2\pi}\int_{0}^{\pi} \left((f(a+t) - l_1) + (f(a-t) - l_2)\right) D_n(t)\,dt \end{aligned}$$

and the proof is completed as before since both $(f(a+t) - l_1)/t$ and $(f(a-t) - l_2)/t$ are bounded near $t = 0$. ■

Example 7.17

(i) Let $f(x) = (\pi - x)/2$ on $(0, 2\pi)$, $f(0) = f(2\pi) = 0$. Then

$$Sf(x) = \sum_{n=1}^{\infty} \frac{\sin nx}{n}$$

and the series is convergent with sum $f(x)$ at all points of $[0, 2\pi]$.

(ii) Let $a \in (0, \pi)$ and let

$$f(x) = \begin{cases} 1 & \text{for } -a \leq x \leq a, \\ 0 & \text{elsewhere.} \end{cases}$$

Then

$$Sf(x) = \frac{a}{\pi} + \frac{2}{\pi} \sum_{n=1}^{\infty} \frac{\sin na}{n} \cos nx$$

and the series is convergent at all points of $[-\pi, \pi]$, with sum equal to $f(x)$ except at $\pm a$ where the sum is $1/2$.

For (i), note that f is odd, so that for $n \geq 1$,

$$\begin{aligned} b_n &= \frac{2}{\pi} \int_0^{\pi} f(x) \sin nx \, dx = \frac{1}{\pi} \int_0^{\pi} (\pi - x) \sin nx \, dx \\ &= \frac{1}{n\pi} \left[-(\pi - x) \cos nx \right]_0^{\pi} + \frac{1}{n\pi} \int_0^{\pi} (-1) \cos nx \, dx \\ &= \frac{1}{n} - \frac{1}{n^2 \pi} \left[\sin nx \right]_0^{\pi} = \frac{1}{n}. \end{aligned}$$

Since f is Lipschitz continuous at all points of $(0, 2\pi)$ the series is convergent with sum $(\pi - x)/2$ at each point of the open interval. (We found in Example 6.19 that the series $\sum \sin nx/n$ is convergent; now we have found its sum.) At $x = 0$ or 2π, f is zero and every term of the series is zero, so we have convergence to the correct sum there also. (We could have used (ii) of the theorem, but we leave that for the next part.)

For (ii) the function is even, and

$$a_n = \frac{2}{\pi} \int_0^{\pi} f(x) \cos nx \, dx = \frac{2}{\pi} \int_0^{a} \cos nx \, dx.$$

If $n = 0$, this gives $a_0 = 2a/\pi$, while for $n \geq 1$, $a_n = (2/n\pi) \left[\sin nx \right]_0^a = 2 \sin na / (n\pi)$ which gives the required series. Part (i) of the theorem gives convergence to f at all points except $\pm a$, and part (ii) gives convergence at $\pm a$ with sum equal to the average of the left-hand and right-hand limits, which is $1/2$. Notice that if we put $x = a$ in the series we get

$$\frac{1}{2} = \frac{a}{\pi} + \frac{2}{\pi} \sum_{n=1}^{\infty} \frac{\sin na \cos na}{n} = \frac{a}{\pi} + \frac{1}{\pi} \sum_{n=1}^{\infty} \frac{\sin 2na}{n}$$

which is the result of (i) with $2a = x$. ♦

At this point we have to call a halt and admit that we are only scratching the surface of a very large subject. Without giving further proofs, we mention the following results which are on the same kind of level, and which can be found in any of the standard books on Fourier theory.

A. In the theorem just proved, if the Lipschitz condition holds uniformly on $[0, 2\pi]$, i.e. there is some M such that $|f(x+h) - f(x)| \leq M|h|$ for all x, h then the Fourier series is uniformly convergent on $[0, 2\pi]$.

B. Functions such as $f(x) = \sqrt{|x|}$ or $1/\ln(|x|/2\pi)$ on $[-\pi, \pi]$ are not Lipschitz continuous at 0, but are continuous, and are monotone on $[0, \pi]$. For such functions there is a theorem which says that if f is monotone on $(a, b) \subset [0, 2\pi]$ (and for instance regulated on the rest of the interval) then at each point $c \in (a, b)$ the Fourier series is convergent with sum $(f(c+) + f(c-))/2$. If in addition f is continuous at c, then the sum of the series is $f(c)$, and if f is continuous on (a, b) then the series is uniformly convergent on $[a + \delta, b - \delta]$ for every $\delta > 0$. This theorem is independent of our Theorem 7.16 – the reader is invited to find suitable examples to justify this assertion.

C. The function in Example 7.17(i) has a discontinuity at the origin with $f(0+) = \pi/2$ and $f(0) = 0$. Its Fourier series has the remarkable property that the partial sums 'overshoot' the amount of the discontinuity. More precisely, if m_n denotes the maximum of $S_n f$ on $[0, \pi]$, then m_n tends to a limit strictly greater than $\pi/2$ as $n \to \infty$. This is an example of the 'Gibbs' phenomenon' which is universal in the sense that it holds at all such discontinuities of piecewise monotone functions.

7.2 Fourier Integrals

The preceding section on Fourier series provided a way of representing periodic phenomena as a sum of simple exponentials. It is natural to want to use similar methods for non-periodic phenomena such as earthquakes or sonic booms. This leads to the Fourier transform for functions on $\mathbb{R}$.

Before giving a formal definition, we shall outline a heuristic and entirely non-rigorous argument which shows how the form of the Fourier transform may be inferred from the Fourier series by a limiting process.

For a function f with period $2a$ (which we shall soon suppose is large), we have the formulae

$$c_n = \frac{1}{2a}\int_{-a}^{a} f(x)\, e^{-in\pi x/a} dx, \tag{7.16}$$

$$f(x) = \sum_{-\infty}^{\infty} c_n e^{in\pi x/a} \tag{7.17}$$

which connect f with its Fourier coefficients. Rewrite (7.16) in the form

$$2ac_n = \int_{-a}^{a} f(x)\, e^{-in\pi x/a} dx$$

and suppose that n, a both $\to \infty$ in such a way that $n/a \to 2y$ for some real y (the choice of the factor 2 here leads to the symmetric form of the equations (7.19) below). Then write $g(y)$ for the result:

$$g(y) = \lim (2ac_n) = \int_{-\infty}^{\infty} f(x)\, e^{-2\pi ixy} dx.$$

Notice that the integral certainly exists for all y since $\int_{-\infty}^{\infty} |f|$ is finite.

Then (7.17) becomes

$$f(x) = \sum_{-\infty}^{\infty} c_n e^{in\pi x/a} = \frac{1}{2a} \sum_{-\infty}^{\infty} 2ac_n e^{in\pi x/a} \tag{7.18}$$

and we have to find the limit of this as $n, a \to \infty$ with $n/a \to 2y$. Write $y_n = n/(2a)$ so that (7.18) becomes

$$f(x) = \sum_{-\infty}^{\infty} (y_{n+1} - y_n)\, g(y_n)\, e^{2\pi ixy_n}.$$

This sum is an approximation to the integral $\int_{-\infty}^{\infty} g(y) e^{2\pi ixy} dy$, using $g(y_n)$ $e^{2\pi ixy_n}$ in place of $g(y)e^{2\pi ixy}$ on the interval (y_n, y_{n+1}), and hence it may reasonably be expected to tend to $\int_{-\infty}^{\infty} g(y)\, e^{2\pi ixy} dy$ as $n, a \to \infty$. This leads to the pair of symmetric formulae

$$g(y) = \int_{-\infty}^{\infty} f(x)\, e^{-2\pi ixy} dx, \quad f(x) = \int_{-\infty}^{\infty} g(y)\, e^{2\pi ixy} dx \tag{7.19}$$

which define the Fourier transform, Definition 7.18 below. Of course at this stage we have proved nothing, and any attempt to make the above reasoning rigorous would be particularly messy. Instead we simply use (7.19) as a starting point, and investigate the consequences.

The scale factor of 2, introduced in this reasoning, when $n/a \to 2y$ is arbitrary, but leads to the symmetric formulae given. Other choices in common use are

$$\begin{aligned} g_1(y) &= \int_{-\infty}^{\infty} f(x)\, e^{-ixy} dx, \quad f(x) = \frac{1}{2\pi} \int_{-\infty}^{\infty} g_1(y)\, e^{ixy} dx, \text{ or} \\ g_2(y) &= \frac{1}{\sqrt{2\pi}} \int_{-\infty}^{\infty} f(x)\, e^{-ixy} dx, \quad f(x) = \frac{1}{\sqrt{2\pi}} \int_{-\infty}^{\infty} g_2(y)\, e^{ixy} dx \end{aligned}$$

in place of (7.19).

For the purposes of this section (and to avoid clumsy repetition) 'integrable on $\mathbb{R}$' will mean 'regulated on $\mathbb{R}$, and $\int_{-\infty}^{\infty} |f(x)|\, dx$ finite'.

Definition 7.18

(i) Let f be integrable on $\mathbb{R}$. The Fourier transforms $\mathcal{F}f$ $\mathcal{F}^{-1}f$ are defined by

$$\mathcal{F}f(y) = \int_{-\infty}^{\infty} f(x)\, e^{-2\pi ixy} dx, \quad \mathcal{F}^{-1}f(y) = \int_{-\infty}^{\infty} f(x)\, e^{2\pi ixy} dx.$$

(ii) If f is even then

$$\mathcal{F}f(y) = 2\int_0^{\infty} f(x)\cos(2\pi xy)\, dx = \mathcal{F}^{-1}f(y);$$

if f is odd then

$$\mathcal{F}f(y) = -2i\int_0^{\infty} f(x)\sin(2\pi xy)\, dx = -\mathcal{F}^{-1}f(y).$$

(iii) The reflection operator R is defined by $Rf(x) = f(-x)$. With this notation

$$R\mathcal{F}f(y) = \mathcal{F}Rf(y) = \mathcal{F}^{-1}f(y).$$

At the moment the notation is purely formal; it will be a little while before we can show that under suitable conditions, $\mathcal{F}$ and $\mathcal{F}^{-1}$ are mutually inverse. When this is done, the inversion formulae (7.19) can be written simply as $\mathcal{F}^2 = R$. The notation $\hat{f}$ for $\mathcal{F}f$ is common and avoids calligraphy, but can cause confusion when combined with other operators. Before going further we need some examples.

Example 7.19

For $a > 0$ we have the following Fourier transforms.

(i) If $f = \chi_{[-a,a]}$, $\mathcal{F}f(y) = (\sin(2\pi ay))/(\pi y)$ if $y \neq 0$, $= 2a$ if $y = 0$.

(ii) If $f(x) = e^{-a|x|}$, $\mathcal{F}f(y) = 2a/\left(a^2 + 4\pi^2y^2\right)$.

(iii) If $f(x) = e^{-ax^2}$, $\mathcal{F}f(y) = \sqrt{\pi/a}e^{-\pi^2y^2/a}$. In particular, $e^{-\pi x^2}$ is its own Fourier transform.

These are direct calculations from the definition. For instance in (i),

$$\begin{aligned}
\mathcal{F}f(y) &= \int_{-\infty}^{\infty} \chi_{[-a,a]} e^{-2\pi ixy} dx = \int_{-a}^{a} e^{-2\pi ixy} dx \\
&= \frac{-1}{2\pi iy}\left[e^{-2\pi ixy}\right]_{-a}^{a} = \frac{\sin(2\pi ay)}{\pi y} \qquad \text{if } y \neq 0, \\
&= \int_{-a}^{a} dx = 2a \qquad \text{if } y = 0.
\end{aligned}$$

For (ii)

$$\begin{aligned} \mathcal{F}f(y) &= \int_{-\infty}^{\infty} e^{-a|x|}e^{-2\pi ixy}dx = \int_{0}^{\infty} e^{-ax}e^{-2\pi ixy}dx + \int_{-\infty}^{0} e^{ax}e^{-2\pi ixy}dx \\ &= \frac{1}{a+2\pi iy} + \frac{1}{a-2\pi iy} = \frac{2a}{a^2+4\pi^2y^2}. \end{aligned}$$

For (iii) we begin with the observation that for real y,

$$g(y) = \int_{-\infty}^{\infty} e^{-a(x+iy)^2}dx$$

is independent of y. To see this, differentiate under the integral sign (justified below by dominated convergence) to get

$$g'(y) = -2ia\int_{-\infty}^{\infty}(x+iy)\,e^{-a(x+iy)^2}dx = -i\left[e^{-a(x+iy)^2}\right]_{-\infty}^{\infty} = 0$$

and so g is constant. Hence

$$\begin{aligned} \mathcal{F}f(y) &= \int_{-\infty}^{\infty} e^{-ax^2}e^{-2\pi ixy}dx = e^{-\pi^2y^2/a}\int_{-\infty}^{\infty} e^{-a(x+i\pi y/a)^2}dx \\ &= e^{-\pi^2y^2/a}\int_{-\infty}^{\infty} e^{-ax^2}dx = e^{-\pi^2y^2/a}\sqrt{\frac{\pi}{a}} \end{aligned}$$

using the result of Example 5.13. It remains to justify the differentiation above, i.e. to show that

$$\lim_{h\to 0}\frac{g(y+h)-g(y)}{h} = -2ia\int_{-\infty}^{\infty}(x+iy)\,e^{-a(x+iy)^2}dx.$$

We have

$$\begin{aligned} \frac{g(y+h)-g(y)}{h} &= \frac{1}{h}\left(\int_{-\infty}^{\infty} e^{-a(x+i(y+h))^2}dx - \int_{-\infty}^{\infty} e^{-a(x+iy)^2}dx\right) \\ &= \int_{-\infty}^{\infty} e^{-a(x+iy)^2}\left[\frac{1}{h}\left(e^{-a\left(2ihx-2hy-h^2\right)}-1\right)\right]dx, \end{aligned}$$

and the expression in square brackets has the properties (i) it tends to $-2ia\,(x+iy)$ as $h\to 0$, and (ii) it is bounded by $(2|x|+2|y|+|h|)e^{2|hx|+|hy|+h^2}$ since $|e^z-1|\le|z|\,e^{|z|}$ for all complex z (Exercise 24, Chapter 6). But since $a>0$ it follows that $\int_{-\infty}^{\infty}e^{-ax^2+cx}dx$ is convergent for any c and so the conditions of the Dominated Convergence Theorem 6.67 (continuous version) are satisfied and $g'(y)$ has the value stated. ♦

The Fourier transform has the following useful properties.

Proposition 7.20

Let f be integrable on $\mathbb{R}$. Then its Fourier transform is bounded, continuous, and tends to zero at infinity.

Proof

The boundedness is immediate: for all y, $|\mathcal{F}f(y)| \leq \int_{-\infty}^{\infty} |f|$. For continuity, note first that for real y, $\left|e^{iy} - 1\right| = \left|\int_0^y e^{it} dt\right| \leq \left|\int_0^y dt\right| = |y|$ and hence $\left|e^{i(x+y)} - e^{ix}\right| \leq |y|$ for all real x, y. Then given $\varepsilon > 0$, there is some $A > 0$ such that $\int_{|x| \geq A} |f| < \varepsilon$. Hence we can write

$$\begin{aligned}
|\mathcal{F}f(y+h) - \mathcal{F}f(y)| &= \left|\int_{-\infty}^{\infty} f(x) \left(e^{-2\pi i x(y+h)} - e^{-2\pi i x y}\right) dx\right| \\
&\leq 2\int_{|x|\geq A} |f(x)|\, dx + \int_{-A}^{A} |f(x)|\, 2\pi\, |xh|\, dx \\
&\leq 2\varepsilon + 2\pi\, |h| \int_{-A}^{A} |x f(x)|\, dx
\end{aligned}$$

which can be made $< 3\varepsilon$ for $|h|$ small enough, and continuity is proved.

To show that $\mathcal{F}f(y) \to 0$ as $|y| \to \infty$, we put $(x + 1/(2y))$ for x in the integral to get

$$\begin{aligned}
\mathcal{F}f(y) &= \int_{-\infty}^{\infty} f(x)\, e^{-2\pi i x y} dx = -\int_{-\infty}^{\infty} f\left(x + \frac{1}{2y}\right) e^{-2\pi i x y} dx \\
&= \frac{1}{2}\int_{-\infty}^{\infty} \left(f(x) - f\left(x + \frac{1}{2y}\right)\right) e^{-2\pi i x y} dx,
\end{aligned}$$

and so

$$|\mathcal{F}f(y)| \leq \frac{1}{2}\int_{-\infty}^{\infty} \left|f(x) - f\left(x + \frac{1}{2y}\right)\right| dx.$$

But we showed in Theorem 5.9 that $\int_{-\infty}^{\infty} |f(x) - f(x+h)|\, dx \to 0$ as $h \to 0$ which is what we want. ■

The proof of inversion will go through a number of steps. To begin with

Lemma 7.21

Let f be integrable on $\mathbb{R}$. Then for any real a,

$$\int_{-a}^{a} \mathcal{F}f(y)\, dy = \int_{-\infty}^{\infty} \frac{\sin(2\pi a x)}{\pi x} f(x)\, dx \tag{7.20}$$

$$= \int_0^\infty \frac{\sin(2\pi a x)}{\pi x} (f(x) + f(-x))\, dx.$$

Proof

Write $g = \mathcal{F}f$, and define functions h, j by

$$h(a) = \int_{-a}^{a} g(y)\, dy, \quad j(a) = \int_{-\infty}^{\infty} \frac{\sin(2\pi a x)}{\pi x} f(x)\, dx.$$

We shall show that $h'(a) = j'(a)$. From the fundamental theorem of calculus,

$$\begin{aligned} h'(a) &= g(a) + g(-a) = \int_{-\infty}^{\infty} f(x) \left(e^{-2\pi i a x} + e^{2\pi i a x}\right) dx \\ &= 2\int_{-\infty}^{\infty} f(x) \cos(2\pi a x)\, dx \end{aligned}$$

and it remains to find j'. For this, consider

$$\begin{aligned} \frac{j(a+t) - j(a)}{t} &= \int_{-\infty}^{\infty} \frac{\sin(2\pi(a+t)x) - \sin(2\pi a x)}{2\pi x t} f(x)\, dx \\ &= 2\int_{-\infty}^{\infty} \frac{\sin(\pi t x)\cos(\pi(2a+t)x)}{\pi x t} f(x)\, dx. \end{aligned}$$

Since $|f(x)|$ is integrable and $\sin(\pi t x)/(\pi t x) \le 1$ for all (real) x, t and $\to 1$ as $t \to 0$, it follows by dominated convergence (Theorem 6.67) that

$$j'(a) = 2\int_{-\infty}^{\infty} \cos(2\pi a x) f(x)\, dx$$

and so $h'(a) = j'(a)$. Since both $h(0) = j(0) = 0$ the result follows. ∎

Note that this would be a 'one-liner' if we used Fubini's theorem (Appendix A), since then we could simply say

$$\begin{aligned} \int_{-a}^{a} \mathcal{F}f(y)\, dy &= \int_{-a}^{a}\int_{-\infty}^{\infty} f(x) e^{-2\pi i x y} dx\, dy = \int_{-\infty}^{\infty} f(x) \int_{-a}^{a} e^{-2\pi i x y} dy\, dx \\ &= \int_{-\infty}^{\infty} f(x) \left[\frac{e^{-2\pi i x y}}{-2\pi i x}\right]_{-a}^{a} dx = \int_{-\infty}^{\infty} f(x) \frac{\sin(2\pi a x)}{\pi x} dx. \end{aligned}$$

Taking a limit as $a \to \infty$ in (7.20) gives

Theorem 7.22

If f is integrable on $\mathbb{R}$ and Lipschitz continuous at x then

$$\lim_{a\to\infty}\int_{-a}^{a}\mathcal{F}f(y)\,e^{2\pi ixy}dy = f(x).$$

If f is not continuous, but has right- and left-hand limits l_1, l_2 at x such that for some $M, \delta > 0$ and $0 < h < \delta$, both $|f(x+h) - f(x)|$ and $|f(x-h) - f(x)|$ are $\le Mh$, then

$$\lim_{a\to\infty}\int_{-a}^{a}\mathcal{F}f(y)\,e^{2\pi ixy}dy = \frac{l_1+l_2}{2}.$$

With Fubini's theorem we can get a theorem (Exercise 6) in which the Lipschitz continuity is replaced by ordinary continuity. In practice however it seems to make little difference; most commonly encountered functions are Lipschitz continuous. Notice that the result does not require that $\mathcal{F}f$ is integrable.

Proof

We begin with the case when $x = 0$. From the lemma,

$$\int_{-a}^{a}\mathcal{F}f(y)\,dy = \int_{-\infty}^{\infty} f(x)\frac{\sin(2\pi ax)}{\pi x}dx$$

and so

$$\int_{-a}^{a}\mathcal{F}f(y)\,dy - f(0) = \int_{-\infty}^{\infty}(f(x)-f(0))\frac{\sin(2\pi ax)}{\pi x}dx \tag{7.21}$$

using the result of Example 5.7(i). Since f is Lipschitz continuous at 0, there are M, δ such that $|f(x) - f(0)| \le M|x|$ for $|x| \le \delta$. Hence $|f(x) - f(0)|/|x|$ is integrable over $(-\delta, \delta)$, and hence integrable over $\mathbb{R}$ since $|f|$ is. Thus the integral on the right of (7.22) is the Fourier integral of the integrable function $(f(x) - f(0))/x$ and so tends to zero by the last part of Proposition 7.20. This proves the result when f is continuous. The reader is given the job of modifying the proof in the case when f is discontinuous, using the final statement in Lemma 7.21.

For $x_1 \ne 0$, consider $f_1(x) = f(x + x_1)$, when

$$\begin{aligned}\mathcal{F}f_1(y) &= \int_{-\infty}^{\infty} e^{-2\pi ixy}f_1(x)\,dx = \int_{-\infty}^{\infty} e^{-2\pi i(t-x_1)y}f(t)\,dt\\ &= e^{2\pi ix_1y}\mathcal{F}f(y).\end{aligned}$$

Applying the above reasoning to f_1 we find

$$\begin{aligned} f(x_1) &= f_1(0) = \lim_{a\to\infty} \int_{-a}^{a} \mathcal{F}f_1(y)\,dy \\ &= \lim_{a\to\infty} \int_{-a}^{a} e^{2\pi i x_1 y} \mathcal{F}f(y)\,dy \end{aligned}$$

which is what we want. ■

Since we have not required the Fourier transform to be integrable, we have $\lim_{a\to\infty} \int_{-a}^{a} \mathcal{F}f(y)\, e^{2\pi i x y} dy$, not $\int_{-\infty}^{\infty} \mathcal{F}f(y)\,dy$.

Example 7.23

In Example 7.19(i) $f = \chi_{[-a,a]}$, and $\mathcal{F}f(y) = (\sin(2\pi a y))/(\pi y)$ if $y \neq 0$, $=2a$ if $y = 0$. The theorem now allows us to deduce that

$$\lim_{b\to\infty} \int_{-b}^{b} \frac{\sin(2\pi a y)}{\pi y} e^{2\pi i x y} dy = \begin{cases} 1 & \text{for } -a < x < a, \\ 1/2 & \text{for } x = \pm a \\ 0 & \text{for } |x| > a. \end{cases}$$ ♦

If the Fourier transform *is* integrable, then we can replace $\lim \int_{-a}^{a} \mathcal{F}f(y)\,dy$ by $\int_{-\infty}^{\infty} \mathcal{F}f(y)\,dy$ to get the inversion theorem

Theorem 7.24

If both f and $\mathcal{F}f$ are integrable, and f is Lipschitz continuous on $\mathbb{R}$, then for all $x \in \mathbb{R}$,

$$f(x) = \int_{-\infty}^{\infty} \mathcal{F}f(y)\, e^{2\pi i x y} dy.$$

The proof is immediate since if $\mathcal{F}f$ is integrable then $\int_{-a}^{a} \mathcal{F}f(y)\, e^{2\pi i x y} dy \to \int_{-\infty}^{\infty} \mathcal{F}f(y)\, e^{2\pi i x y} dy$ as $a \to \infty$. In other words, $\mathcal{F}^{-1}\mathcal{F}f = f$ for functions satisfying the conditions of the theorem, so $\mathcal{F}^{-1}$ and $\mathcal{F}$ are indeed inverse operators for such functions. We shall see when we come to the study of distributions in Section 7.3 that we can operate much more freely with both inversion and differentiation.

Example 7.25

In Example 7.19(ii) $f(x) = e^{-a|x|}$, $\mathcal{F}f(y) = 2a/\left(a^2 + 4\pi^2 y^2\right)$. Then for $a > 0$, and all real x,

$$\int_{-\infty}^{\infty} \frac{2a}{a^2 + 4\pi^2 y} e^{2\pi i x y} dy = e^{-a|x|}.$$

This follows since $\mathcal{F}f$ is integrable; the result is not easy to prove directly. ♦

There is an interesting relation between differentiation and multiplication of $f(x)$ by x. For instance

Proposition 7.26

(i) Suppose that f is differentiable, f' is continuous, and that both f, f' are integrable. Then $\mathcal{F}f'(y) = 2\pi i y \mathcal{F}f(y)$.

(ii) Suppose that both f,g are integrable, where $g(x) = xf(x)$. Then $\mathcal{F}f(y)$ is differentiable, and

$$\frac{d}{dy}\mathcal{F}f(y) = -2\pi i \mathcal{F}g(y).$$

Proof

(i) Since f' is continuous and integrable, we have $f(x) = \int_{-\infty}^{x} f'(t)\,dt$ and so $f(x) \to 0$ as $x \to -\infty$. Also $f(x) \to \int_{-\infty}^{\infty} f'(t)\,dt$ as $x \to \infty$ and this limit must be zero, otherwise f would not be integrable. Hence we can integrate by parts to obtain

$$\begin{aligned}
\mathcal{F}f'(y) &= \int_{-\infty}^{\infty} f'(x)\,e^{-2\pi i x y}dx \\
&= \left[f(x)\,e^{-2\pi i x y}\right]_{-\infty}^{\infty} + 2\pi i y \int_{-\infty}^{\infty} f(x)\,e^{-2\pi i x y}dx \\
&= 2\pi i y \mathcal{F}f(y)
\end{aligned}$$

since we just showed that the integrated term is zero.

Part (ii) can be proved by differentiation under the integral sign (justified as usual by dominated convergence), or, more ambitiously, deduced from (i) using the inversion theorem. The details are left to the reader in Exercise 12. ■

Notice that the first part of the proposition concerns $\mathcal{F}(f')$, and the second $(\mathcal{F}f)'$.

This result illustrates a general property of Fourier transforms (which we shall not attempt to formulate precisely) that smoothness properties of a function are reflected in properties of $\mathcal{F}f$ at ∞ and vice versa; in this example the differentiability of $\mathcal{F}f$ (local smoothness) is reflected in the integrability of $xf(x)$ (slow growth of f at infinity).

Example 7.27

Let $f(x) = \operatorname{sech}(ax)$ for $a > 0$. Then

$$\mathcal{F}f(y) = (\pi/a)\operatorname{sech}\left(\pi^2 y/a\right),$$

and applying part (i) of the proposition gives

$$\mathcal{F}f'(y) = \mathcal{F}\left(-\frac{a\sinh(ax)}{\cosh^2(ax)}\right) = \frac{2\pi^2 iy}{a\cosh(\pi^2 y/a)}.$$

To find the Fourier integral, we compute

$$\begin{aligned}
\int_{-\infty}^{\infty}\frac{e^{-2\pi ixy}}{\cosh(ax)}dx &= 2\int_{-\infty}^{\infty}\frac{e^{-2\pi ixy}}{e^{ax}+e^{-ax}}dx = 2\int_{-\infty}^{\infty}\frac{e^{(a-2\pi iy)x}}{e^{2ax}+1}dx\\
&= \frac{1}{a}\int_0^{\infty}\frac{t^{(1/2-\pi iy/a)}}{t+1}dt\\
&= \frac{\pi}{a}\Gamma(1/2-\pi iy/a)\,\Gamma(1/2+\pi iy/a)\\
&= \frac{\pi}{a\sin(\pi(1/2-\pi iy/a))} = \frac{\pi}{a\cosh(\pi^2 y/a)}
\end{aligned}$$

where we have put $e^{ax} = t$, and used the result of Exercise 6, Chapter 5 to evaluate the integral and Example 6.77(ii) for the product of Gamma functions. Notice in particular that the case $a = \pi$ shows that $\operatorname{sech}(\pi x)$ is invariant under the Fourier transform. ♦

The last result in this section is Parseval's formula which as well as being important in its own right, will be needed in Section 7.3 on distributions. Its proof uses Fubini's theorem in an essential way – the reader may take this on trust, or consult the appendix for a proof.

Theorem 7.28 (Parseval's Equation)

Let f, g be continuous integrable functions, such that $\mathcal{F}f$ and $\mathcal{F}g$ are also integrable . Then

$$\int_{-\infty}^{\infty} f(x)\,\mathcal{F}g(x)\,dx = \int_{-\infty}^{\infty} g(y)\,\mathcal{F}f(y)\,dy \tag{7.22}$$

and

$$\int_{-\infty}^{\infty} |f(x)|^2\,dx = \int_{-\infty}^{\infty} |\mathcal{F}f(y)|^2\,dy. \tag{7.23}$$

Proof

The continuity of $\mathcal{F}f$ and $\mathcal{F}g$ is automatic from Proposition 7.20. Consider the double integral

$$\int_{-\infty}^{\infty}\int_{-\infty}^{\infty} f(x)\, g(y)\, e^{-2\pi ixy}dx\, dy.$$

The integral surely exists since the integrand is continuous with respect to both variables, and $\int_{-\infty}^{\infty}\int_{-\infty}^{\infty} |f(x)|\,|g(y)|\, dx\, dy$ is finite. Hence by Fubini's theorem (Appendix A) the value may be found by taking the repeated integral in either order. But

$$\int_{-\infty}^{\infty}\left(\int_{-\infty}^{\infty} f(x)\, e^{-2\pi ixy}dx\right) g(y)\, dy = \int_{-\infty}^{\infty} g(y)\, \mathcal{F}f(y)\, dy$$

and

$$\int_{-\infty}^{\infty}\left(\int_{-\infty}^{\infty} g(y)\, e^{-2\pi ixy}dy\right) f(x)\, dx = \int_{-\infty}^{\infty} f(x)\, \mathcal{F}g(x)\, dx$$

as required.

The second result is the special case in which $g(y) = (\mathcal{F}f(y))^{-}$ where the bar denotes complex conjugation. Then

$$\begin{aligned} g(y) &= \left[\int_{-\infty}^{\infty} f(x)\, e^{-2\pi ixy}dx\right]^{-} \\ &= \int_{-\infty}^{\infty} \bar{f}(x)\, e^{2\pi ixy}dx = \mathcal{F}^{-1}\left(\bar{f}\right)(y). \end{aligned}$$

The conditions of the theorem are enough to ensure that $\bar{f}$ satisfies the inversion theorem, so since $g = \mathcal{F}^{-1}\left(\bar{f}\right)$, it follows that $\bar{f} = \mathcal{F}g$. Substituting for g in (7.22) gives

$$\int_{-\infty}^{\infty} f(x)\, \bar{f}(x)\, dx = \int_{-\infty}^{\infty} \mathcal{F}f(y)\, (\mathcal{F}f(y))^{-}\, dy$$

which is what we want. ■

Example 7.29

From Example 7.19(ii) with $a > 0$,

$$\int_{-\infty}^{\infty} e^{-2a|x|}dx = \frac{1}{a} = \int_{-\infty}^{\infty} \frac{4a^2}{\left(a^2 + 4\pi^2 y^2\right)^2}dy. \ \blacklozenge$$

7.3 Distributions

The definitions of two of the basic processes of analysis, those of differentiation in Chapter 3, and of taking Fourier transforms in this chapter, were subject to what one might consider undesirable restrictions. In the case of differentiation, continuous functions need not be differentiable, and a function which is once differentiable need not be twice differentiable; similarly for Fourier integrals, all kinds of 'nice' functions, $\cos x$ for instance, do not possess Fourier transforms. The theory of distributions, or generalised functions as they are also known, is a way of getting round these difficulties; continuous functions become infinitely differentiable, and the scope of the theory of Fourier integrals is greatly widened. There is a price to be paid however, in that we must enlarge our view of what kind of objects to consider. So far we have considered 'ordinary' functions as mappings from numbers to numbers:

$$\text{number} \longrightarrow \boxed{\text{function}} \longrightarrow \text{number}.$$

But a distribution is a mapping from functions to numbers:

$$\text{function} \longrightarrow \boxed{\text{distribution}} \longrightarrow \text{number}.$$

To illustrate this, in advance of the formal definitions below, consider the following example.

Example 7.30

The Dirac delta-function δ is the distribution whose value at a function f defined on $\mathbb{R}$ is given by $\delta(f) = f(0)$. More generally, for $a \in \mathbb{R}$, δ_a is the translation of δ given by $\delta_a(f) = f(a)$. ♦

Notice again, for emphasis, that the input is a function f, and the output is a number (real or complex).

To begin with, we define the functions which we shall use as inputs – a more restricted class than the arbitrary functions which appeared in the example.

Definition 7.31

(i) A test function ϕ is an infinitely differentiable real or complex valued function on $\mathbb{R}$, for which each derivative tends to zero at infinity faster than any power of x. Formally, for any integers $k, n \geq 0$,

$$x^n \phi^{(k)}(x) \to 0 \qquad \text{as } |x| \to \infty.$$

(ii) A sequence (ϕ_m) of test functions is null (tends to zero) if for all $n, k \geq 0$, $x^n \phi_m^{(k)}(x)$ tends uniformly to zero on $\mathbb{R}$ as $m \to \infty$.

We use Greek letters ϕ, ψ etc. for test functions, and denote the set of all test functions by Φ.

Example 7.32

(i) For any $a > 0$, $\operatorname{sech} ax = 1/\cosh ax$ is a test function.

(ii) For any $a > 0$, e^{-ax^2} is a test function.

(iii) For any real a, and $h > 0$,

$$\phi(x) = \begin{cases} \exp\left(1/\left((x-a)^2 - h^2\right)\right) & \text{for } a - h \leq x \leq a + h, \\ 0 & \text{elsewhere} \end{cases}$$

is a test function which is zero outside the interval $(a - h, a + h)$.

(iv) $1/\left(1 + x^2\right)$ is not a test function.

For (i), it is easily verified by induction that the nth derivative of $\operatorname{sech}(ax)$ is of the form $\sinh(ax)\, p_n(\cosh(ax)) \cosh^{-n-1}(ax)$ if n is odd, or $q_n(\cosh(ax))$ $\cosh^{-n-1}(ax)$ if n is even, where p_n, q_n are polynomials of degree at most n. Thus the nth derivatives tend to zero at infinity like $e^{-a|x|}$, which is faster than any power of x. For (ii) similarly, the nth derivative is of the form e^{-ax^2} multiplied by a polynomial in x so $x^k f^{(n)}(x) \to 0$ as $|x| \to \infty$. For (iii) the nth derivative is $\exp\left(1/\left((x-a)^2 - h^2\right)\right)$ multiplied by a rational function of x and so tends to zero as $x \to a \pm h$.

Part (iv) is interesting in that $1/\left(1 + x^2\right)$ and all its derivatives tend to zero at infinity, but not faster than any power of x. ♦

Many other names (smooth, good, rapidly decreasing, etc.) for test functions are common in the literature of distribution theory, so the reader should be careful to check definitions when looking elsewhere for results.

Proposition 7.33

(i) All sums and products of test functions are again test functions. (In particular, Φ is a complex vector space.)

(ii) All derivatives of test functions are test functions.

Proof

This is immediate from Definition 7.31. ∎

We shall see later in Proposition 7.48 that all Fourier transforms of test functions are again test functions. The set Φ of test functions is *very* well behaved.

Now we come to the main definition of the section.

Definition 7.34

A distribution (a generalised function) is a function $T : \Phi \to \mathbb{C}$ which is

(i) linear: $T(a\phi + b\psi) = aT\phi + bT\psi$ for all $a, b \in \mathbb{C}$ and $\phi, \psi \in \Phi$, and

(ii) continuous: if $\phi_n \to 0$ in Φ then $T\phi_n \to 0$ in $\mathbb{C}$.

We denote the set of all distributions by Δ. Δ is usually called the space of tempered distributions, but we shall not consider any other kind.

Example 7.35

(i) The delta function $\delta_a : \phi \to \phi(a)$ is a distribution.

(ii) Given any function f which is (improperly) integrable over every subinterval $[a, b] \subset \mathbb{R}$ (Definition 5.1) and bounded on $\mathbb{R}$ by some power of $(1 + |x|)$, the mapping $f \to \Lambda f$ given by

$$\Lambda f(\phi) = \int_{-\infty}^{\infty} f(x)\,\phi(x)dx \tag{7.24}$$

is a distribution.

Part (ii) is important in that it allows us to regard most 'ordinary' functions f, as special kinds of distribution, using the mapping $f \to \Lambda f$. For instance $f(x) = x^2$ determines a distribution (P^2 in Definition 7.40 below) but $g(x) = e^x$ does not, since it is not bounded by any power of $(1 + |x|)$.

In both cases the linearity is evident and we have to show continuity. In (i), if $\phi_n \to 0$ then in particular $\phi_n(a) \to 0$ for all real a, so $\delta_a(\phi_n) \to 0$ as required. In (ii), suppose that $|f(x)| \leq M(1 + |x|)^k$ for all x. Since for any given m, $x^m\phi_n(x) \to 0$ uniformly on $\mathbb{R}$ as $n \to \infty$, there is $K > 0$ such that $(1 + |x|)^{k+2}|\phi_n(x)| \leq K$ for all n, x, so that $(1 + |x|)^k|\phi_n(x)|$ tends uniformly to zero and is dominated by $K(1 + |x|)^{-2}$. It follows that

$$\int_{-\infty}^{\infty} f(x)\,\phi_n(x)dx = \int_{-\infty}^{\infty} \frac{f(x)}{(1 + |x|)^k}(1 + |x|)^k\phi_n(x)\,dx \to 0$$

as $n \to \infty$ which is what we want. ♦

Part (ii) of this example shows us the direction in which we should proceed. The distribution Λf corresponds to the ordinary function f, and we shall define differentiation and Fourier transforms for distributions in order to respect this correspondence, as can be seen from Proposition 7.37(ii) and Theorem 7.50(iii) below. Not all distributions are of the form Λf however; the delta function is an example of one which is not.

We can now define differentiation for distributions.

Definition 7.36

Let T be a distribution. Its derivative DT is the distribution given by

$$DT(\phi) = -T(\phi'). \tag{7.25}$$

Notice that DT is again a distribution since if (ϕ_n) is a null sequence of distributions, then so is (ϕ_n'). The motivation for the definition comes from the formula for integration by parts; if f, f' are continuous and bounded by powers of $(1+|x|)$, then

$$\int_{-\infty}^{\infty} f'(x)\,\phi(x)\,dx = -\int_{-\infty}^{\infty} f(x)\,\phi'(x)\,dx$$

since the integrated term $[f(x)\,\phi(x)]_{-\infty}^{\infty}$ is zero.

Proposition 7.37

(i) Every distribution is infinitely differentiable with $D^k T(\phi) = (-1)^k T(\phi^{(k)})$. In particular, $\delta^{(k)}(\phi) = (-1)^k \phi^{(k)}(0)$.

(ii) If f and f' are continuous and bounded by powers of $(1+|x|)$, then $\Lambda f' = D\Lambda f$.

Proof

For (i), apply (7.25) repeatedly. For (ii)

$$\begin{aligned}
\Lambda f'(\phi) &= \int_{-\infty}^{\infty} f'(x)\,\phi(x)\,dx && \text{from (7.24)}\\
&= -\int_{-\infty}^{\infty} f(x)\,\phi'(x)\,dx && \text{by integration by parts}\\
&= -\Lambda f(\phi') && \text{from (7.24) again}
\end{aligned}$$

$$= D\Lambda f(\phi) \qquad \text{from the definition (7.25).} \blacksquare$$

The result of (ii) is emphatically not true in general.

Example 7.38

Let $H(x) = 1$ for $x \geq 1$, $= 0$ elsewhere (this is Heaviside's function). Then $H' = 0$ for all $x \neq 0$, so $\Lambda H' = 0$, but $D\Lambda H = \delta$.

This follows from the fact that for any test function ϕ,

$$\begin{aligned} D\Lambda H(\phi) &= -\Lambda H(\phi') = -\int_{-\infty}^{\infty} H(x)\phi'(x)\,dx \\ &= -\int_{0}^{\infty} \phi'(x)\,dx = -[\phi(x)]_0^{\infty} = \phi(0) = \delta(\phi). \end{aligned}$$

The reader should put in the reasons for each step here, as we did in the proof of Proposition 7.37. ♦

One of the most striking results in the theory of distributions is that it extends the classical theory in the most economical way which makes every continuous function infinitely differentiable. More explicitly

Theorem 7.39

For every distribution T, there is a continuous function f, bounded by a power of $1 + |x|$, and an integer k such that $T = D^k \Lambda f$.

For instance, we already noted that the Dirac distribution δ is not of the form Λf for any f. However, it follows from Example 7.38 that if $f(x) = x$ for $x \geq 0$, $= 0$ for $x < 0$ then f is continuous and $\delta = D^2 \Lambda f$.

It is important to realise that in this result we are only talking about what in the wider theory are called tempered distributions – in the general case the result is only true locally – every distribution is equal on each bounded interval to a derivative of a continuous function. Since we shall not make use of this result, and since the proof is on a much deeper level than anything else we have done, we shall simply refer the reader to [13] or [9] for a proof (this is the converse of our motto – see the proof of Proposition 1.6). Instead we look at the way in which powers of x fit into the framework of distribution theory. To maintain the distinction between f and Λf we have to introduce new notation, for instance P^2 for the generalised function which corresponds to the ordinary function $x \to x^2$ on $(0, \infty)$ (Λx^2 would not do since since x^2 is not the name of a function!).

Definition 7.40 (Powers on $(0, \infty)$)

(i) For $a \geq 0$, $p_a(x) = x^a$ is continuous on $[0, \infty)$; for $-1 < a < 0$, p_a is continuous on $(0, \infty)$ and improperly integrable on $(0, 1)$. In either case we define P^a by

$$P^a(\phi) = \Lambda p_a(\phi) = \int_0^\infty x^a \phi(x)\, dx. \tag{7.26}$$

In particular $P^0 = \Lambda H$ where H is Heaviside's function from Example 7.38.

(ii) For $a < -1$, not an integer, we define

$$P^a = \frac{1}{(a+1)(a+2)\cdots(a+n)} D^n P^{a+n}, \tag{7.27}$$

where n is an integer with $n + a > -1$ so that we can use (7.26).

(iii) For negative integer powers,

$$\begin{aligned} P^{-1} &= D\Lambda \ln, \\ P^{-k-1} &= \frac{(-1)^k}{k!} D^k P^{-1}, \qquad k = 1, 2, \ldots. \end{aligned} \tag{7.28}$$

This definition calls for a great deal of comment! Firstly all P^a are defined in terms of known objects, namely Λp^a and $\Lambda \ln$ and their derivatives, and so they are all distributions. Secondly there is a problem of consistency in (ii) in that the value of n is not given – we leave it to the reader (Exercise 11) to show that the value does not depend on the choice of n. Then we should realise that the exponents are purely formal – since we have no multiplication for distributions, the distribution P^2 is in no sense the square of the distribution P. But most importantly we have to find out to what extent these distributions actually correspond to the powers which they are supposed to represent. We do this by calculating $P^a(\phi)$ for several interesting special values of a.

Example 7.41

(i) $P^{-3/2}(\phi) = \int_0^\infty (\phi(x) - \phi(0)) / x^{3/2}\, dx$.

(ii) $P^{-1}(\phi) = \int_0^1 (\phi(x) - \phi(0)) / x\, dx + \int_1^\infty \phi(x) / x\, dx$.

(iii)

$$P^{-2}(\phi) = \int_0^1 \frac{\phi(x) - \phi(0) - x\phi'(0)}{x^2} dx + \int_1^\infty \frac{\phi(x)}{x^2} dx - \phi(0) - \phi'(0).$$

To deduce (i), use (7.26) and (7.27) as follows.

$$\begin{aligned} P^{-3/2}(\phi) &= -2DP^{-1/2}(\phi) = 2P^{-1/2}(\phi') \\ &= 2\int_0^\infty \frac{\phi'(x)}{\sqrt{x}} dx = 2\left[\frac{\phi(x) - \phi(0)}{\sqrt{x}}\right]_0^\infty + \int_0^\infty \frac{\phi(x) - \phi(0)}{x^{3/2}} dx \end{aligned}$$

which is the required result since the integrated term is zero. For (ii) use (7.28) to write

$$\begin{aligned} P^{-1}(\phi) &= D\Lambda \ln(\phi) = -\Lambda \ln(\phi') \\ &= -\int_0^\infty \ln x\, \phi'(x)\, dx = -\int_0^1 \ln x\, \phi'(x)\, dx - \int_1^\infty \ln x\, \phi'(x)\, dx \\ &= -\left[(\phi(x) - \phi(0)) \ln x\right]_0^1 + \int_0^1 \frac{\phi(x) - \phi(0)}{x} dx \\ &\quad - \left[\phi(x) \ln x\right]_1^\infty + \int_1^\infty \frac{\phi(x)}{x} dx \end{aligned}$$

and again the integrated terms vanish. For (iii)

$$\begin{aligned} P^{-2}(\phi) &= -DP^{-1}(\phi) = P^{-1}(\phi') \\ &= \int_0^1 \frac{\phi'(x) - \phi'(0)}{x} dx + \int_1^\infty \frac{\phi'(x)}{x} dx \qquad \text{from (ii)} \\ &= \left[\frac{\phi(x) - \phi(0) - x\phi'(0)}{x}\right]_0^1 + \int_0^1 \frac{\phi(x) - \phi(0) - x\phi'(0)}{x^2} dx \\ &\quad + \left[\frac{\phi(x)}{x}\right]_1^\infty + \int_1^\infty \frac{\phi(x)}{x^2} dx \\ &= \int_0^1 \frac{\phi(x) - \phi(0) - x\phi'(0)}{x^2} dx + \int_1^\infty \frac{\phi(x)}{x^2} dx - \phi(0) - \phi'(0) \end{aligned}$$

where the Taylor expansion of ϕ about 0 shows that the stated limits and integrals exist. ♦

Part (ii) of this example shows up a possible misunderstanding: it is not the case that $P^{-a} \to P^{-1}$ as $a \to 1$ (where convergence of operators is considered pointwise; $T_n \to T$ means $T_n(\phi) \to T(\phi)$ for all test functions ϕ). The correct result is as follows.

Example 7.42

$$\lim_{\varepsilon \to 0+} \left\{ P^{-1+\varepsilon} - \frac{1}{\varepsilon}\delta \right\} = P^{-1}.$$

To see this, consider any test function ϕ and $\varepsilon > 0$. Then

$$\begin{aligned}
P^{-1+\varepsilon}(\phi) &= \int_0^\infty \frac{\phi(t)}{t^{1-\varepsilon}} dt \\
&= \int_0^1 \frac{\phi(t) - \phi(0)}{t^{1-\varepsilon}} dt + \int_0^1 \frac{\phi(0)}{t^{1-\varepsilon}} dt + \int_1^\infty \frac{\phi(t)}{t^{1-\varepsilon}} dt \\
&= \int_0^1 \frac{\phi(t) - \phi(0)}{t^{1-\varepsilon}} dt + \frac{\phi(0)}{\varepsilon} + \int_1^\infty \frac{\phi(t)}{t^{1-\varepsilon}} dt \\
P^{-1+\varepsilon}(\phi) - \frac{1}{\varepsilon}\delta(\phi) &= \int_0^1 \frac{\phi(t) - \phi(0)}{t^{1-\varepsilon}} dt + \int_1^\infty \frac{\phi(t)}{t^{1-\varepsilon}} dt
\end{aligned}$$

and the right-hand side tends to $P^{-1}(\phi)$ using the result of Example 7.41(ii). The limit on the left at -1 is Exercise 13. ♦

In order to define powers on $\mathbb{R}$, we need the reflection operator R given for functions by $R\phi(x) = \phi(-x)$, and for distributions by $RT(\phi) = T(R\phi)$.

Definition 7.43

For all real a,

$$E^a = P^a + RP^a, \quad O^a = P^a - RP^a.$$

Informally, E^a corresponds to the ath power, $|x|^a$ of $|x|$ on $(-\infty, \infty)$, while O^a corresponds to $|x|^a \operatorname{sgn} x$. ($E$ and O are chosen to suggest even and odd functions, respectively.)

The evaluation of $E^a(\phi)$, $O^a(\phi)$ for $a = -1, -3/2, -2$, as in Example 7.41 above, is Exercise 14.

Another useful operation is the multiplication of a distribution by a function. The class of functions which are suitable here are called slowly increasing – like test functions, they are infinitely differentiable, but now the condition at infinity is that they (and their derivatives) are bounded by powers of $1 + |x|$. More precisely we have the following definition.

Definition 7.44

A slowly increasing function f on $\mathbb{R}$ is one which is infinitely differentiable and such that for each derivative $f^{(k)}$ there are $M, n \geq 0$ such that $\left|f^{(k)}(x)\right| \leq M(1+|x|)^n$ for all $x \in \mathbb{R}$.

Example 7.45

Any polynomial is slowly increasing. Any infinitely differentiable periodic function is slowly increasing. ♦

The product of a distribution by a slowly increasing function is now defined as follows.

Definition 7.46

Given a distribution T and a slowly increasing function f, the product fT is the distribution defined by

$$fT(\phi) = T(f\phi).$$

The reader is left to check that $f\phi$ is again a test function, and to verify such identities as $(fT)' = f'T + fT'$, and $f\Lambda g = \Lambda(fg)$ under suitable conditions. Note that the operation does not work the other way round: Tf is not defined. Similarly there is no definition of either multiplication or composition for distributions since $T\phi U\phi$ is not linear in ϕ, while $T(U\phi)$ is undefined.

Example 7.47

Let $p(x) = x$ for all $x \in \mathbb{R}$. Then for all real a, $pE^a = O^{a+1}$, $pO^a = E^{a+1}$, $p\delta = 0$.

The first two statements are left to the reader to verify in Exercise 15. For $p\delta$, consider

$$p\delta(\phi) = \delta(p\phi) = p(0)\phi(0) = 0.$$

This shows that, just as for the multiplication of ordinary functions, we may have 'zero-divisors', i.e. non-zero objects whose product is zero. ♦

Distributions show their greatest interest and importance when we consider Fourier transforms. This is because, as mentioned earlier, the class of test functions remains fixed under Fourier transforms.

Proposition 7.48

For every test function $\phi \in \Phi$, the Fourier transform $\mathcal{F}\phi$ is again a test function and satisfies the inversion formula

$$\phi(x) = \int_{-\infty}^{\infty} \mathcal{F}\phi(y)\, e^{2\pi ixy} dy.$$

Proof

This is Proposition 7.26 applied repeatedly. Since ϕ goes to zero at infinity faster than any power of $|x|$, all of $\phi(x)$, $x\phi(x)$, $x^n\phi(x)$ are integrable, and hence $\mathcal{F}\phi$ has derivatives of all orders. Again, since ϕ has derivatives of all orders, then for any n, $y^n\mathcal{F}\phi(y) = \text{Const.}\mathcal{F}\phi^{(n)}(y)$ tends to zero by Proposition 7.20. Thus ϕ is a test function, and satisfies the conditions for the inversion Theorem 7.24. ■

The beauty of this result is that we can now extend the Fourier transform to distributions, using Parseval's equation (Theorem 7.28) for motivation.

Definition 7.49

For any distribution T, its Fourier transform is defined by

$$\mathcal{F}T(\phi) = T(\mathcal{F}\phi),$$
$$\mathcal{F}^{-1}T(\phi) = T(\mathcal{F}^{-1}\phi).$$

In order to be sure that this is a sensible definition, we have to check that $\mathcal{F}T$ really is a distribution (i.e. that it is linear and continuous, as Definition 7.34 requires), that the definition is consistent ($\mathcal{F}\Lambda f = \Lambda\mathcal{F}f$) and that it satisfies the inversion theorem, $\mathcal{F}^{-1}\mathcal{F}T = T$. This, and other useful properties of the Fourier transform are listed next.

Theorem 7.50

(i) For all $T \in \Delta$, $\mathcal{F}T \in \Delta$,

(ii) for all $\phi \in \Phi$, $\mathcal{F}\Lambda\phi = \Lambda\mathcal{F}\phi$,

(iii) for all $T \in \Delta$, $\mathcal{F}^{-1}\mathcal{F}T = T$,

(iv) $\mathcal{F}DT = 2\pi ip\mathcal{F}(T)$ and $D\mathcal{F}T = -2\pi i\mathcal{F}(pT)$, where $p(x) = x$ as usual,

(v) if $T_n \to T$ as $n \to \infty$ (i.e. $T_n(\phi) \to T(\phi)$ for all ϕ as $n \to \infty$, as in Example 7.42) then also $\mathcal{F}T_n \to \mathcal{F}T$.

Proof

(i) Linearity is immediate; as usual it is the continuity which requires the effort. (Note that we should also have done this in Definition 7.46 above – i.e. we should have shown for instance that if $\phi_n \to 0$ then $fT(\phi_n) \to 0$ also. Full marks (and supply the proof!) if you noticed.)

Suppose then that $\phi_n \to 0$ as $n \to \infty$; we have to show that $\mathcal{F}T(\phi_n) \to 0$ also. From the definition of $\mathcal{F}T$, and the continuity of T, it is enough to show that $\mathcal{F}\phi_n \to 0$ in Φ. Definition 7.31 shows that if $\phi_n \to 0$ in Φ, then ϕ_n converges uniformly to zero on any finite interval of $\mathbb{R}$, and is bounded at infinity, say $|\phi_n(x)| \le K(1+|x|)^{-2}$ for some K and $x \le A$. Hence since $|\mathcal{F}\phi_n(y)| \le \int_{-\infty}^{\infty} |\phi_n|$, $\mathcal{F}\phi_n \to 0$ uniformly on $\mathbb{R}$. But the same argument applies to all products of derivatives of ϕ_n by powers of x, and so $\mathcal{F}\phi_n \to 0$ in Φ also.

Fortunately the other parts just require checking that the pieces fit together properly. For instance in (ii), given any $\psi \in \Phi$,

$$\begin{aligned}\mathcal{F}\Lambda\phi(\psi) &= \Lambda\phi(\mathcal{F}\psi) = \int_{-\infty}^{\infty} \phi(x)\,\mathcal{F}\psi(x)\,dx \\ &= \int_{-\infty}^{\infty} \mathcal{F}\phi(x)\,\psi(x)\,dx = \Lambda(\mathcal{F}\phi)(\psi)\end{aligned}$$

by Parseval's equation. Part (iii) follows from the inversion theorem for test functions, and for (iv),

$$\begin{aligned}\mathcal{F}DT(\phi) &= DT(\mathcal{F}\phi) = -T(D\mathcal{F}\phi) = T(2\pi i\mathcal{F}(p\phi)) \\ &= 2\pi i\mathcal{F}T(p\phi) = 2\pi i p\mathcal{F}T(\phi)\end{aligned}$$

where the crucial step uses Proposition 7.26(ii). Part (v) is left to the reader. ■

Example 7.51

$\mathcal{F}\delta = \Lambda 1 = E^0$ and for integer $n \ge 1$, $\mathcal{F}\left(\delta^{(n)}\right) = (2\pi i)^n \Lambda p^n$ (or $(2\pi i)^n X^n$ in the notation of Definition 7.55 below).

The first statement is immediate since for any ϕ, $\mathcal{F}\delta(\phi) = \delta(\mathcal{F}\phi) = \mathcal{F}\phi(0) = \int_{-\infty}^{\infty} \phi(t)\,dt = \Lambda 1(\phi)$ and the second follows from part (iv) of Theorem 7.50. ♦

Note that the inversion theorem $\mathcal{F}^{-1}\mathcal{F} = I$ (for test functions or distributions) can be written equivalently as $\mathcal{F}^2 = R$, and hence as $\mathcal{F}^2 = \pm I$ when the input is even or odd respectively.

To round off the section, we shall find the Fourier transforms of the distributions E^a, O^a. This requires some preparation.

Lemma 7.52

For complex x, c with $\Re c > 0$ and $\Re x > 0$,

$$\int_0^\infty e^{-ct} t^{x-1} dt = c^{-x} \Gamma(x).$$

The result is also true for $x \in (0,1)$ and $\Re c = 0$, $c \neq 0$.

Proof

This looks as if it should be easy. After all, we can put $ct = u$ and get

$$\int_0^\infty e^{-ct} t^{x-1} dt = \int_0^\infty e^{-u} (u/c)^{x-1} du/c = c^{-x} \int_0^\infty e^{-u} u^{x-1} du = c^{-x} \Gamma(x)$$

and we appear to have finished. The trouble with this argument is that it replaces $t \in [0, \infty)$ with $u = ct$ and if c is not real, then the integral with respect to u is not over $[0, \infty)$ as the notation misleadingly suggests. We have to be a little more devious. The simple-minded argument above does however give us one step in the right direction: if we put $c = |c| v$ say with $|v| = 1$, $v = e^{i\alpha}$, $-\pi/2 \leq \alpha \leq \pi/2$, then the substitution $|c| t = u$ is legitimate, and gives

$$\int_0^\infty e^{-ct} t^{x-1} dt = \int_0^\infty e^{-|c| vt} t^{x-1} dt = |c|^{-x} \int_0^\infty e^{-vu} u^{x-1} du$$

and we have reduced the problem to the case in which $|v| = 1$.

Suppose first that $\Re c > 0$ so that $|\alpha| < \pi/2$. Consider the two integrals

$$I_1(r) = \int_0^r e^{-t} t^{x-1} dt,$$
$$I_2(r) = v^x \int_0^r e^{-vt} t^{x-1} dt;$$

our result will follow if we can show that they have the same limit as $r \to \infty$, so we need an expression for their difference. Consider

$$I_3(r) = \int_0^\alpha e^{-re^{i\theta}} \left(re^{i\theta}\right)^{x-1} ire^{i\theta} d\theta = ir^x \int_0^\alpha e^{-re^{i\theta}} e^{ix\theta} d\theta.$$

(The form of this integral will be no surprise to those who know some complex integration, but otherwise can be taken 'out of the hat' and no harm will be done.) We shall show that $I_1 + I_3 = I_2$, and that $I_3 \to 0$ as $r \to \infty$ which will give what we want. Consider the derivatives with respect to r. We have

$$I_1'(r) = e^{-r} r^{x-1},$$
$$I_2'(r) = v^x e^{-vr} r^{x-1}$$

from the Fundamental Theorem of Calculus. For I_3 we have

$$I_3'(r) = ixr^{x-1}\int_0^\alpha e^{-re^{i\theta}}e^{ix\theta}d\theta + ir^x\int_0^\alpha \left(-e^{i\theta}\right)e^{-re^{i\theta}}e^{ix\theta}d\theta$$

where the reader should (a) consider exactly which properties of complex powers of the real variable r are required to show that $(r^x)' = xr^{x-1}$, and (b) justify the differentiation under the integral sign, as in the appendix, Theorem A.8. This gives

$$\begin{aligned} I_3'(r) &= \frac{x}{r}I_3(r) + ir^x\left\{\left[\frac{e^{ix\theta}}{ir}e^{-re^{i\theta}}\right]_0^\alpha - \frac{x}{r}\int_0^\alpha e^{-re^{i\theta}}e^{ix\theta}d\theta\right\} \\ &= \frac{x}{r}I_3(r) + r^{x-1}\left\{e^{ix\alpha}e^{-re^{i\alpha}} - e^{-r}\right\} - \frac{x}{r}I_3(r) \\ &= I_2'(r) - I_1'(r) \end{aligned}$$

after integration by parts. Thus $I_1 + I_3 = I_2$+Constant, and since all three integrals are zero at $r = 0$, it follows that $I_1 + I_3 = I_2$. Showing that $I_3(r) \to 0$ as $r \to \infty$ is easy since

$$|I_3(r)| \leq \alpha r^{\Re x}e^{-r\cos\alpha}e^{|\alpha\Im x|}$$

and $|\alpha| < \pi/2$. Thus letting $r \to \infty$ gives the required result when $\Re c > 0$.

For the final step when $c = ic_2 \neq 0$, we know already (Example 5.6) that the integral exists for $0 < x < 1$. Hence from Exercise 8, Chapter 5,

$$\begin{aligned} c^{-x}\Gamma(x) &= \lim_{c_1\to 0+}(c_1 + ic_2)^{-x}\Gamma(x) \\ &= \lim_{c_1\to 0+}\int_0^\infty e^{-(c_1+ic_2)t}t^{x-1}dt = \int_0^\infty e^{-ic_2t}t^{x-1}dt \end{aligned}$$

as required. ■

The particular case when $x = 1/2$ gives the value of the Fresnel integrals from Example 5.7.

Corollary 7.53

$\int_0^\infty \cos x^2 dx = \int_0^\infty \sin x^2 dx = \sqrt{\pi/8}$.

Putting $x^2 = y \geq 0$ gives $\int_0^\infty \exp(ix^2)\,dx = \int_0^\infty \exp(iy)/(2\sqrt{y})\,dy = (-i)^{-1/2}\Gamma(1/2)/2 = (1+i)\sqrt{\pi}/(2\sqrt{2})$. ♦

Lemma 7.52 is what we need to find the Fourier transforms of the distributions E^a, O^a, P^a when a is not an integer.

Theorem 7.54

For real non-integer values of a,

$$\begin{aligned}
\mathcal{F}E^a &= \frac{-2\sin(a\pi/2)\,\Gamma(a+1)}{(2\pi)^{a+1}}E^{-a-1}, \\
\mathcal{F}O^a &= \frac{-2i\cos(a\pi/2)\,\Gamma(a+1)}{(2\pi)^{a+1}}O^{-a-1}, \\
\mathcal{F}P^a &= (\mathcal{F}E^a + \mathcal{F}O^a)/2.
\end{aligned}$$

Proof

We begin with the case $-1 < a < 0$, and then show how the other non-integer values can be deduced from it. For $-1 < a < 0$ we have from Lemma 7.52 that

$$\begin{aligned}
\int_0^\infty x^a e^{-2\pi i x y}dx &= (2\pi i y)^{-a-1}\,\Gamma(a+1) \\
&= \begin{cases} (2\pi|y|)^{-a-1} e^{-i\pi(a+1)/2}\Gamma(a+1) & \text{if } y>0 \\ (2\pi|y|)^{-a-1} e^{i\pi(a+1)/2}\Gamma(a+1) & \text{if } y<0 \end{cases} \\
&= \frac{\Gamma(a+1)}{(2\pi|y|)^{a+1}}\left\{\cos\frac{(a+1)\pi}{2} - i\sin\frac{(a+1)\pi}{2}\operatorname{sgn} y\right\}.
\end{aligned}$$

In terms of distributions this can be written

$$\begin{aligned}
\mathcal{F}P^a &= \frac{\Gamma(a+1)}{(2\pi)^{a+1}}\left\{\cos\frac{(a+1)\pi}{2}E^{-a-1} - i\sin\frac{(a+1)\pi}{2}O^{-a-1}\right\} \\
&= \frac{\Gamma(a+1)}{(2\pi)^{a+1}}\left\{-\sin\frac{a\pi}{2}E^{-a-1} - i\cos\frac{a\pi}{2}O^{-a-1}\right\}.
\end{aligned}$$

This gives the required results for $\mathcal{F}E^a = \mathcal{F}(P^a + RP^a)$ and $\mathcal{F}O^a = \mathcal{F}(P^a - RP^a)$ when $-1 < a < 0$. For other non-integer values, we use Theorem 7.50(iv) and Exercise 15 to move from a to $a \pm 1$. For instance

$$\begin{aligned}
\mathcal{F}E^{a+1} &= \mathcal{F}(pO^a) = \frac{-1}{2\pi i}D\mathcal{F}O^a \\
&= \frac{-1}{2\pi i}D\left(\frac{-2i\cos(a\pi/2)\,\Gamma(a+1)}{(2\pi)^{a+1}}O^{-a-1}\right) \\
&= \frac{1}{\pi}\frac{\cos(a\pi/2)\,\Gamma(a+1)}{(2\pi)^{a+1}}(-a-1)E^{-a-2} \\
&= \frac{-2\sin((a+1)\pi/2)\,\Gamma(a+2)}{(2\pi)^{a+2}}E^{-a-2}
\end{aligned}$$

which is what we want with $a+1$ in place of a. A similar calculation starting from $\mathcal{F}E^{a-1} = \mathcal{F}(DO^a)/a$ gives the result with $a-1$ for a, and the corresponding results for $\mathcal{F}O^a$ follow in the same way. Repetition of the process then gives the result for all non-integer values of a. ∎

For integer powers, it is convenient to use a slightly modified notation.

Definition 7.55

For integers n, let $X^n = P^n + (-1)^n RP^n$ and $Y^n = P^n - (-1)^n RP^n$. Informally X^n corresponds to x^n on $\mathbb{R}$, and Y^n to $x^n \operatorname{sgn} x$. $X^n = E^n$ if n is even, $= O^n$ if n is odd; $Y^n = O^n$ if n is even, $= E^n$ if n is odd.

We need a preliminary result on limits of distributions, before getting to the main result.

Lemma 7.56

(i) $\lim_{\varepsilon\to 0+} O^{-1+\varepsilon} = O^{-1}$,

(ii) $\lim_{\varepsilon\to 0} \left(E^\varepsilon - E^0\right)/\varepsilon = L$, where $L\phi = \int_{-\infty}^{\infty} \ln|t|\,\phi(t)\,dt$, $L = \Lambda \ln + R\Lambda \ln$.

Proof

(i) is immediate since

$$\begin{aligned}\lim_{\varepsilon\to 0+} O^{-1+\varepsilon}(\phi) &= \lim_{\varepsilon\to 0+} \int_0^\infty t^{-1+\varepsilon}(\phi(t) - \phi(-t))\,dt \\ &= \int_0^\infty t^{-1}(\phi(t) - \phi(-t))\,dt = O^{-1}(\phi)\end{aligned}$$

using dominated convergence. For (ii),

$$\begin{aligned}\frac{E^\varepsilon - E^0}{\varepsilon}(\phi) &= \frac{1}{\varepsilon}\int_{-\infty}^\infty \left(|t|^\varepsilon - 1\right)\phi(t)\,dt \\ &\to \int_{-\infty}^\infty \ln|t|\,\phi(t)\,dt = L(\phi)\,. \ \blacksquare\end{aligned}$$

Theorem 7.57

(i) For integers $n \geq 0$,

$$\mathcal{F}(X^n) = \frac{1}{(2\pi i)^n}\delta^{(n)},$$
$$\mathcal{F}(Y^n) = \frac{2\,n!}{(2\pi i)^{n+1}}X^{-n-1}.$$

(ii) For integers $k > 0$ and L as in Lemma 7.56,

$$\mathcal{F}(X^{-k}) = \frac{(-2\pi i)^k}{2(k-1)!}Y^{k-1},$$
$$\mathcal{F}(Y^{-k}) = \frac{-2(-2\pi i)^{k-1}}{(k-1)!}\left(p^{k-1}L + (\gamma + \ln(2\pi))\,X^{k-1}\right).$$

Proof

The first part of (i) follows by applying the inversion theorem to Example 7.51. For the second relation, we have from the lemma that $X^{-1} = O^{-1} = \lim_{\varepsilon\to 0+} O^{-1+\varepsilon}$ and hence by using Theorem 7.50(v)

$$\begin{aligned}\mathcal{F}(X^{-1}) &= \lim_{\varepsilon\to 0+}\mathcal{F}(O^{-1+\varepsilon})\\ &= \lim_{\varepsilon\to 0+}\frac{-2i\cos((\varepsilon-1)\pi/2)\,\Gamma(\varepsilon)}{(2\pi)^{\varepsilon}}O^{-\varepsilon}\\ &= -2i\lim_{\varepsilon\to 0+}\sin(\varepsilon\pi/2)\,\Gamma(\varepsilon+1)/\varepsilon O^{-\varepsilon}\\ &= -\pi i O^0 = -\pi i Y^0.\end{aligned}$$

This gives both the second part of (i) when $n = 0$ (after applying the inversion theorem) and the first part of (ii) when $k = 1$. The cases $n \geq 1$ and $k \geq 2$ now follow by repeated application of the rules $\mathcal{F}DT = 2\pi i p\mathcal{F}T$ and $D\mathcal{F}T = -2\pi i\mathcal{F}(pT)$ as we did in the corresponding part of the proof of Theorem 7.54. This leaves the second part of (ii) for which we have to work a little harder. We know that $E^{-1} = \lim_{\varepsilon\to 0+}(E^{-1+\varepsilon} - 2\delta/\varepsilon)$ (compare Example 7.42, or Exercise 15). Also

$$\begin{aligned}\mathcal{F}(E^{-1+\varepsilon} - 2\delta/\varepsilon) &= \frac{-2\sin((\varepsilon-1)\pi/2)\,\Gamma(\varepsilon)}{(2\pi)^{\varepsilon}}E^{-\varepsilon} - \frac{2}{\varepsilon}E^0\\ &= \frac{2\cos(\varepsilon\pi/2)\,\Gamma(\varepsilon+1)}{(2\pi)^{\varepsilon}\varepsilon}E^{-\varepsilon} - \frac{2}{\varepsilon}E^0\end{aligned}$$

$$\begin{aligned}
&= \frac{2\cos(\varepsilon\pi/2)\,\Gamma(\varepsilon+1)}{(2\pi)^{\varepsilon}}\,\frac{E^{-\varepsilon}-E^{0}}{\varepsilon} \\
&\quad + \frac{2}{\varepsilon}\left(\frac{\cos(\varepsilon\pi/2)\,\Gamma(\varepsilon+1)}{(2\pi)^{\varepsilon}} - 1\right)E^{0} \\
&\to -2L + 2f'(0)\,E^{0}
\end{aligned}$$

where $f(x) = \cos(x\pi/2)\,\Gamma(x+1)\,(2\pi)^{-x}$. But for $f'(0)$ we have

$$\begin{aligned}
\frac{f'(x)}{f(x)} &= -\frac{\pi\sin(x\pi/2)}{\cos(x\pi/2)} + \frac{\Gamma'(x+1)}{\Gamma(x+1)} - \ln(2\pi), \\
f'(0) &= \Gamma'(1) - \ln(2\pi) = -\gamma - \ln(2\pi).
\end{aligned}$$

Hence $\mathcal{F}(Y^{-1}) = \mathcal{F}(E^{-1}) = -2L - 2(\gamma + \ln(2\pi))\,E^{0}$ which gives the second part of (ii) when $k = 1$. The result for $k \geq 2$ follows as before from repeated application of $\mathcal{F}DT = 2\pi ip\mathcal{F}T$. ∎

The two volumes of Schwarz [13] are still the basic reference for the general theory. Alternatively the slightly off-beat treatments in [6] or [9] may be preferred and are at least in English! See Rudin [12] for the theory in $\mathbb{R}^n$ and for applications to differential equations.

7.4 Asymptotics

Once we have succeeded in answering the simplest questions of analysis – does a given sequence converge and what is the value of its limit, we come to a question of more practical utility – at what rate is the limit attained? To take an extreme example, the series $\sum(-1)^n/\ln n$ is convergent, but when $n = 10^6$ the terms are only a little less than 1/10. To determine its sum in this way to an error of less than 0.01 would require around 10^{43} terms. Clearly such series are of no use for calculation.

Asymptotics is that part of analysis which deals with these related questions – how fast does a process converge, and how should we find numerical values effectively. As such it is a rather loosely structured collection of results and methods which we shall simply illustrate by means of examples. For a more systematic exposition, see for instance [18].

To begin, we look at ways of finding the rates of convergence of series, and of estimating their sums. For instance given an alternating series whose terms are decreasing, the difference between its sum and the nth partial sum is at most the $(n+1)^{\text{st}}$ term (this is Theorem 6.16). For series with positive terms, the proof of convergence will usually provide some sort of estimate for the error, though this may not be the best possible. Consider the following example.

Example 7.58

(i) Show that $\sum_{m=n}^{\infty} m2^{-m} < (9/8)(2/3)^{n-3}$,

(ii) $|\sum_{m=n}^{\infty} m \sin(\pi\sqrt{m})\, 2^{-m}| < (9/8)(2/3)^{n-3}$.

We know that the series is convergent by the ratio test, since with $a_m = m2^{-m}$, $a_{m+1}/a_m = (m+1)/(2m) \to 1/2$ as $m \to \infty$. To get an explicit estimate for the remainder, we have $a_{m+1}/a_m = (m+1)/(2m) \leq 2/3$ for $m \geq 3$, and so $a_{m+3}/a_3 \leq (2/3)^m$. Then for $n \geq 3$, $\sum_{m=n}^{\infty} a_m \leq a_3 \sum_{m=n}^{\infty} (2/3)^{m-3} = (9/8)(2/3)^{n-3}$ as required. The second result is immediate from the first. ♦

For a more interesting example, consider the problem of finding the sum of the series $\sum_1^{\infty} 1/n^2$, say correct to 10 decimal places. This has some historical significance, since Euler found the sum of the series numerically before being led to the exact value $\pi^2/6$ which we found in Example 7.8(ii).

Example 7.59

Estimate the sum $\sum_{m=20}^{\infty} 1/m^2$ and hence find the sum $\sum_1^{\infty} 1/n^2$ with an error $< 10^{-10}$.

A crude estimate for the tail can be obtained by simply saying $\sum_{20}^{\infty} 1/m^2 \leq \int_{19}^{\infty} t^{-2} dt = 1/19 = 0.052...$ as in the proof of the integral test (Theorem 6.13) but this is not much use if we want higher accuracy. Instead we take $k = 7$, $f(x) = x^{-2}$ in the Euler–Maclaurin formula,

$$\begin{aligned}\sum_{r=m}^{\infty} f(r) &= \int_m^{\infty} f(t)\, dt + \frac{1}{2} f(m) \\ &\quad - \sum_{2 \leq 2j \leq 6} \frac{B_{2j}}{(2j)!} f^{(2j-1)}(m) + \frac{1}{7!} \int_m^{\infty} \tilde{B}_7(t) f^{(7)}(t)\, dt\end{aligned}$$

which is Exercise 3. This gives

$$\begin{aligned}\int_{20}^{\infty} \frac{dt}{t^2} &= 0.05 \\ \frac{1}{2} f(20) &= \frac{1}{2.20^2} = 0.001\,25 \\ -\frac{B_2}{2} f'(20) &= \frac{1}{6.20^3} = 0.000\,020\,833\,33\end{aligned}$$

$$-\frac{B_4}{4!}f^{(3)}(20) = -\frac{1}{30.20^5} = -0.000\,000\,010\,42$$
$$-\frac{B_6}{6!}f^{(5)}(20) = \frac{1}{42.20^7} = 1.86 \times 10^{-11}.$$

Adding these to $\sum_1^{19} 1/m^2 = 1.593\,663\,243\,90$ (found from one's calculator) gives $1.644\,934\,066\,84$. The error term is estimated by

$$\left|\frac{1}{7!}\int_{20}^{\infty} \tilde{B}_7(t)\, f^{(7)}(t)\, dt\right| \leq \frac{1}{\pi^7(2^6-1)}\left|f^{(6)}(20)\right| = \frac{7!}{\pi^7(2^6-1)\,20^8} = 1.03 \times 10^{-12}$$

using the result of Exercise 4. Hence the calculated value of $1.644\,934\,066\,84$ is correct to the required accuracy, in agreement with the known value $\pi^2/6 = 1.644\,934\,066\,85$. ♦

Another frequently occurring problem is to estimate the behaviour of an integral of the form $\int_a^b (f)^n g$ as $n \to \infty$, where f has a maximum value at some point c say of the interval $[a, b]$, and the integral behaves essentially like $[f(c)]^n n^k$ where the exponent k depends on the detailed nature of f and g near c. We build up to this result via a number of related examples and theorems.

Example 7.60

(i) For $a > 0$, $n\int_0^{\infty} e^{-ant}dt \to 1/a$ as $n \to \infty$.

(ii) For $a > 0$ and $0 < d \leq 1/a$, $n\int_0^d (1-at)^n\, dt \to 1/a$ as $n \to \infty$.

In (i), $n\int_0^{\infty} e^{-ant}dt = \left[-e^{-ant}\right]_0^{\infty}/a = [1-0]/a$, and in (ii)

$$n\int_0^d (1-at)^n\, dt = \frac{-n}{a(n+1)}\left[(1-at)^{n+1}\right]_0^d \to 1/a. \ ♦$$

The similarity between these examples is deliberate; in each case we have a decreasing function $f(t)$ with $f(0) = 1$ and $f'(0) = -a$. So the next proposition should not be quite so much of a surprise.

Proposition 7.61

Suppose that for some $d > 0$, the function f is positive, continuous and decreasing on $[0, d]$ with $f(0) = 1$ and let f be differentiable on the right at 0 with $f'_+(0) = -a < 0$. Then

$$\lim_{n\to\infty} n\int_0^d [f(x)]^n\, dx = \frac{1}{a} = \frac{-1}{f'_+(0)}.$$

Proof

The idea of the proof is simple; we choose $\delta > 0$ and divide the interval $[0, d]$ into $[0, \delta]$ and $[\delta, d]$. On the first we approximate f by $1 - ax$, while the contribution of the second is shown to be negligible.

Suppose then that $\varepsilon > 0$ is given with $\varepsilon < a$, and we choose $\delta > 0$ so that

$$0 \leq 1 - (a + \varepsilon) x < f(x) < 1 - (a - \varepsilon) x$$

on $[0, \delta]$. Then

$$n \int_0^{\delta} [1 - (a + \varepsilon) x]^n \, dx < n \int_0^{\delta} [f(x)]^n \, dx < n \int_0^{\delta} [1 - (a - \varepsilon) x]^n \, dx$$

and the first and last terms here tend to $1/(a + \varepsilon)$ and $1/(a - \varepsilon)$ as in the example above.

For the rest of the interval we have $0 \leq n \int_{\delta}^{d} [f(x)]^n \, dx \leq n (d - \delta) f(\delta)^n \to 0$ as $n \to \infty$ since $0 \leq f(\delta) < 1$. Hence for large enough n,

$$\frac{1}{a + 2\varepsilon} < n \int_0^{d} [f(x)]^n \, dx < \frac{1}{a - 2\varepsilon}$$

and the result follows. ■

The result is true (with almost the same proof) if instead of continuous and decreasing, we require only that f is regulated, and has the property that for each $\delta > 0$, f has a greatest upper bound on $[\delta, d]$ which is strictly less that 1, but we shall not need this sort of generality. More usefully, it can be applied when the maximum is at either end of a general interval $[a, b]$, when the limit is $-1/f'_+(a)$ or $1/f'_-(b)$ respectively. If the maximum is at an interior point with non-zero right- and left-hand derivatives, the result is simply the sum of the two contributions from the right and left. A further useful observation is that, although we have thought of n as an integer, the result, and that of the theorem below, are equally valid when $n = x$ say, is a real variable tending to infinity.

Example 7.62

$n 2^n \int_0^{\pi/6} \sin^n x \, dx \to 1/\sqrt{3}$ as $n \to \infty$.

Here $f(x) = 2 \sin x$ is continuous and increasing on $[0, \pi/6]$ with $f(\pi/6) = 1$, $f'(\pi/6) = \sqrt{3}$. Hence the limit is $1/f'(\pi/6) = 1/\sqrt{3}$ as stated. ♦

Proposition 7.61 is such a nice result that it is natural to want to extend it in several directions. For instance we might have a function with a maximum which was not equal to 1, and there might be other functions present under the integral sign. Most interestingly, the function might have a maximum of a different form than that specified in the proposition – for instance it might have a zero derivative as when $f(x) = \cos x$ or e^{-x^2} at $x = 0$, or an undefined derivative, as when $f(x) = 1 - \sqrt{x}$.

The following theorem covers most cases.

Theorem 7.63

Suppose that for some $d > 0$, the function f is positive, continuous and decreasing on $[0, d]$ with

$$f(x) = l - (a + \eta)x^p,$$

where l, p, a are positive constants and $\eta \to 0$ as $x \to 0+$. Let g be regulated on $[0, d]$, continuous on the right at 0 with $g(0) \neq 0$, and let $b > 0$. Then

$$n^{b/p} l^{-n} \int_0^d x^{b-1} g(x) [f(x)]^n \, dx \to \frac{g(0)}{p} \left(\frac{l}{a}\right)^{b/p} \Gamma(b/p) \qquad \text{as } n \to \infty.$$

The result can be written

$$\int_0^d x^{b-1} g(x) [f(x)]^n \, dx \sim \frac{g(0)}{p} l^n \left(\frac{l}{na}\right)^{b/p} \Gamma(b/p) \qquad \text{as } n \to \infty \qquad (7.29)$$

where $f \sim g$ means $f/g \to 1$.

Proof

Despite its somewhat messy appearance, this is just an elaboration of the proposition above with $g(x)$ and x^{b-1} thrown in for good measure. We outline the proof, in the belief that the reader is by now good at filling in the details. Given $\varepsilon > 0$ we choose $\delta > 0$ such that $g(0) - \varepsilon < g(x) < g(0) + \varepsilon$ and

$$l - (a + \varepsilon)x^p < f(x) < l - (a - \varepsilon)x^p$$

on $[0, \delta]$. On this interval (the contribution from the rest is negligible as before) we approximate $\int_0^\delta x^{b-1} g(x) [f(x)]^n \, dx$ first by

$$l^n (g(0) \pm \varepsilon) \int_0^\delta x^{b-1} \left[1 - \frac{a \pm \varepsilon}{l} x^p\right]^n dx$$

and then by

$$l^n (g(0) \pm \varepsilon) \int_0^\delta x^{b-1} \exp\left(-n \frac{a \pm \varepsilon}{l} x^p\right) dx.$$

Putting $x^p = lt/n\,(a \pm \varepsilon)$ gives the Gamma integral

$$\frac{(g(0) \pm \varepsilon)\, l^n}{p} \left(\frac{l}{n\,(a \pm \varepsilon)}\right)^{b/p} \int_0^\infty t^{b/p-1} e^{-t} dt = \frac{(g(0) \pm \varepsilon)\, l^n}{p} \left(\frac{l}{n\,(a \pm \varepsilon)}\right)^{b/p} \Gamma\,(b/p)$$

where again the contribution from the interval $[\delta, \infty)$ can be neglected. This gives the result since ε is arbitrary. ∎

Having made essential use of the Gamma function in proving this theorem, it is good to be able to turn the tables and apply the result to tell us more about the Gamma function, namely Stirling's famous approximation to $\Gamma\,(x)$ for large x.

Example 7.64

$$\frac{\Gamma\,(x)}{e^{-x} x^{x-1/2}} \to \sqrt{2\pi} \qquad \text{as } x \to \infty.$$

To see this, consider the integral for $x\Gamma\,(x) = \Gamma\,(x+1) = \int_0^\infty t^x e^{-t} dt$ which we found in Theorem 5.12. The integrand has a maximum at $t = x$, so we consider

$$\begin{aligned} \int_x^\infty t^x e^{-t} dt &= \int_0^\infty (t+x)^x\, e^{-t-x} dt = x^x e^{-x} \int_0^\infty \left(1 + \frac{t}{x}\right)^x e^{-t} dt \\ &= x^{x+1} e^{-x} \int_0^\infty \left((1+u)\, e^{-u}\right)^x du \end{aligned}$$

on putting $t = xu$. The Taylor expansion of $(1+u)\,e^{-u}$ about $u = 0$ is $(1+u)\left(1 - u + u^2/2 + \cdots\right) = 1 - u^2/2 + \cdots$ and so the theorem is satisfied with $g\,(x) = 1$, $b = l = 1$, $p = 2$, $a = 1/2$. Hence from the theorem,

$$\int_0^\infty \left((1+u)\, e^{-u}\right)^x du \sim \frac{1}{2} \left(\frac{2}{x}\right)^{1/2} \Gamma\,(1/2) = x^{-1/2} \left(\frac{\pi}{2}\right)^{1/2},$$

and so

$$\int_x^\infty t^x e^{-t} dt \sim x^{x+1/2} e^{-x} \left(\frac{\pi}{2}\right)^{1/2}.$$

The integral over $[0,x]$ gives the same contribution, and we end up with

$$\Gamma(x) = \frac{1}{x}\Gamma(x+1) \sim 2x^{x-1/2}e^{-x}\left(\frac{\pi}{2}\right)^{1/2} = x^{x-1/2}e^{-x}\sqrt{2\pi}$$

which is what we want. ♦

It is possible to refine the method to obtain not just the principal term but to find asymptotic expansions, of which we quote the following without proof:

$$e^{x}x^{-x+1/2}\Gamma(x) = \sqrt{2\pi}\left(1+\frac{\theta}{12n}\right) \qquad \text{where } 0<\theta<1.$$

For details and proofs, see [16, Section 2.5].

Our last investigation concerns sequences defined by iteration. As promised in the introduction, we consider the iterated sine, but the method is quite general, and the reader is invited to apply it to other similar sequences. The result is that for $0 < x_0 < \pi$, and $(x_n)_{n\geq 1}$ defined by $x_{n+1} = \sin x_n$ then $\sqrt{n}x_n \to \sqrt{3}$ as $n \to \infty$. Since $x_1 = \sin x_0 \leq 1 < \pi/2$ and $0 < \sin x < x$ for $0 < x \leq \pi/2$ it is clear that the sequence is decreasing for $n \geq 1$ and $x_n \to 0$ as $n \to \infty$. We can argue informally that if $x_n \sim an^b$ (see above for the notation – this simply means that $x_n/\left(an^b\right) \to 1$ as $n \to \infty$) then from the recurrence, $x_{n+1} = \sin x_n = x_n - x_n^3/6 +$ higher terms, or $a\left(n+1\right)^b = an^b\left(1 - a^2n^{2b}/6 + \cdots\right)$. This gives $\left(1+1/n\right)^b = 1 + b/n + \cdots = 1 - a^2n^{2b}/6 + \cdots$ and equating the second terms gives $2b = -1$ and $b = -a^2/6$, i.e. $b = -1/2$ and $a = \sqrt{3}$ as predicted. Of course this argument proves nothing since we have neglected the higher terms completely, but it does make the result that $x_n \sim \sqrt{3/n}$ at least plausible. We have to make the argument precise by replacing the $\sim$ relation with some inequalities.

Lemma 7.65

For any $\varepsilon > 0$ there is some δ with

(i) $\qquad 1 + 2x \leq (1-x)^{-2} \leq 1 + (2+\varepsilon)x$

and

(ii) $\qquad x - x^3/6 \leq \sin x \leq x - (1/6 - \varepsilon)x^3$

for $0 \leq x < \delta$.

Proof

The estimates (without ε) $1+2x$ and $x-x^3/6$ are the beginnings of the convergent Taylor series of the functions on the left. Thus for instance in (i), $(1-x)^{-2} = 1+2x+\sum_2^\infty (n+1)\,x^n \geq 1+2x$, and if $0 \leq x < 1/2$ then $\sum_2^\infty (n+1)\,x^n \leq x^2 \sum_2^\infty (n+1)\,2^{2-n} = Cx^2$ (the exact value of C is irrelevant) which is $\leq \varepsilon x$ if $0 \leq x < \varepsilon/C$. The reader is left to prove (ii) similarly using Exercise 26, Chapter 6. ■

Example 7.66

Let $x_0 \in (0,\pi)$ and for $n \geq 0$, $x_{n+1} = \sin x_n$. Then $\sqrt{n}x_n \to \sqrt{3}$ as $n \to \infty$.

We have already noted that x_n decreases to 0 as $n \to \infty$, so we can suppose that n is large enough ($n \geq N$ say) to make $x_n^2/6 < \delta$, where δ is as in the lemma. Then

$$\begin{aligned}\frac{1}{x_{n+1}^2} &= \frac{1}{\sin^2 x_n} \leq \frac{1}{x_n^2\,(1-x_n^2/6)^2} \leq \frac{1+(2+\varepsilon)\,x_n^2/6}{x_n^2} \\ &= \frac{1}{x_n^2} + \frac{2+\varepsilon}{6}.\end{aligned}$$

Applying this repeatedly gives

$$\frac{1}{x_{N+k}^2} \leq \frac{1}{x_N^2} + k\frac{2+\varepsilon}{6}$$

and so for large enough k,

$$\frac{1}{(N+k)\,x_{N+k}^2} \leq \frac{2+2\varepsilon}{6}$$

which gives one half of the result. For an estimate in the other direction we have similarly

$$\begin{aligned}\frac{1}{x_{n+1}^2} &= \frac{1}{\sin^2 x_n} \geq \frac{1}{x_n^2\,(1-(1/6-\varepsilon)\,x_n^2)^2} \geq \frac{1+2\,(1/6-\varepsilon)\,x_n^2}{x_n^2} \\ &= \frac{1}{x_n^2} + 2\left(\frac{1}{6}-\varepsilon\right)\end{aligned}$$

and the rest of the argument is as before. ♦

The reader is invited to turn this example into a theorem by considering more generally $x_{n+1} = f\,(x_n)$ where f is increasing on $[0,d)$ and is given there by $f\,(x) = x-(a+\eta)\,x^b$ where $a>0$, $b>1$ and $\eta \to 0$ as $x \to 0^+$ and a refinement of the method gives further terms in the expansion. In fact the field of asymptotics is so large and varied that at this point it is better to stop and leave the reader to follow his or her own interests.

EXERCISES

1. Find the Fourier series and investigate their convergence for the functions given by (i) $1 - |x|/a$ on $[-a, a]$, else zero, where $0 < a < \pi$, (ii) $\cos(x/a)$ on $[-\pi a/2, \pi a/2]$, else zero, where $0 < a < 2$.

2. If we let $r \to 1$ in (6.15) we get

$$\sum_{n=1}^{\infty} \frac{1}{n} \cos nx = -\ln(2\sin(x/2)) \qquad \text{for } 0 < x < 2\pi.$$

What is required to justify the limiting process here?

3.* (i) Let $B_n(x)$ be the Bernoulli polynomials from Exercise 9, Chapter 6. Show that for any f with f, f' continuous on $[0,1]$,

$$\frac{f(0) + f(1)}{2} = \int_0^1 f(t)\, dt + \int_0^1 f'(t)\, B_1(t)\, dt.$$

Integrate the last term repeatedly by parts to get the Euler–Maclaurin formula on $[0,1]$

$$\frac{f(0)+f(1)}{2} = \int_0^1 f(t)\, dt + \sum_{2 \le 2k < n} \frac{B_{2k}}{(2k)!} \left[f^{(2k-1)}(1) - f^{(2k-1)}(0) \right] + \frac{(-1)^n}{n!} \int_0^1 f^{(n)}(t)\, B_n(t)\, dt, \tag{7.30}$$

assuming that all derivatives up to $f^{(n)}$ are continuous.

(ii) Let $\tilde{B}_n(x)$ be the Bernoulli functions, which are the functions on $\mathbb{R}$ with period 1 whose restrictions to $[0,1]$ are equal to $B_n(x)$. Replace $[0,1]$ in (7.30) by $[j, j+1]$ and sum over values $j = m, m+1, \ldots, n-1$ to get the Euler–Maclaurin formula

$$\sum_{r=m}^{n} f(r) = \int_m^n f(t)\, dt + \frac{1}{2}(f(m) + f(n)) + \sum_{2 \le 2j \le k} \frac{B_{2j}}{(2j)!} \left(f^{(2j-1)}(n) - f^{(2j-1)}(m) \right) + \frac{(-1)^{k-1}}{k!} \int_m^n \tilde{B}_k(t)\, f^{(k)}(t)\, dt.$$

We can replace n by ∞ here provided that f and its derivatives tend to zero fast enough to make the integrals converge, as in Example 7.59.

4. Derive by successive integration, starting with the case $k = 1$, the Fourier series for the Bernoulli functions

$$\tilde{B}_k(x) = -2\frac{k!}{(2\pi)^k}\sum_{n=1}^{\infty} n^{-k}\sin(2n\pi x - (k-1)\pi/2).$$

Deduce that for all real x, $\left|\tilde{B}_k(x)\right| \leq k!/\left(\pi^k\left(2^{k-1}-1\right)\right)$ and that for k even, $\left|\tilde{B}_k(x)\right| \leq B_k$.

5.** Give a necessary and sufficient condition on f that its Fourier series should be absolutely convergent.

6. Deduce from Perseval's equation that

$$\int_{-a}^{a}\left(1-\frac{|y|}{a}\right)\mathcal{F}f(y)\,dy = \frac{1}{a\pi^2}\int_{-\infty}^{\infty} f(x)\frac{\sin^2(\pi a x)}{x^2}dx.$$

Deduce the following form of the inversion theorem which uses ordinary continuity in place of Lipschitz continuity.

Theorem 7.67

Let both $f, \mathcal{F}f$ be continuous and integrable. Then for all real x,

$$f(x) = \int_{-\infty}^{\infty}\mathcal{F}f(y)\,e^{2\pi i x y}dy.$$

7. Show that $f(x) = 1 - x^2/a^2$ for $-a \leq x \leq a$, $= 0$ elsewhere, satisfies the conditions of the inversion theorem, and find $\mathcal{F}f(y)$ explicitly.

8.* Primitives of Distributions.

   (i) Let ϕ_0 be a fixed test function with $\int_{-\infty}^{\infty}\phi_0 \neq 0$. Given any test function ϕ, choose c with $\int_{-\infty}^{\infty}(\phi + c\phi_0) = 0$ and let $\psi(t) = \int_{-\infty}^{t}(\phi(t) + c\phi_0(t))\,dt$. Show (a) that ψ is a test function and (b) that if for any constant k and any distribution T, we define U by $U\phi = k\int_{-\infty}^{\infty}\phi - T\psi = kE^0\phi - T\psi$, then $U' = T$.

   (ii) Show that if $U_1' = U_2' = T$ then $U_1 - U_2 = kE^0$ for some constant k.

9. Find explicit expressions, analogous to those in Example 7.41, for $E^a(\phi)$, $O^a(\phi)$ for $-1 \geq a \geq -2$.

10. Extend the result of Example 7.41(iii) to find a formula for $P^{-k}(\phi)$ for any positive integer k.

11. Show that for any $a > 0$, $DP^a = aP^{a-1}$, and deduce the consistency of Definition 7.40(ii). Then show that $DP^a = aP^{a-1}$ for all $a \neq 0$.

12. Complete the proof of part (ii) of Proposition 7.26.

13. Find the result corresponding to Example 7.42 for $\lim_{\varepsilon \to 0-} P^{-1+\varepsilon}$.

14. Show that if $\phi(0) > 0$, $\phi'(0) = 0$, $0 \leq \phi(x) \leq \phi(0)$ on $[0, 1]$ and $\phi(x) = 0$ on $[1, \infty)$ then $P^{-2}(\phi) < 0$. Thus P^2 does not satisfy $P^2(\phi) \geq 0$ when $\phi \geq 0$.

15. (i) Show that for real $a \neq 0$, $DE^a = aO^{a-1}$, $DO^a = aE^{a+1}$. Also $DE^0 = 0$, $DO^0 = 2\delta$.

    (ii) Let $p(x) = x$ as in the text. Then for all real a, $pE^{a-1} = O^a$ and $pO^{a-1} = E^a$. (Take care with $a = 0$.)

    (iii) From Exercise 8 we have that if $DT = 0$ then $T = kE^0$ for some constant k. Deduce, using Fourier transforms, that if $pT = 0$ then $T = k\delta$ for some constant k.

    (iv) Show that $O^{-1} = DL$ where L is as in Lemma 7.56 and find L_1 with $E^{-1} = DL_1$.

    (v) Show that $O^0 = \lim_{\varepsilon \to 0} O^{\varepsilon}$, $O^{-1} = \lim_{\varepsilon \to 0+} O^{-1+\varepsilon}$, $E^{-1} = \lim_{\varepsilon \to 0+} \left(E^{-1+\varepsilon} - 2\delta/\varepsilon\right)$.

16. Show that for integer n,

$$\int_0^{2\pi} e^{t \cos x} \cos nx \, dx \sim e^t \sqrt{\frac{2\pi}{t}} \qquad \text{as } t \to +\infty.$$

17.* Given $0 < b_0 < a_0$, let $M = M(a_0, b_0)$ be the arithmetic-geometric mean (AGM) of a_0, b_0 as in Exercise 11, Chapter 1.

    (i) Show that for all n, $a_{n+1} - b_{n+1} \leq (a_n - b_n)/2$, and that if we choose d, N such that for $n \geq N$, $0 < d \leq b_n < a_n$ and $a_n - b_n \leq 8d$, then for $n \geq N$, $a_{n+1} - b_{n+1} \leq (a_n - b_n)^2/(8d)$, showing that the rate of convergence is similar to that of Newton's method. (However, unlike Newton's method, the AGM is *not* self-correcting, since an error in calculation gives the mean of a new pair of starting values. Thus to get a result to a given accuracy, each stage of the calculation must be to that accuracy.)

    (ii) The common limit M is given explicitly in terms of a_0, b_0 by means of an elliptic integral:

$$\frac{\pi}{2M} = \int_0^{\pi/2} \frac{dt}{\sqrt{a_0^2 \cos^2 t + b_0^2 \sin^2 t}}.$$

Prove this as follows. Denote the integral on the right by $G(a_0, b_0)$ and show that

$$G(a_0, b_0) = \int_0^\infty \frac{dx}{\sqrt{(1+x^2)(a_0^2 + b_0^2 x^2)}}.$$

Make the substitutions $b_0 x^2 = a_0 y^2$, $v = 2y/(1-y^2)$ to show that

$$G(a_0, b_0) = G(a_1, b_1) = \cdots = G(a_n, b_n) = \cdots = G(M, M) = \frac{\pi}{2M}.$$

18. Given $x_0 > 0$, define (x_n) by $x_{n+1} = \ln(1 + x_n)$ for $n \geq 0$. Show that $nx_n \to 2$ as $n \to \infty$.

19. Justify the following alternative method for finding the value of the Fresnel integrals, Corollary 7.53.

 (i) Let

$$I(\lambda) = \int_0^A \frac{\sin(\lambda^2(1+x^2))}{1+x^2} dx.$$

 Differentiate with respect to λ to get

$$\begin{aligned} I'(\lambda) &= 2\lambda \int_0^A \cos(\lambda^2(1+x^2))\, dx \\ &= 2\cos\lambda^2 \int_0^{A\lambda} \cos t^2 dt - 2\sin\lambda^2 \int_0^{A\lambda} \sin t^2 dt, \\ I(\lambda) &= 2\int_0^\lambda \left\{ \cos u^2 \int_0^{Au} \cos t^2 dt - \sin u^2 \int_0^{Au} \sin t^2 dt \right\} du. \end{aligned}$$

 Now let $A \to \infty$ to get

$$\int_0^\infty \frac{\sin(\lambda^2(1+x^2))}{1+x^2} dx = 2C \int_0^\lambda \cos u^2 du - 2S \int_0^\lambda \sin u^2 du, \tag{7.31}$$

 where

$$C = \int_0^\infty \cos t^2 dt,$$
$$S = \int_0^\infty \sin t^2 dt.$$

(ii) Starting similarly from

$$J(\lambda) = \int_0^A \frac{\cos\left(\lambda^2\left(1+x^2\right)\right)}{1+x^2}dx,$$

show that

$$\int_0^\infty \frac{\cos\left(\lambda^2\left(1+x^2\right)\right)}{1+x^2}dx = \frac{\pi}{2} - 2C\int_0^\lambda \sin u^2 du - 2S\int_0^\lambda \cos u^2 du. \tag{7.32}$$

(iii) The integrals on the left of (7.31) and (7.32) are (slightly disguised) Fourier integrals, and so tend to zero as $\lambda \to \infty$ by Proposition 7.20. This gives

$$0 = 2C^2 - 2S^2,$$
$$0 = \frac{\pi}{2} - 4CS$$

from which $C = S = \sqrt{\pi/8}$ follows.

(I am indebted to Dr. Philip Heywood who showed me this, which was set as a final honours examination question to undergraduates in Edinburgh in 1950.)

A
Fubini's Theorem

Throughout this book we have concentrated on functions of a single real (or occasionally complex) variable, other quantities such as the a in $\cos(ax)$ (Example 7.9) having the role of parameters. However in several places the variables have a more equal role, notably when differentiating under the integral sign in the proof of Lemma 7.52 or interchanging the order of integration to obtain Parseval's equation, Theorem 7.28. In order to justify these results we have to look at functions of several real variables. We sketch as much of the theory as we need, and refer the reader to [10] for a more systematic development.

To begin with notation, we reserve p, q for points in $\mathbb{R}^2$ with for instance $p = (x, y)$ where $x, y \in \mathbb{R}$. (Much of the appendix could equally be developed in $\mathbb{R}^n$, but two dimensions will be sufficient for our needs and will save on suffixes.) A function f is then (as in the introduction to Chapter 2) a rule which, for each p in some subset of $\mathbb{R}^2$ assigns a real or complex number $f(p)$, the value of f at p. We have to extend the concepts of continuity and differentiability from Chapters 2 and 3, and for this we need the notion of distance in $\mathbb{R}^2$.

Definition A.1

Given points $p = (x, y)$, $q = (u, v) \in \mathbb{R}^2$, the distance $d(p, q) = |p - q|$ is defined as the positive square root

$$|p - q| = \sqrt{(x-u)^2 + (y-v)^2}.$$

Definition A.2

A (real or complex valued) function f defined on $E \subset \mathbb{R}^2$ is said to be continuous on E if for each $p \in E$ and each $\varepsilon > 0$ there is some $\delta > 0$ (depending on both p and ε) such that

$$|f(p) - f(q)| < \varepsilon \qquad \text{for all } q \in E \text{ with } |p - q| < \delta.$$

Another way of stating the definition is to say that with $p = (x, y)$, then $f(x + h, y + k) \to f(x, y)$ as $(h, k) \to (0, 0)$ from which it follows at once that f must be continuous with respect to each variable separately, as considered below. The dependence of δ on p turns out to be a disadvantage, and we immediately consider a stronger property.

Definition A.3

A (real or complex valued) function f defined on $E \subset \mathbb{R}^2$ is said to be uniformly continuous on E if for each $\varepsilon > 0$ there is some $\delta > 0$ (depending on ε only) such that

$$|f(p) - f(q)| < \varepsilon \qquad \text{for all } p, q \in E \text{ with } |p - q| < \delta.$$

It is perhaps surprising that a function which is continuous on a set of the form $[a, b] \times [c, d]$ (a closed rectangle in $\mathbb{R}^2$) must be uniformly continuous there. We could have proved the corresponding result in Chapter 2, but chose on a first acquaintance not to go beyond the results of Section 2.4. The analogues of those results are also true for functions on a closed rectangle, with essentially the same proofs.

Theorem A.4

Let $E = [a, b] \times [c, d]$ and f be continuous on E. Then f is uniformly continuous on E.

Proof

Suppose that f is not uniformly continuous on E. Then for some $\varepsilon > 0$ and every $\delta > 0$ there are points p, q of E with $|p - q| < \delta$ and $|f(p) - f(q)| \geq \varepsilon$. (This tests the reader's fluency with quantifiers!) Hence for each integer $n \geq 1$, we can choose p_n, q_n with $|p_n - q_n| < 1/n$ and $|f(p_n) - f(q_n)| \geq \varepsilon$. The sequence (p_n) is bounded, since it lies in $[a, b]$, and so by Corollary 1.29 has a convergent subsequence, say (p_{n_j}) whose limit, x say, is in $[a, b]$. Then for this subsequence,

(q_{n_j}) has again a convergent subsequence whose limit, y say, is in $[c, d]$. To avoid an excess of subscripts, now discard all terms not in the subsequence of (q_{n_j}) and relabel them as (p_r, q_r). This sequence of points has the properties that $(p_r, q_r) \in E$, $p_r \to x, q_r \to y$, $|p_r - q_r| \to 0$ as $r \to \infty$, and $|f(p_r) - f(q_r)| \geq \varepsilon$ for all r. But f is supposed continuous at $u = (x, y) \in E$, so there is some $\delta > 0$ such that $|f(u) - f(v)| < \varepsilon/2$ for all $v \in E$ with $|v - u| < \delta$. Take r large enough so that $|p_r - x|, |q_r - y| < \delta$, when it follows that

$$\begin{aligned} |f(p_r) - f(q_r)| &\leq |f(p_r) - f(u)| + |f(u) - f(q_r)| \\ &< \varepsilon/2 + \varepsilon/2 = \varepsilon \end{aligned}$$

and we have a contradiction. ∎

Despite its similarity to Definition 2.9, the new definition of continuity conceals some subtleties. For instance it is natural to consider one of the variables x, y as fixed and allow the other to vary, thus giving the notion of continuity in each variable separately. However this is weaker than Definition A.2 as the next example shows.

Example A.5

Let $f(x, y) = xy/(x^2 + y^2)$ if $(x, y) \neq (0, 0)$, $f(0, 0) = 0$. Then f is continuous in each variable separately at all points of $\mathbb{R}^2$, but is not continuous at $(0, 0)$.

To see this, note that $f(x, y) = 0$ whenever either x or $y = 0$, and so f is continuous in each variable separately at $(0, 0)$. At other points it is continuous in each variable separately since xy and $x^2 + y^2$ are continuous and non-zero there. However at points where $x = y$ we have $f(x, y) = 1/2$, and since such points may be arbitrarily close to $(0, 0)$ where $f(0, 0) = 0$ it follows that f is not continuous there. ♦

A similar subtlety arises when we come to differentiation. It is again natural to consider fixing one of the variables, and differentiate with respect to the other, leading to the notion of partial derivatives as in the next definition.

Definition A.6

The limit

$$\lim_{h \to 0} \frac{f(x + h, y) - f(x, y)}{h},$$

if it exists, is called the first partial derivative of f at (x, y) (the partial derivative with respect to the first variable) and is denoted variously by

$$\frac{\partial f}{\partial x}(x,y), \ f_x(x,y), \ f_1(x,y), \text{ or } D_1 f(x,y).$$

Similarly the limit

$$\lim_{k \to 0} \frac{f(x, y+k) - f(x,y)}{k},$$

is the second partial derivative (the partial derivative with respect to the second variable – not to be confused with $\partial^2 f / \partial y^2$) and denoted $\partial f / \partial y (x, y)$, etc.

However the mere existence of partial derivatives at a point does not even imply continuity in the sense of Definition A.2 since for instance Example A.5 gives a function with partial derivatives at all points (in particular $f_1(0,0) = f_2(0,0) = 0$). For the definition of differentiability, we have, as in Definition A.2, to allow both variables to vary together.

Definition A.7

The function f is differentiable at (x, y) if there exist constants a, b and a function $\varepsilon(h, k)$ such that for all sufficiently small h, k, $f(x+h, y+k)$ is defined and satisfies

$$f(x+h, y+k) = f(x,y) + ah + bk + \varepsilon(h,k)\sqrt{h^2 + k^2}$$

where $\varepsilon(h,k) \to 0$ as $(h,k) \to (0,0)$ (equivalently ε is continuous at $(0,0)$ with $\varepsilon(0,0) = 0$.)

It follows at once from the definition that a, b are the partial derivatives f_1, f_2 at (x, y), and it also follows that if f is differentiable then it is continuous. Geometrically, Definition A.7 implies the existence of a tangent plane at (x, y) in the same way as Definition 3.1 implies the existence of a tangent line. Since we need only to prove the facts required in Chapter 7 we shall not develop the theory further, but proceed at once to the main results.

Theorem A.8

Let f and f_2 (the second partial derivative, $\partial f / \partial y$) be continuous on a rectangle $[a, b] \times (c, d)$ where $a, b \in \mathbb{R}$ and c, d may be finite or infinite. Then the functions ϕ, ψ given by

$$\phi(y) = \int_a^b f(x,y)\,dx, \quad \psi(y) = \int_a^b f_2(x,y)\,dx$$

exist on (c, d) and ϕ is differentiable there with $\phi' = \psi$.

Proof

The integrals exist since f, f_2 are continuous with respect to x. Given $y \in (c,d)$ we want to show that $\phi'(y) = \psi(y)$ for all $y \in (c,d)$. Choose a bounded interval $[c_1, d_1] \subset (c,d)$ with $c_1 < y < d_1$. Then f and f_2 are continuous on $E = [a,b] \times [c_1, d_1]$ so by Theorem A.4 they are uniformly continuous there. Hence given $\varepsilon > 0$ there is some $\delta > 0$ with $|f_2(p) - f_2(q)| < \varepsilon$ when $p, q \in E$ and $|p - q| < \delta$. Then for $k \neq 0$ and $y + k \in [c_1, d_1]$ we have

$$\begin{aligned}\frac{\phi(y+k) - \phi(y)}{k} - \psi(y) &= \int_a^b \left\{ \frac{f(x, y+k) - f(x,y)}{k} - f_2(x,y) \right\} dx \\ &= \int_a^b \{ f_2(x, y+k') - f_2(x,y) \} \, dx\end{aligned}$$

where $0 < k'/k < 1$, using the Mean Value Theorem (for functions of one variable). Hence if $|k| < \delta$, then

$$\left| \frac{\phi(y+k) - \phi(y)}{k} - \psi(y) \right| \leq \int_a^b \varepsilon \, dx = \varepsilon (b-a)$$

and the result follows. ∎

This version of the theorem is not the strongest known – it can with some effort be extended to cover the case in which $[a,b]$ is replaced by an unbounded interval – but it is easy to establish, and sufficient to prove Lemma 7.52 for instance.

Finally we want Fubini's theorem which shows that under reasonable conditions, the value of a repeated integral is independent of the order in which the integrations are carried out.

Theorem A.9 (Fubini)

(i) Let f be continuous on $E = [a,b] \times [c,d]$. Then $h(x,y) = \int_c^y f(x,v) \, dv$ is continuous on E and

$$\int_c^d \left(\int_a^b f(u,v) \, du \right) dv = \int_a^b h(u,d) \, du = \int_a^b \left(\int_c^d f(u,v) \, dv \right) du. \tag{A.1}$$

(ii) If f is continuous on $I \times J$ where I, J are intervals in R which may be finite or infinite, and if f is positive on E then the integrals in (A.1) are either all infinite, or all finite and equal.

(iii) If f is continuous on $I \times J$ where I, J are intervals in R which may be finite or infinite, and if f is such that any one of the intervals in (A.1) exists with $|f|$ in place of f, then all are finite and equal.

Proof

We shall prove only (i); the other parts follow exactly as for double series in Proposition 6.30 and Theorem 6.31.

The fact that h, as defined by $h(x,y) = \int_c^y f(x,v)\,dv$, is continuous on E is immediate from the uniform continuity of f – the details are left to the reader. Clearly $\partial h/\partial y = f(x,y)$ from the Fundamental Theorem of Calculus.

For $y \in [c,d]$, let

$$\begin{aligned} F(y) &= \int_a^b h(x,y)\,dx = \int_a^b \left(\int_c^y f(x,v)\,dv \right) dx, \\ G(y) &= \int_c^y \left(\int_a^b f(u,v)\,du \right) dv. \end{aligned}$$

Again F, G are continuous functions on E, and we want to show that they are in fact equal. But $F'(y) = \int_a^b \partial h/\partial y\,(u,y)\,du = \int_a^b f(u,y)\,du$ by Theorem A.8 above, and $G'(y) = \int_a^b f(u,y)\,du$ from the Fundamental Theorem of Calculus. Hence $F(y) - G(y)$ must be constant, and since $F(c) = G(c)$ the constant must be zero. Hence in particular $F(d) = G(d)$ which is the required result. ■

We finish with an example to illustrate the failure of (iii) of the theorem when the integral of $|f|$ is infinite.

Example A.10

Let $f(x,y) = \sin(y-x)$ if $x, y \geq 0$, $|y-x| \leq \pi$, $= 0$ otherwise. Then

$$\int_0^\infty \left(\int_0^\infty f(x,y)\,dy \right) dx = \pi = -\int_0^\infty \left(\int_0^\infty f(x,y)\,dy \right) dx.$$

The integral on the left is

$$\begin{aligned} &\int_0^\pi \left(\int_0^{x+\pi} \sin(y-x)\,dy \right) dx + \int_\pi^\infty \left(\int_{x-\pi}^{x+\pi} \sin(y-x)\,dy \right) dx \\ &= -\int_0^\pi \left([\cos(y-x)]_0^{x+\pi} \right) dx - \int_\pi^\infty \left([\cos(y-x)]_{x-\pi}^{x+\pi} \right) dx \\ &= -\int_0^\pi (\cos\pi - \cos x)\,dx = \pi \end{aligned}$$

and the integral on the right is the negative of this, on interchanging the roles of x, y. ♦

B
Hints and Solutions for Exercises

It should go without saying that these hints and (partial) solutions are not intended to be read in parallel with the text; consult them only *after* making a serious attempt to solve the question independently. For more on this, and on the philosophy of 'learning through problem solving,' see the preface to [8].

Chapter 1

1. If $x = a$, $y = -b$ with $a > b > 0$ then $|x + y| = a - b < a$ while if $b > a > 0$ then $|x + y| = b - a < b$. If x, y have the same sign, then this is also the sign of $x + y$.

2. $$s_{n+1} - s_n = \frac{ad - bc}{(cn + d)(c(n + 1) + d)}$$
   so the sequence is increasing if and only if $ad - bc > 0$.

3. Any s_n is an upper bound for S and so by Axiom 1.11, S has a least upper bound l say, with $l \le s_n$ for all n. Thus l is a lower bound for (s_n). If $l' > l$ then some $s_m < l'$, since otherwise l' would be a lower bound for (s_n) and thus an element of S, contrary to the definition of l as the least upper bound of S. Thus l' is not a lower bound so l must be the greatest.

4. For some A, $\left|a_0 + a_1 n + \cdots + a_{k-1}n^{k-1}\right| \le An^{k-1}$ and hence $s_n \ge a_k n^k - An^{k-1}$ which (i) is > 0 for $n > A/a_k$ so the sequence is bounded below,

and (ii) is $> a_k n^k/2$ for $n > 2A/a_k$ and so the sequence is not bounded above.

5. The limits are 3/4, 1/2, 1, 2/3 respectively.

6. Note that $||s_n| - |l|| \leq |s_n - l|$. The converse is false; consider $((-1)^n)$.

7. If $|s_n| \leq m$ and $\varepsilon > 0$, then there is N such that $|t_n| < \varepsilon$ if $n \geq N$ and so $|s_n t_n| < m\varepsilon$.

8. If $v_0 = 0$, $v_{n+1} = \sqrt{v_n + 2}$ then $2 - v_{n+1} = 2 - \sqrt{v_n + 2} = (2 - v_n)/(2 + \sqrt{v_n + 2}) < (2 - v_n)/3$, so $0 < 2 - v_n < (2 - v_0)/3^n$ as required.

9. Let r be the greatest lower bound of $\left(r_n^{1/n}\right)$; we show $r_n^{1/n} \to r$ as $n \to \infty$. Given $\varepsilon > 0$, there is some m with $r_m^{1/m} < r + \varepsilon$. For any n, write $n = km + t$ for integers k, t with $0 \leq t \leq m - 1$. Then $r_n \leq r_{km} r_t \leq r_m^k r_t \leq (r + \varepsilon)^{km} r_t$, and $r_n^{1/n} \leq (r + \varepsilon)^{km/n} r_t^{1/n}$. But $km/n \to 1$ and $r_t^{1/n}$ both $\to 1$ as $n \to \infty$, so $r_n^{1/n} \leq r + 2\varepsilon$ for large enough n, as required.

10. The sequence is increasing by induction, and bounded above by the positive root of the equation $x^2 = ax + b$.

11. Both a_{n+1}, b_{n+1} are strictly between a_n and b_n, and $a_{n+1} - b_{n+1} = \left(\sqrt{a_n} - \sqrt{b_n}\right)^2/2 > 0$.

12. Suppose $s_n \to l$ and $t_n = (s_1 + s_2 + \cdots + s_n)/n$. Given $\varepsilon > 0$, take m with $|s_n - l| < \varepsilon$ for $n \geq m$. Then $t_n - l = (s_1 - l + s_2 - l + \cdots + s_n - l)/n$, so for $n \geq m$,

$$\begin{aligned} |t_n - l| &\leq \frac{|s_1 - l| + \cdots + |s_{m-1} - l|}{n} + \frac{|s_m - l| + \cdots + |s_n - l|}{n} \\ &\leq \frac{\text{Const.}}{n} + \frac{n - m + 1}{n}\varepsilon < 2\varepsilon \end{aligned}$$

for n large enough.

The example $s_n = (-1)^n$ for which $t_n \to 1/2$ shows that the converse fails.

13. Let $n^{1/n} = 1 + x_n$ where $x_n > 0$. Then by the Binomial Theorem, $n = (1 + x_n)^n > 1 + nx_n + n(n-1)x_n^2/2 > n(n-1)x_n^2/2$, so $x_n^2 < 2/(n-1) \to 0$ as $n \to \infty$.

14. For $k \geq 0$, $|s_{N+k}| \leq c^k |s_N| \to 0$ as $k \to \infty$.

15. Suppose for some $x \in [0, 2\pi]$, that $f_n(x) = \cos nx \to l$ as $n \to \infty$. Then $f_{n+1}(x) + f_{n-1}(x) = 2\cos nx \sin x$ and taking a limit gives $2l = 2l\cos x$, so either $l = 0$, or $\cos x = 1$. But $f_{2n}(x) = \cos 2nx = 2\cos^2 nx - 1$ and so l must satisfy $l = 2l^2 - 1$ and in particular $l \neq 0$. Hence for convergence we

must have $\cos x = 1$, $x = 2k\pi$ for integer k when $f_n(x) = 1$ for all n and so (f_n) converges. Hence $\cos nx$ has a limit if and only if $\cos x = 1$, $x = 2k\pi$.

If $g_n(x) = \sin nx \to m$ as $n \to \infty$, then $g_{n+1}(x) - g_{n-1}(x) = 2\cos nx \sin x \to 0$ so either $\sin x = 0$, or $\cos nx \to 0$ which we just found is impossible. Hence $\sin nx$ has a limit if and only if $\sin x = 0$, $x = k\pi$.

16. The condition implies that the sequence is bounded, and hence by Corollary 1.29 it has a convergent subsequence with limit l say. Given $\varepsilon > 0$ and N with $s_n > s_m - \varepsilon$ for $n > m > N$, let $n \to \infty$ while restricting s_n to the subsequence. This gives $l \geq s_m - \varepsilon$, or $s_m \leq l + \varepsilon$ for $m > N$. Since the subsequence tends to l we can choose s_m in the subsequence with $m > N$ and $s_m > l - \varepsilon$ and hence for $n > m > N$, $s_n > s_m - \varepsilon > l - 2\varepsilon$. Combining these shows that the whole sequence tends to l.

17. Immediate by induction.

18. If for some $\delta > 0$, $t_n \geq \delta > 0$, then $s_n t_n \geq s_n \delta \to \infty$. For examples to show the failure of the converse, take for instance $s_n = n^2$, $t_n = 1/n$, $1/n^2$, $1/n^3$.

19. For $x_0 = 2$, $x_{n+1} = 2 - 1/x_n$, the sequence is decreasing and > 1 by induction, and the value follows from taking limits on both sides. The same argument works for any $c > 2$. What happens if $c < 2$?

20. (i) We have $x_1 = 1$, $x_2 - x_1 = 1/2$, $x_4 - x_2 > 1/2, \ldots, x_{2^k} - x_{2^{k-1}} > 1/2$ so $x_{2^k} > 1 + k/2$ follows by addition.

    (ii) $y_{n+1} - y_n = 1/(2n+1) + 1/(2n+2) - 1/(n+1) = 1/((2n+1)(2n+2)) > 0$ and $y_n > n/(2n) = 1/2$.

21. This follows by induction, since $n_1 \geq 1$, and $n_{k+1} > n_k \geq k$.

22. If $|s_n - l| < \varepsilon$ for $n \geq N$ then this also holds for those s_n which are in the subsequence.

23. (i) The result is obvious if $L = \infty$, so suppose L finite, and let (s_{n_k}) be the subsequence used in the proof of Theorem 1.27 which converges to L. Then the way in which (s_{n_k}) was constructed shows that if $n_k \leq j \leq n_{k+1}$ then $s_j \leq s_{n_k}$ when (s_{n_k}) is decreasing, $s_j \leq s_{n_{k+1}}$ when (s_{n_k}) is increasing. Hence any limit of terms s_j must be $\leq \lim s_{n_k} = L$.

    (ii) That $s_n > L - \varepsilon$ for infinitely many n follows by putting $n = n_k$. Also $s_n < L + \varepsilon$ if $n = n_k$ and k is large enough, and the inequalities noted in (i) show that then $s_j < L + \varepsilon$ if $j \geq n_k$.

    (iii) When (s_{n_k}) is increasing, $\sup_{m \geq n} s_m$ is eventually constant $= L$. If (s_{n_k}) is decreasing, then each $\sup_{m \geq n} s_m$ is equal to some peak point.

24. If the result is false, then from (iii) of the previous exercise, there is some N such that for all $n \geq N$

$$\left(\frac{a_1 + a_{n+1}}{a_n}\right)^n < \left(1 + \frac{1}{n}\right)^n,$$

equivalently

$$\frac{a_1}{n+1} < \frac{a_n}{n} - \frac{a_{n+1}}{n+1} \qquad \text{for } n \geq N.$$

Adding these inequalities gives

$$a_1 \left(\frac{1}{N+1} + \cdots + \frac{1}{n}\right) < \frac{a_N}{N} - \frac{a_{n+1}}{n+1}$$

which is impossible since it gives an upper bound for $1/(N+1) + \cdots + 1/n$ which we know tends to infinity.

(This striking result is from [7, Volume 1, pages 79–83], which gives an attractive discussion of the general approach to problem solving.)

25. The iterated limits are both 1, but the repeated limit does not exist since for instance the terms are zero when $m = n$.

26. For instance $5/7$ occurs in the $7^{\text{th}}, 14^{\text{th}}$, $7k^{\text{th}}$ block of terms, while $20 + 5/7$ occurs in the $7k^{\text{th}}$ block when $7k \geq 21$. Generally a fraction p/q in lowest terms occurs in the qk^{th} blocks when $qk > p/q$.

27. This exercise should bear a health warning! Verify by hand or machine that the result is true up to at least 30, and plot the number of iterations required to return to 1. Try another block of initial values around say 100. (Be warned that this has been done already up to many millions!) Try to explain why there is a tendency for sets of adjacent initial values to have equal numbers of iterations to return to zero. There is at the time of writing no convincing reason, heuristic or otherwise, why the result should be true in general.

Chapter 2

1. Suppose that f is increasing and has jumps at x, y with $x < y$ and $I = (f(x-), f(x+))$, $J = (f(y-), f(y+))$. Then if $x < u < v < y$ then $f(u) \leq f(v)$ and so letting $u \to x+$, $v \to v-$, $f(x+) \leq f(y-)$. Thus I, J are disjoint. There can be only countably many such intervals since each contains a rational number by Theorem 2.39 and these numbers must be distinct.

2. If $x \neq 1$ then divide by $x-1$ to get $f(x) = x^4 + x^3 + x^2 + x - 5 \to -1$ as $x \to 1$. Hence k should be 1.

3. Let f be increasing, and suppose f maps (a,b) onto (c,d). Let $x \in (a,b)$ and let $\varepsilon > 0$ where $(f(x) - \varepsilon, f(x) + \varepsilon) \subset (c,d)$. Then there are u, v with $u < x < v$, $f(u) = f(x) - \varepsilon$, $f(v) = f(x) + \varepsilon$ and so f maps (u,v) onto $(f(x) - \varepsilon, f(x) + \varepsilon)$ which proves continuity.

4. One way is proved as Proposition 2.18. For the other suppose that f is not continuous at c. Then there is some $\varepsilon > 0$ such that for every $\delta > 0$ there is some x with $|x - c| < \delta$ and $|f(x) - f(c)| \geq \varepsilon$. For each $n \geq 1$, choose x_n with $|x_n - c| < 1/n$ and $|f(x_n) - f(c)| \geq \varepsilon$. Then $x_n \to c$ but $f(x_n) \nrightarrow f(c)$ which gives the converse.

5. Take $\varepsilon = |f(c)|/2$ in the definition of continuity.

6. Let $(a_n), (b_n), c$ be as in the proof of the theorem, so that $a_n \to c-$, $b_n \to c+$. Since f is regulated at c there is some $\delta > 0$ such that $|f(x) - f(c-)| < 1$ on $(c - \delta, c)$ and $|f(x) - f(c+)| < 1$ on $(c, c + \delta)$, so f is bounded on $(c - \delta, c + \delta)$ and the proof is completed as before.

7. (a_n) is increasing, (b_n) is decreasing, and $b_n - a_n \to 0$.

8. For $\alpha > 0$, $|x^\alpha \sin(1/x)| \leq |x|^\alpha \to 0$ as $|x| \to 0$. For $x > \pi$, $\sin(1/x) \leq 1/x$ (Proposition 6.49(ii)) and so $|x^\alpha \sin(1/x)| \leq |x|^{\alpha - 1} \to 0$ as $x \to \infty$.

9. $|x^2 - a^2| \leq |x - a|\,|x + a|$. However $\left|\sqrt{|x|} - 0\right|$ is not $\leq M|x|$ on $(0,1)$ for any M.

10. If $x < u < y < v$, then $f(x) \leq f(u) \leq f(y) \leq f(v)$. Letting $u \to x$, $v \to y$ then $f(x) \leq f(x+) \leq f(y) \leq f(y+)$ so $f(x+)$ is monotone. If f is continuous on the right, then for any $\varepsilon > 0$, there is some $\delta > 0$ with $0 \leq f(y) - f(x) < \varepsilon$ when $0 \leq y - x < \delta$. Then $0 \leq f(y+) - f(x+) \leq \varepsilon$ when $0 \leq y - x < \delta$ follows as in the proof of the previous statement.

11. Let $g(x) = f(x + 1/2) - f(x)$. Then g is continuous and $g(0) + g(1/2) = 0$ since $f(0) = f(1)$. Thus either $g(0) = g(1/2) = 0$, when we are done, or $g(0)$ and $g(1/2)$ have opposite signs and the result follows from the Intermediate Value Theorem. A similar argument, with $g(x) = f(x + 1/3) - f(x)$ gives the result with $1/3$ (or generally $1/n$) in place of $1/2$.

   To show the result false for $g(x) = f(x + c) - f(x)$ where $1/3 < c < 1/2$, define f as follows. Take some d with $1/3 < d < c$ and let $f(0) = f(d) = f(1/2) = f(1 - d) = f(1) = 0$. Then define f on $[1/2, 1 - d]$ to be continuous and strictly positive on $(1/2, 1 - d)$. Let m be an upper bound for f on $[1/2, 1 - d]$. Define f on $[0, d]$ to be continuous and strictly positive on $(0, d)$ with the additional requirement that $f(x) > m$ on $[1/2 - d, 1 - 2d]$

(not forgetting to verify that $[1/2 - d, 1 - 2d] \subset (0, d)$). For other values of x, let $f(x) = -f(1-x)$. Show that this f does the trick, and then extend the argument to cover other values of $c \neq 1/n$.

12. Let's answer the second question first. Borrow from the proof of Theorem 1.27 the idea of a peak point, so $x \in [a, b]$ is a peak point if $f(x) \geq f(y)$ for all $y \geq x$. Thus G is the set of all x which are not peak points. Then if we imagine the sun rising from a distant point in the direction of positive x, G consists of the points which are in shadow and the result says that f has equal values at the end points of the intervals of shadow.

    Suppose then that $(c, d) \subset G$, $c, d \notin G$, so c, d are peak points. Then $f(c) \geq f(d)$ since c is a peak point. If $f(c) > f(d)$, f attains a maximum on $[c, d]$ at some point $k < d$ which must be a peak point since d is a peak point. Hence $k \notin G$ which contradicts $(c, d) \subset G$. This leaves $f(c) = f(d)$ as the only possibility.

13. (i) Clearly $T_0(x) = 1$, $T_1(x) = x$, $T_2(x) = 2x^2 - 1$. Since $\cos(n+1)\theta + \cos(n-1)\theta = 2\cos n\theta \cos\theta$ we have $T_{n+1}(x) + T_{n-1}(x) = 2xT_n(x)$ which defines T_n inductively with the required properties. Notice that in addition, as $x = \cos\theta$ decreases from 1 to -1, θ increases from 0 to π and so $\cos n\theta$ changes monotonically from 1 to -1 and back n times, finishing at $(-1)^n$ when $x = -1$.

    (ii) If p is a polynomial with the given properties, then the graph of p must intersect the graph of T_n at least n times on the interval $[-1, 1]$ (once on each branch from -1 to 1). Thus since $p - T_n$ has degree $n-1$ with n zeros, it must be identically zero. (The reader should check that the argument works when some of the zeros coincide.)

    (iii) If q has degree n, leading coefficient 1, and $|q(x)| \leq 2^{1-n}$ on $[-1, 1]$ then $q = 2^{1-n}T_n$ from (ii); equivalently if $q \neq 2^{1-n}T_n$, then $|q(x)| > 2^{1-n}$ for some $x \in [-1, 1]$.

14.
$$f(x) = \begin{cases} x & \text{for } x \in \mathbb{Q}, \\ x + 1/2 & \text{for } 0 < x < 1/2,\ x \notin \mathbb{Q}, \\ x - 1/2 & \text{for } 1/2 < x < 1,\ x \notin \mathbb{Q}. \end{cases}$$

15. (i) Rearranging as $x = (5 - x)^{1/3}$ gives $x = 1.51598$.

    (ii) Rearranging as $x = \tan^{-1} x + \pi$, gives $x = 4.49341$.

16. For any $n \geq 0$, $|x_{n+1} - d| = |f(x_n) - f(d)| \leq c|x_n - d|$.

17. If $f(x) \to l$ then $f(x) - f(y) \to 0$ is immediate. Suppose conversely that $f(x) - f(y) \to 0$ as $x, y \to \infty$. Then if $x_n \to \infty$, $f(x_n)$ is a Cauchy

sequence with a limit l, say, and the limit is independent of the choice of (x_n) since if also $y_n \to \infty$ we consider $(x_1, y_1, \ldots, x_n, y_n, \ldots)$. The rest of the argument is as in Exercise 4 above.

Chapter 3

1. The given functions are (a) continuous on (i) $[0, \infty)$, (ii) $\mathbb{R}$, (iii) $[0, \infty)$, (iv) $[-1, 1]$, and (b) differentiable on (i) $(0, \infty)$, (ii) $\mathbb{R} - \pi\mathbb{Z}$, (iii) $[0, \infty)$ (including on the right at 0), (iv) $(-1, 1)$.

2. Suppose $f'(c) < t < f'(d)$ and consider $g(x) = f(x) - tx$. Then g is differentiable on $[c, d]$ with $g'(c) < 0 < g'(d)$ so g must attain a local minimum at some $e \in (c, d)$ at which $g'(c) = 0$.

3. (i) Suppose f is differentiable at t, $f'(t) = k$ and $x < t < y$. Then $(f(t) - f(x)) / (t - x) \to k$ as $x \to t-$ and $(f(y) - f(t)) / (y - t) \to k$ as $y \to t+$. Also

$$\begin{aligned} \frac{f(y) - f(x)}{y - x} &= \frac{t - x}{y - x} \frac{f(t) - f(x)}{t - x} + \frac{y - t}{y - x} \frac{f(y) - f(t)}{t - x} \\ &= \alpha \frac{f(t) - f(x)}{t - x} + \beta \frac{f(y) - f(t)}{t - x} \end{aligned}$$

where $\alpha, \beta > 0$ and $\alpha + \beta = 1$. Hence $(f(y) - f(x)) / (y - x)$ lies between $(f(t) - f(x)) / (t - x)$ and $(f(y) - f(t)) / (y - t)$ and so must $\to k$ too.

(ii) Let $x_n = 1/(n\pi)$ so $f(x_n) = (-1)^n / (n^2\pi^2)$. Show that

$$(-1)^n (f(x_n) - f(x_{n+1})) / (x_n - x_{n+1}) \to 2/\pi.$$

4. (i) Given $\varepsilon > 0$ there is some $\delta > 0$ such that $|f'(t) - l| < \varepsilon$ for $a < t < a + \delta$. Then for each $x \in (a, a + \delta)$ there is some $t \in (a, x)$ such that $(f(x) - f(a)) / (x - a) = f'(t)$ by the Mean Value Theorem. Hence for each $x \in (a, a + \delta)$, $|(f(x) - f(a)) / (x - a) - l| < \varepsilon$ as required.

(ii) If not, then for some $p > m$, $f(b) - f(a) \geq p(b - a)$ (or $< -p(b - a)$ in which case the argument is similar). Consider lines $y = px + c$ parallel to the chord on $[a, b]$ and choose the least c for which the line intersects the curve. Let t be the greatest value of x which lies both on the line and the graph of f. Then $f'_+(t) \geq p$ which is a contradiction.

(iii) The reader is left to modify the proof of (i) to cover this case.

5. (i) The tangent at (t, t^2) is $y - t^2 = 2t(x - t)$ and this passes through $(0, -1)$ if $t = \pm 1$ but never through $(0, 1)$.

   (ii) The normal at (t, t^2) is $2t(y - t^2) + (x - t) = 0$ and this passes through $(0, -1)$ if $t = 0$ only, through $(0, 1/2)$ if $t = 0$ (triple solution) and through $(0, 4)$ if $t = 0, \pm\sqrt{7/2}$.

   (iii) The tangent at (t, e^t) is $y - e^t = e^t(x - t)$ which passes through the origin if $t = 1$.

6. $f(x) = a^\alpha (1 - x)^\beta$ is continuous on the right at 0 if $\alpha > 0$ and on the left at 1 if $\beta > 0$, and is always differentiable on $(0, 1)$.

7. For $f(x) = x^2$, $f'(c) = 2c = (f(b) - f(a)) / (b - a) = (b + a)/2$, so $\lim_{b \to a} (c - a) / (b - a) = 1/2$. In fact the limit is $1/2$ for any function f and any point at which f'' is continuous and non-zero, by considering the Taylor expansions of f and f' about a.

8. Suppose $|f'(x)| \le m$ on $(a, b]$. Then for $a < x < b$, $|f(x) - f(b)| \le m|x - b| \le m(b - a)$ and so f is bounded on $(a, b]$.

9. Suppose that f is defined and satisfies (3.20) on $[a, b]$ – otherwise choose $[a', b'] \subset (a, b)$ and apply the solution on $[a', b']$.

   (i) The statement in (a) is equivalent to (3.20). For (b), suppose $x < u < t$, and $f(u)$ is below the chord joining $(t, f(t))$ to $(y, f(y))$. Then $f(t)$ is above the chord joining $(u, f(u))$ to $(y, f(y))$ which contradicts (3.20).

   (ii) Continuity follows from (i) since for any $x \in (a, b)$ the graph of f on (x, b) is between the the chord joining $(a, f(a))$ to $(x, f(x))$ and the chord joining $(x, f(x))$ to $(b, f(b))$. Also (3.20) shows that $(f(t) - f(x)) / (t - x)$ is decreasing as $t \to x+$ and bounded below by the slope of the chord joining $(a, f(a))$ to $(x, f(x))$ so the right-hand derivative exists and is finite. A similar argument works on the left and shows that $f'_-(x) \le f'_+(x)$.

   (iii) Choose u, v with $x < u < v < y$. Then $(f(u) - f(x))/(u - x) \le (f(y) - f(v))/(y - v)$ and so $f'_+(x) \le f'_-(y)$ by taking limits. Hence if f is not differentiable at any points x, y then the intervals $(f'_-(x), f'_+(x))$ and $(f'_-(y), f'_+(y))$ are disjoint and non-empty, and so can be at most countable in number. (Compare the solution to Exercise 1, Chapter 2.)

   (iv) is immediate from the Mean Value Theorem.

10.* (i) If f is not strictly increasing on (a, b) then there is $(c, d) \subset (a, b)$ with $f(c) \ge f(d)$. Then f cannot be constant on (c, d) since then f'_+ would be zero on (c, d) which is an uncountable set. Hence either $f(c) > f(d)$

or, if $f(c) = f(d)$ then there is some $e \in (c, d)$ with $f(e) \neq f(c)$; in any case we have points p, q from among c, d, e with $p < q$ and $f(p) > f(q)$.

Now consider, for each $y \in (f(q), f(p))$ (there are uncountably many such values of y), the greatest x with $f(x) = f(y)$. For such a point, $f(x) \geq f(x')$ for $x < x' < q$ and so $f'_+(x) \leq 0$. This gives uncountably many points with $f'_+ \leq 0$ and a contradiction.

(ii) Apply (i) to $f(x) + kx$ for arbitrary $k > 0$.

11. Replace x by $y/3$ to get a cubic with both f and f' having integer roots.

12. One needs to generate an infinite family of solutions f to the previous exercise, depending on a parameter which is chosen to make the roots of $\int f$ also integers. But this has not been done.

13. The construction of $(x_n), (y_n)$ shows that $y_n < y_{n+1} < x_{n+1} < x_n$ so that $y_n \to l$ for some l and we know that $x_n \to d$ where $f(d) = 0$. Taking a limit in (3.19) gives $l = d$ and so $y_n < d < x_n$ for all n. We leave it to the reader to show that if $f' \geq m > 0$ on (a, b) then $x_n - y_{n+1} \leq f(x_n)/m$ which gives convergence of (y_n) to d at the same rate as (x_n).

14. The definition of x_{n+1} gives $f(x_n)/(x_n - x_{n+1}) = f'(x_n)$ and $f(x_n)/(x_n - d) = f'(u)$ for some $u \in (d, x_n)$ by the Mean Value Theorem. Hence $(x_{n+1} - d)/(x_n - x_{n+1}) = f'(x_n)/f'(u) \leq f'(x_n)/f'(y_n)$.

15. Let $f_n(x) = nx(1-x)^n$. Then $f_n(x) \to 0$ for each $x \in [0,1]$ but f_n attains its maximum at $x_n = 1/(n+1)$ and $f_n(x_n) = (n/(n+1))^n \to 1/e \neq 0$.

16. $f(x) = x^2(x-5)(x-9)$ is zero at $0, 5, 9$ and is > 0 on $(-\infty, 0)$, $(0, 5)$ and $(9, \infty)$, and < 0 on $(5, 9)$. In particular it has a local minimum at 0. $f'(x) = 2x(x-3)(2x-15)$ and so since there must be a local maximum on $(0, 5)$, it is at $x = 3$, and since there must be a local minimum on $(5, 9)$, it is at $x = 15/2$ and is a global minimum. There is no global maximum.

17. If $a, b > 0$ then $f(x) = x^a(1-x)^b$ is continuous on $[0, 1]$, positive and differentiable on $(0, 1)$ with $f(0) = f(1) = 0$. The only stationary point is at $x = a/(a+b)$ which must be a maximum. If $a, b < 0$ then $f \to \infty$ as $x \to 0, 1$ with a unique minimum at $a/(a+b)$. If $ab < 0$ then f' has fixed sign on $(0, 1)$.

18. With $f(x) = (\sin x)/x$, $f'(x) = (x\cos x - \sin x)/x^2$ which is zero when $\tan x = x$ which happens exactly once in each interval $(n\pi, (n+1/2)\pi)$, $n \geq 1$. Similarly $f''(x) = 0$ when $\tan x = 2x/(2-x^2)$ which happens exactly once in each interval $((n-1/2)\pi, n\pi)$, $n \geq 0$.

19. The statement is the same except for replacing $x \to a$ by $x \to \infty$, and the

proof is got by replacing $0 < |x-a| < \delta$ by $x > A$ (so x is 'near infinity') and making the right verbal changes.

20.* One obvious example is when f is linear, $f(x) = ax + b$ when $(f(x+h) - f(x))/h = a$ for all x, h. The limit does not exist for polynomials of degree greater than 1, or for trigonometric or exponential functions, a^x with $a > 1$. The limit does exist, and is equal to zero, when $f(x) = \ln x$.

What is the relationship between the 'derivative at infinity' and $\lim_{x\to\infty} f'(x)$?

21. If $g = f_1 f_2$ then by the product rule, Theorem 3.14,

$$\frac{g'}{g} = \frac{f_1'}{f_1} + \frac{f_2'}{f_2}$$

which is the basis for an induction: if $g = f_1 f_2 \cdots f_n f_{n+1}$ then

$$\begin{aligned}\frac{g'}{g} &= \frac{(f_1 f_2 \cdots f_n)'}{f_1 f_2 \cdots f_n} + \frac{f_{n+1}'}{f_{n+1}} \\ &= \frac{f_1'}{f_1} + \frac{f_2'}{f_2} + \cdots + \frac{f_n'}{f_n} + \frac{f_{n+1}'}{f_{n+1}}.\end{aligned}$$

Chapter 4

1. (i) Put $x = \sin\theta$ to get $\int_0^1 \sqrt{1-x^2}dx = \int_0^{\pi/2} \cos^2\theta d\theta = \int_0^{\pi/2}(1 - \cos 2\theta)/2\, d\theta = \pi/4$ (showing that the calculus and geometrical values for the area of a quarter-circle are the same). (ii) $\int_1^2 \ln t\,/t\, dt = \left[\ln^2 t/2\right]_1^2 = \left(\ln^2 2\right)/2$. (iii) $\int_0^{\pi/4} \tan u\, du = \int_{1/\sqrt{2}}^1 dv/v$ (put $v = \cos u$) $= \ln\left(\sqrt{2}\right) = (\ln 2)/2$. (iv) Integrate by parts twice to get $\left(\pi^2 - 2b^2\right) b^{-3} \sin(b\pi) + 2\pi b^{-2}\cos(b\pi)$. (v) Put $\tan(x/2) = t$, $dx = 2dt/\left(1+t^2\right)$, $\cos x = \left(1-t^2\right)/\left(1+t^2\right)$ to get

$$\begin{aligned}\int_0^{2\pi} \frac{dx}{a+\cos x} &= 2\int_0^{\pi} \frac{dx}{a+\cos x} = 4\int_0^{\infty} \frac{dt}{a(1+t^2)+1-t^2} \\ &= \frac{4}{a-1}\int_0^{\infty} \frac{dt}{t^2+b^2}, \qquad \text{where } b^2 = \frac{a+1}{a-1} \\ &= \frac{4}{(a-1)b}\left[\tan^{-1}(t/b)\right]_0^{\infty} = \frac{2\pi}{\sqrt{a^2-1}}.\end{aligned}$$

(Of course in the last example we have been making free with results from elementary calculus which will be justified in the next two chapters.)

2. A regulated function can only have jump discontinuities, i.e. points at which the right- and left-hand limits are unequal, either to each other or to the value of the function. But Exercise 2, Chapter 3 shows that if f is differentiable on $[a,b]$ then f' assumes on (a,b) every value between $f'(a)$ and $f'(b)$ (irrespective of whether f' is continuous) and so cannot have jump discontinuities.

   It is worth noting that the situation in quite different for one-sided derivatives: $f(x) = |x|$ has a right-derivative at every point, equal to -1 for $x < 0$ and $+1$ for $x \geq 0$ – see Exercise 6 below.

3. These are immediate since by putting $x = -t$ we have $\int_{-a}^{0} f(x)\,dx = \int_0^a f(t)\,dt = \int_0^a f(x)\,dx$ if f is even, $\int_{-a}^{0} f(x)\,dx = -\int_0^a f(t)\,dt = -\int_0^a f(x)\,dx$ if f is odd.

4.* For (i), $f(t) = x_1 + t(x_2 - x_1)$, $g(t) = y_1 + t(y_2 - y_1)$ for $0 \leq t \leq 1$, and for (ii), $f(t) = r\cos t$, $g(t) = r\sin t$ for $0 \leq t \leq 2\pi$.

   For the cycloid we have $f'(t) = r(1 - \cos t)$, $g'(t) = r\sin t$ so the length is

$$\begin{aligned} r^2 \int_0^{2\pi} \sqrt{(1-\cos t)^2 + \sin^2 t}\,dt &= r^2 \int_0^{2\pi} \sqrt{2 - 2\cos t}\,dt \\ &= 4r^2 \int_0^{2\pi} \sin(t/2)\,dt = 8r^2. \end{aligned}$$

5. (i) For $n \geq 2$,

$$\begin{aligned} I_n &= \int_0^{\pi/2} \sin^n x\,dx = \int_0^{\pi/2} \sin^{n-1} x \sin x\,dx \\ &= \left[-\cos x \sin^{n-1} x\right]_0^{\pi/2} + \int_0^{\pi/2} \cos x\,(n-1)\sin^{n-2} x \cos x\,dx \\ &= (n-1)\int_0^{\pi/2} \sin^{n-2} x\left(1 - \sin^2 x\right)dx = (n-1)\left(I_{n-2} - I_n\right) \end{aligned}$$

   as required. Then (4.11) is equivalent to $\sqrt{n}I_n = \sqrt{n-2}I_{n-2}(n-1)/\sqrt{n(n-2)}$ and (ii) follows since $n - 1 \geq \sqrt{n(n-2)}$.

   Since $I_n \geq I_{n+1}$ we have $\sqrt{n}I_n \geq \sqrt{n+1}I_{n+1}\sqrt{n/(n+1)}$ and so taking limits with n even gives $a \geq b$, and with n odd gives $b \geq a$. Multiplying the expressions in (4.11) gives $I_{2n}I_{2n+1} = \pi/(2(2n+1))$, or $\sqrt{2n}I_{2n}\sqrt{2n+1}I_{2n+1} = (\pi/2)\sqrt{2n/(2n+1)}$ and hence $a = b = \sqrt{\pi/2}$.

6. Let $g(x) = \int_a^x f'_+$. We know from Theorem 4.13 that g is continuous on $[a,b]$ and has a right-hand derivative g'_+ which is equal to f'_+ on $[a,b)$,

using Exercise 4(iii), Chapter 3. Hence we have to show that if f, g are continuous on $[a, b]$ and $g'_+ = f'_+$ then $f = g$. Exercise 10(ii), Chapter 3 is what is required.

The reader is now challenged to use the full strength of that exercise to prove the following:

Theorem B.1

Let f be continuous on $[a, b]$ and suppose that f'_+ exists on $[a, b] - D$ for some countable set D. Define $g(x) = f'_+(x)$ on $[a, b] - D$, $g(x) = 0$ on D, and suppose that g is regulated. Then

$$\int_a^b g = f(b) - f(a).$$

7. $\int_a^b f = \int_a^b f^+ - \int_a^b f^-$ while $\int_a^b |f| = \int_a^b f^+ + \int_a^b f^-$.

Chapter 5

1. If $c < a' < a < b$ then $\int_c^{a'} f + \int_{a'}^b f = \int_c^{a'} f + \int_{a'}^a f + \int_a^b f = \int_c^a f + \int_a^b f$.

2. For $0 < a < 1$, $\int_0^a \left(1 - t^2\right)^{-1/2} dt = \int_0^{\sin^{-1} a} dx = \sin^{-1} a$, putting $t = \sin x$. Now let $a \to 1-$.

   Similarly

$$\begin{aligned} \int_0^a t\left(1 - t^2\right)^{-1/2} dt &= \int_0^{\sin^{-1} a} \sin x dx = [-\cos x]_0^{\sin^{-1} a} \\ &= 1 - \sqrt{1 - a^2} = a^2 / \left(1 + \sqrt{1 - a^2}\right) \to 1 \end{aligned}$$

   as $a \to 1-$.

   We already have from Example 4.18 the integrals $\int e^{-ax} \sin(bx) dx = e^{-ax}(-a \sin(bx) - b \cos(bx))/(a^2 + b^2)$, $\int e^{-ax} \cos(bx) dx = e^{-ax}(-a \cos(bx) + b \sin(bx))/(a^2 + b^2)$, so putting in limits $0, c$ and letting $c \to \infty$ gives $\int_0^\infty e^{-ax} \sin(bx) dx = b/\left(a^2 + b^2\right)$, $\int_0^\infty e^{-ax} \cos(bx) dx = a/\left(a^2 + b^2\right)$.

3. Let $4b - a^2 = t^2$, $t > 0$. Then (putting in the infinite limits directly) we

have

$$\begin{aligned}\int_{-\infty}^{\infty}\frac{dx}{x^2+ax+b} &= 4\int_{-\infty}^{\infty}\frac{dx}{4x^2+4ax+a^2+t^2}\\ &= 2\int_{-\infty}^{\infty}\frac{dy}{y^2+t^2}=\frac{2}{t}\left[\tan^{-1}\frac{y}{t}\right]_{-\infty}^{\infty}\\ &= \frac{2\pi}{t}=\frac{\pi}{\sqrt{b-a^2/4}}.\end{aligned}$$

The partial fractions (to be verified!) are

$$\frac{1}{x^4+1}=\frac{1}{2\sqrt{2}}\left[\frac{x+\sqrt{2}}{x^2+\sqrt{2}x+1}-\frac{x-\sqrt{2}}{x^2-\sqrt{2}x+1}\right]$$

and so the integral is

$$\frac{1}{4\sqrt{2}}\left[\ln\left(\frac{x^2+\sqrt{2}x+1}{x^2-\sqrt{2}x+1}\right)\right]_{-\infty}^{\infty}+\frac{1}{4}\left[\int_{-\infty}^{\infty}\frac{dx}{x^2+\sqrt{2}x+1}\right.$$

$$\left.+\int_{-\infty}^{\infty}\frac{dx}{x^2-\sqrt{2}x+1}\right]=\frac{\pi}{4}\frac{2}{\sqrt{1-1/2}}=\frac{\pi}{\sqrt{2}}.$$

The method for $\int_{-\infty}^{\infty} x^2/\left(x^4+1\right)dx$ is similar, and gives the same answer, $\pi/\sqrt{2}$.

4. Since $\int_b^\infty g$ is finite, $\int_y^z g \to 0$ as $y, z \to \infty$, so $\left|\int_y^z f\right| \to 0$ also. Now use Exercise 17, Chapter 2.

5. The function given by $f(x) = n$ on $[n, n+n^{-3})$, $n = 1, 2, 3, \ldots$, $f(x) = 0$ otherwise, is unbounded and $\int_{-\infty}^{\infty}|f| = \sum_1^\infty n^{-2}$ which is finite.

   The function given by $f(x) = n^2$ on $[n, n+n^{-3}/2)$, $= -n^2$ on $[n + n^{-3}/2, n+n^{-3})$ is integrable (improperly) since $\sum(-1)^n/n$ is convergent, but $\int_{-\infty}^{\infty}|f| = \sum 1/n$ diverges.

6. Put $t^4 = u$ (more generally $t^b = u$) to reduce to a Beta-integral. For the integral over $(0,\infty)$ put $t^b = u$, then $u/(1-u) = v$.

7. $\int_0^a \sin(\sqrt{x})\,dx = 2\sqrt{a}\cos(\sqrt{a}) - \sin(\sqrt{a})$ which is unbounded as $a \to \infty$, but if $a = (2n+1/2)^2\pi^2$, the integral has the constant value -1.

8. We have to show that if $\int_0^a f \to l$ as $a \to \infty$ then subject to the given conditions on g, $\int_0^\infty f(x)\,g(tx)\,dx \to l$ as $t \to 0+$. Suppose that $\varepsilon > 0$ is given, and that $A > 0$ is such that $\left|\int_a^b f\right| < \varepsilon$ for $A \le a < b$. We first show that

$$\left|\int_a^b fg\right| < \varepsilon g(a) \qquad \text{for } A \le a < b. \tag{B.1}$$

Note that it is enough to show (B.1) when f, g are step functions (why is this?), and hence that it is enough to prove the analogue for sums: if $|\sum_1^n a_r| \leq M$ for $1 \leq n \leq N$ and (b_r) is decreasing and positive, then for $1 \leq n \leq N$,

$$\left|\sum_1^n a_r b_r\right| \leq b_1 M. \tag{B.2}$$

But on rearranging we have with $A_n = \sum_1^n a_r$,

$$\begin{aligned}\sum_1^n a_r b_r &= A_1 b_1 + (A_2 - A_1) b_2 + \cdots + (A_n - A_{n-1}) b_n \\ &= A_1 (b_1 - b_2) + \cdots + A_{n-1}(b_{n-1} - b_n) + A_n b_n, \\ \left|\sum_1^n a_r b_r\right| &\leq M (b_1 - b_2 + \cdots + b_{n-1} - b_n + b_n) = M b_1\end{aligned}$$

as required. Now returning to the proof of the main result, let A be as in (B.1), and let $\delta > 0$ be such that $g(0) - g(tA) < \varepsilon/A$ if $0 < t < \delta$. Then

$$\left|\int_0^A f(x) g(tx)\, dx - \int_0^A f(x)\, dx\right| \leq A(\varepsilon/A)$$

which is enough, since we already know that $\int_A^\infty f$ and $\int_A^\infty fg$ are small.

For the examples we have the values $\int_0^\infty e^{-tx} \sin(bx)\, dx = b/(t^2 + b^2)$, $\int_0^\infty e^{-tx} \cos(bx)\, dx = t/(t^2 + b^2)$ from Exercise 2 above. Hence letting $t \to 0+$ gives the $g-$values $\int_0^\infty \cos(bx)\, dx = 0$, $\int_0^\infty \sin(bx)\, dx = 1/b$ for real $b \neq 0$.

9. The idea of the solution is to show that if f does not tend to zero, then using the fact that $f' \to 0$, there are intervals on which $\int f$ does not tend to zero, which contradicts the existence of $\int_0^\infty f$.

   More explicitly, if f does not tend to zero, there is some $\delta > 0$ such that $|f(x_n)| \geq \delta$ for $x_n \to \infty$. Take $\varepsilon = \delta^2/2$; since $f' \to 0$, there is some N such that $|f'(x)| \leq \varepsilon$ if $x \geq N$ and we assume $x_n \geq N$ from now on. Then on the interval $[x_n, x_n + 1/\delta]$ we have $|f(x_n)| \geq \delta$, $|f'(x)| \leq \varepsilon = \delta^2/2$ and so $|f(x)| \geq \delta/2$. Hence

$$\left|\int_{x_n}^{x_n + 1/\delta} f\right| \geq (\delta/2)(1/\delta)$$

   does not tend to zero, and so $\int_0^\infty f$ cannot exist by Exercise 4 above.

Chapter 6

1. We have $\sum_{n=0}^{\infty} r^n = (1-r)^{-1}$ for $-1 < r < 1$ from the Binomial Theorem, and so by differentiation (to be justified) $\sum_{n=1}^{\infty} nr^{n-1} = (1-r)^{-2}$ and so $\sum_{n=1}^{\infty} nr^n = r(1-r)^{-2}$. For $k = 2, 3$, etc. differentiate again.

2. For instance if $k = 3$ we have

$$\begin{aligned}\sum_{n=1}^{\infty} \frac{1}{n(n+3)} &= \lim_{N\to\infty} \sum_{n=1}^{N} \frac{1}{n(n+3)} = \frac{1}{3} \lim_{N\to\infty} \sum_{n=1}^{N} \left(\frac{1}{n} - \frac{1}{n+3}\right) \\ &= \frac{1}{3} \lim_{N\to\infty} \left[\left(1 + \frac{1}{2} + \frac{1}{3} + \cdots + \frac{1}{N}\right) - \left(\frac{1}{4} + \frac{1}{5} + \frac{1}{6}\right.\right. \\ &\quad \left.\left. + \cdots + \frac{1}{N+3}\right)\right] \\ &= \frac{1}{3}\left[\left(1 + \frac{1}{2} + \frac{1}{3}\right) - \lim_{N\to\infty}\left(\frac{1}{N+1} + \frac{1}{N+2} + \frac{1}{N+3}\right)\right] \\ &= \frac{1}{3}\left(1 + \frac{1}{2} + \frac{1}{3}\right)\end{aligned}$$

and the general pattern is clear.

3. Use the inequality

$$\int_2^{\infty} f(x)\,dx > \sum_3^{\infty} f(n) > \int_3^{\infty} f(x)\,dx$$

and the given value of the integral.

4. If $m = 2n$ is even, $S \le s_{2n+1} = s_{2n} + b_{2n+1}$, so $S - s_{2n} \le b_{2n+1}$.

5. For each $N = 1, 2, \ldots$, first take enough terms $\sum p_n$ for the partial sum to exceed $2N$, then enough terms $\sum q_n$ for the partial sum to be less than $2N - 1$. The result can be further extended to make the rearrangement have arbitrary limits superior and inferior (Exercise 23, Chapter 1).

6.* $\sum_0^{\infty} (-1)^n$ diverges, but by grouping pairs we can get series $(1-1) + (1-1) + \cdots$ which converges with sum zero, or $1 - (1-1) - (1-1) - \cdots$ with sum 1.

7.* The absolute convergence (to the correct sum) of the Cauchy product is a special case of Theorem 6.31. If $a_n = (-1)^n/\sqrt{n}$ then since $\sqrt{ab} \le (a+b)/2$, $(-1)^n \sum_{r=1}^{n-1} a_r a_{n-r} = \sum_{r=1}^{n-1} 1/\sqrt{r(n-r)} \ge \sum_{r=1}^{n-1} 2/n = 2(n-1)/n$ does not tend to zero.

8.** Some necessary conditions, and some (different!) sufficient conditions are in [4].

9.* The recurrence in (ii) follows from equating powers of t in

$$te^{tx} = \left(e^t - 1\right) \sum_{n=0}^{\infty} \frac{B_n(x)}{n!} t^n. \tag{B.3}$$

This proves that the B_n are polynomials of degree n. Differentiating (B.3) with respect to x gives $B_n'(x) = nB_{n-1}(x)$ and integrating this gives $\int_0^1 B_n(x)\,dx = 0$ for $n \geq 1$.

10.* (i) is immediate from the previous question.

(ii) With $x = 0$ we have

$$\begin{aligned} \frac{t}{e^t - 1} &= -\frac{t}{2} + \sum_{n=0}^{\infty} \frac{B_{2n}}{(2n)!} t^{2n}, \\ \frac{t\left(e^t + 1\right)}{2\left(e^t - 1\right)} &= \sum_{n=0}^{\infty} \frac{B_{2n}}{(2n)!} t^{2n}, \end{aligned}$$

and putting $t = 2iz$ gives the result.

11. Suppose

$$\frac{2e^{tx}}{e^t + 1} = \sum_{n=0}^{\infty} E_n(x) \frac{t^n}{n!}$$

and show, as in the previous question that E_n is a polynomial of degree n and $\eta_n = 2^n E_n(1/2)$ is an integer (the nth Euler number). Then $x = 1/2$, $t = 2iz$ gives

$$\begin{aligned} \frac{2e^{t/2}}{e^t + 1} &= \sum_{n=0}^{\infty} E_n(1/2) \frac{t^n}{n!}, \\ \sec z &= \sum_{n=0}^{\infty} (-1)^n \eta_{2n} \frac{x^{2n}}{(2n)!} \end{aligned}$$

showing incidentally that all η_n with odd n are zero.

12.* (i) If f is positive and summable over A then for each integer $n \geq 1$, the set E_n on which $f(x) \geq 1/n$ must be finite, otherwise $\sum_{E_n} f(x)$ would be infinite. Hence the set E on which $f \neq 0$ is at most countable, and $\sum_E f(x) = \sum_A f(x)$ is finite. Conversely if $\sum_A f(x)$ is finite then f can only be > 0 on a countable set and f is summable over A. The rest follows by considering positive and negative parts (and real and imaginary parts if necessary), as in the proof of Proposition 6.24; in particular if say $g(x)$, $h(x)$ are the positive and negative parts and $\sum g(x)$ is infinite, then choose E as the set on which $g(x) > 0$.

(ii) The fact that the sums are either both finite or both infinite follows since $\sum_A |f(a)|$ is an upper bound for $\sum_{n=1}^{\infty} \left(\sum_{a\in A_n} |f(a)|\right)$ (over finite subset of A_n) and conversely. The equality follows (inevitably?) by considering positive and negative parts.

(iii) For Theorem 6.27, $A = \mathbb{Z}^+$ and each A_n is the one-element set $\{n\}$; for Theorem 6.31, $A = \mathbb{Z}^+ \times \mathbb{Z}^+$ and A_n is $\{n\} \times \mathbb{Z}^+ = \{(n,1),(n,2),\ldots,(n,k),\ldots\}$.

13. Put $\rho = \limsup_{n\to\infty} |a_n|^{1/n}$, including the cases in which $\rho = 0, \infty$.

If $\rho > 0$, then given any $\varepsilon > 0$, $|a_n| > (\rho - \varepsilon)^n$ for infinitely many n. Then if $|z| > 1/(\rho - \varepsilon)$, $|a_n z^n| > 1$ for infinitely many n and so $\sum a_n z^n$ cannot converge. Hence all points at which the series converges must have $|z| \leq 1/(\rho - \varepsilon)$, and so $|z| \leq 1/\rho$ since ε is arbitrary. In particular if $\rho = \infty$, the series converges only at $z = 0$.

If $\rho < \infty$, then given any $\varepsilon > 0$, there is some N such that $|a_n| < (\rho + \varepsilon)^n$ for all $n \geq N$. Hence if $|z| < 1/\rho$ we choose ε such that $|z| < 1/(\rho + 2\varepsilon)$, and so $|a_n z^n| < ((\rho + \varepsilon)/(\rho + 2\varepsilon))^n$ for $n \geq N$, and the series is convergent. In particular if $\rho = 0$, the series converges for all z.

14. Let $s_n = f_1 + f_2 + \cdots + f_n$. The construction shows that the maximum of both s_1 and s_2 is attained at $x = 1/2$ (and other places) when $y = 1/2$. Similarly the maximum of both s_3 and s_4 is attained at $x = 1/2 - 1/8$ (and other places) when $y = 1/2 + 1/8$. Generally, the maximum of both s_{2n-1} and s_{2n} is attained at $x = 1/2 - 1/8 - \cdots - 1/2^{2n-1}$(and other places) when $y = 1/2 + 1/8 + \cdots + 1/2^{2n-1}$. Letting $n \to \infty$ (and anticipating Exercise 27 below), the maximum is at $x = 1/3$ when $y = 2/3$. For the value of f at other points, t say, write $t = \sum_1^\infty t_n/2^n$ with each $t_n \in \{0,1\}$ (i.e. write t in binary) and for $k \geq 1$, let $r_k = \sum_{n=k}^\infty t_n/2^{n-k}$. Show that $f(t) = \sum_1^\infty 2^{1-n} f_1(r_n)$.

15. For $|z| \leq 1$, $|\sin z| \leq \sum_0^\infty |z|^{2n+1}/(2n+1)! \leq |z| \sum_0^\infty 1/(2n+1)! < 2|z|$.

Since the graph of sine is concave down on $(0, \pi/2)$, the graph lies above the chord from $(0,0)$ to $(\pi/2, 1)$ which is what the result states.

If $|z| \leq 1/2$, $|\log(1+z) - z| / |z|^2 = \left|\sum_2^\infty (-1)^{n-1} z^{n-2}/n\right| \leq \sum_2^\infty 2^{2-n}/2$ $= 1$.

16. For instance $\left(a^b\right)^x = \exp\left(x \ln\left(a^b\right)\right) = \exp(bx \ln a) = a^{(bx)}$.

17. This is of course immediate from Example 6.77. Without going this far, the formulae (6.13) show that for instance $\sin z = 0$ when $\exp(iz) = \exp(-iz)$, or $\exp(2iz) = 1$, which we know from Theorem 6.51(iii) is true only when z is a multiple of π.

18.* $\cos(x+iy) = \cos x \cos(iy) - \sin x \sin(iy) = \cos x \cosh y - i \sin x \sinh y$. For tan, rationalise the denominator.

19. Given z, write $z = e^{a+ib}$ with $-\pi < b \leq \pi$ so that $\log z = a + ib$ and $\exp \log z = z$. On the other hand, if $z = x + iy$ then $\exp z = e^{x+iy}$ and so $\log \exp z = x + iy_0$ where $y_0 = y + 2k\pi i$ and k is chosen to make $y_0 \in (-\pi, \pi]$. Thus $\log \exp z = z$ if and only if $y = y_0 \in (-\pi, \pi]$.

20. If $c = u + iv$, $f = g + ih$ then $\int cf = u\int g - v\int h + i\left(u\int h + v\int g\right) = (u+iv)\left(\int g + i\int h\right)$, etc.

    Given f, choose c such that $|c| = 1$ and $\int cf > 0$ (the result is trivial if $\int f = 0$). Then $\left|\int f\right| = \left|\int cf\right| = \int cf = \int \Re(cf) \leq \int |cf| = \int |f|$.

21. For instance

$$\begin{aligned}\sum_{n=1}^{\infty} nr^{n-1} \cos n\theta &= \frac{d}{dr}\left(\frac{1 - r\cos\theta}{1 - 2r\cos\theta + r^2}\right) \\ &= \frac{\cos\theta - 2r + r^2\cos\theta}{\left(1 - 2r\cos\theta + r^2\right)^2}.\end{aligned}$$

    The same results can be obtained by differentiation with respect to θ (as should be checked).

    Letting $r \to 1-$ gives the g–sum $-1/\left(4\sin^2(\theta/2)\right)$ for the (divergent) series $\sum_{n=1}^{\infty} n\cos n\theta$ when $0 < \theta < 2\pi$.

22. Suppose $c_1 + c_2 + \cdots + c_n < 1$, (otherwise the result is trivial) and use induction. For $1 \leq r < n$,

$$\begin{aligned}&(1-c_1)(1-c_2)\cdots(1-c_r)(1-c_{r+1}) \\ &\quad \geq (1 - (c_1 + c_2 + \cdots + c_r))(1 - c_{r+1}) \\ &\quad = 1 - (c_1 + c_2 + \cdots + c_r + c_{r+1}) + (c_1 + c_2 + \cdots + c_r)c_{r+1} \\ &\quad > 1 - (c_1 + c_2 + \cdots + c_r + c_{r+1}).\end{aligned}$$

23. This follows by expanding all products and replacing each term by its absolute value.

24.

$$\begin{aligned}|e^z - 1| &= \left|\sum_1^{\infty} \frac{z^n}{n!}\right| \leq |z| \sum_1^{\infty} \frac{|z|^{n-1}}{n!} = |z| \sum_0^{\infty} \frac{|z|^n}{(n+1)!} \\ &\leq |z| \sum_0^{\infty} \frac{|z|^n}{n!} = |z|\, e^{|z|}.\end{aligned}$$

25.* (i) The binomial series for $\left(1-y^2\right)^{-1/2}$ is uniformly convergent on every interval $(-1+\delta, 1-\delta)$ and so may be integrated termwise to get the series for arcsin on $(-1,1)$. At $y=\pm 1$, the series are convergent by the Alternating Series Test and we can take a limit as $y \to \pm 1$ (anticipating the result of Exercise 28 below).

(ii) The series for arctan follows by integrating $\left(1+y^2\right)^{-1}$ as in (i), or more interestingly, by putting $\theta = \pi/2$ in (6.16).

26. Since $\cos x \le 1$, $\int_0^x \cos t \, dt \le \int_0^x dt$, i.e. $\sin x \le x$. Then $\int_0^x \sin t \, dt \le \int_0^x t \, dt$, $1-\cos x \le x^2/2$. The general result follows by induction.

27. Let m be the maximum of f on $[a,b]$ and let $f(c)=m$. Given $\varepsilon > 0$ there is N such that $|f(x) - f_n(x)| < \varepsilon$ for all $x \in [a,b]$ and $n \ge N$.

Then $f_n(c) > f(c) - \varepsilon$ for $n \ge N$ so $m_n \ge m - \varepsilon$ for $n \ge N$. Also for all $x \in [a,b]$, $f(x) \le f(c) < f(c) + \varepsilon$ and so for $n \ge N$, $f_n(x) \le f(c) + \varepsilon$ since $|f(x) - f_n(x)| < \varepsilon$. Hence $m_n \le m + \varepsilon$ for $n \ge N$ as required.

28. The solution to Exercise 8, Chapter 5 concerning summability of integrals reduces to the case of series which is what we have here.

Since $\sum_0^\infty r^n = (1-r)^{-1}$ and $\sum_1^\infty n r^{n-1} = (1-r)^{-2}$ for $|r| \le 1$, then letting $r \to 1-$ gives $1/2$, $1/4$ for the g-values of $\sum_0^\infty (-1)^n$ and $\sum_1^\infty (-1)^{n-1} n$. Compare also Exercise 21 above.

Chapter 7

1. (i) The function is even, so we can use (7.4) to find (a_n).

$$\begin{aligned}
a_n &= \frac{2}{\pi}\int_0^\pi f(x)\cos nx \, dx = \frac{2}{\pi}\int_0^a \left(1-\frac{x}{a}\right)\cos nx \, dx \\
&= \frac{2}{\pi}\left[x - \frac{x^2}{2}\right]_0^a = \frac{a}{\pi} \qquad \text{if } n = 0 \\
&= \frac{2}{n\pi}\left[\left(1-\frac{x}{a}\right)\sin nx\right]_0^a + \frac{2}{n\pi a}\int_0^a \sin nx \, dx \\
&= \frac{2}{n^2\pi a}(1-\cos na) = \frac{4}{n^2\pi a}\sin^2\left(\frac{na}{2}\right) \qquad \text{if } n \ge 1.
\end{aligned}$$

The Fourier series is absolutely convergent for all x by the M-test, and so its sum is f as required.

(ii) Here again, f is even and

$$\begin{aligned} a_n &= \frac{2}{\pi}\int_0^{\pi a/2} \cos\left(\frac{x}{a}\right)\cos nx\,dx \\ &= \frac{1}{\pi}\int_0^{\pi a/2}\left\{\cos\left(\left(\frac{1}{a}+n\right)x\right)+\cos\left(\left(\frac{1}{a}-n\right)x\right)\right\}dx \\ &= \frac{2a}{\pi\left(1-n^2a^2\right)}\cos\left(\frac{\pi n a}{2}\right) \qquad \text{for all } n \geq 0. \end{aligned}$$

(The reader should think about what happens to the term in which $n = k$ when $a = 1/k$ for some integer k.) Convergence follows as in (i).

2. The easiest way seems to be to use the fact that the series is convergent by Example 6.19 and apply the result of Exercise 28, Chapter 6.

3.* (i) To get started,

$$\begin{aligned} \int_0^1 f'(t)\,B_1(t)\,dt &= \int_0^1 f'(t)\left(t-\frac{1}{2}\right)dt \\ &= \left[f(t)\left(t-\frac{1}{2}\right)\right]_0^1 - \int_0^1 f(t)\,dt \\ &= \frac{f(0)+f(1)}{2} - \int_0^1 f(t)\,dt. \end{aligned}$$

Then the question guides you through the remaining steps.

(ii) From (i),

$$\begin{aligned} \frac{f(j)+f(j+1)}{2} &= \int_j^{j+1} f(t)\,dt + \sum_{2\leq 2k<n}\frac{B_{2k}}{(2k)!}\left[f^{(2k-1)}(j+1)\right. \\ &\quad \left.-f^{(2k-1)}(j)\right] + \frac{(-1)^n}{n!}\int_j^{j+1} f^{(n)}(t)\,\tilde{B}_n(t)\,dt, \end{aligned}$$

and the result follows by adding from $j = m$ to $n-1$.

4. For $0 < t < 2\pi$ we have from Example 7.17,

$$\frac{\pi - t}{2} = \sum_{n=1}^{\infty}\frac{\sin nt}{n}$$

which gives the case $k = 1$, putting $t = 2\pi x$. Then integrate repeatedly, using $\int \sin t\,dt = \sin(t-\pi/2)$.

For any $k \geq 2$,

$$\begin{aligned}
\left|\tilde{B}_k(x)\right| &\leq 2\frac{k!}{(2\pi)^k}\sum_{n=1}^{\infty} n^{-k} \leq 2\frac{k!}{(2\pi)^k}\left(1+2^{-k}+3^{-k}+\cdots\right) \\
&< 2\frac{k!}{(2\pi)^k}\left(1+2^{-k}+\int_2^{\infty} t^{-k}dt\right) \\
&= 2\frac{k!}{(2\pi)^k}\left(1+\frac{k+1}{(k-1)\,2^k}\right) \leq 2\frac{k!}{(2\pi)^k}\left(1+2^{1-k}\right).
\end{aligned}$$

If k is even

$$\begin{aligned}
\tilde{B}_k(x) &= \pm 2\frac{k!}{(2\pi)^k}\sum_{n=1}^{\infty} n^{-k}\cos(2\pi n x), \\
\left|\tilde{B}_k(x)\right| &\leq 2\frac{k!}{(2\pi)^k}\sum_{n=1}^{\infty} n^{-k} = \left|\tilde{B}_k(0)\right|.
\end{aligned}$$

5.** It is certainly necessary for instance that f should be continuous, and sufficient that f, f', f'' should all be continuous since then $|c_n| \leq K/n^2$ for some K. More precisely, it is known that f being Lipschitz continuous of order $\alpha > 1/2$ ($|f(x) - f(y)| \leq M|x-x|^{\alpha}$ for all x, y) is sufficient, but that Lipschitz continuity of order $1/2$ is not – see for instance [19, page 240]. But conditions which are both necessary and sufficient are unknown.

6. With $g(x) = (1 - |x|/a)$ on $[-a, a]$, $= 0$ elsewhere,

$$\begin{aligned}
\mathcal{F}g(y) &= 2\int_0^{\infty} g(x)\cos(2\pi x y)\,dx = 2\int_0^{a}(1-x/a)\cos(2\pi x y)\,dx \\
&= \left[\frac{1}{\pi y}(1-x/a)\sin(2\pi x y)\right]_0^a + \frac{1}{\pi a y}\int_0^a \sin(2\pi x y)\,dx \\
&= \frac{1-\cos(2\pi a y)}{2\pi^2 a y^2} = \frac{\sin^2(\pi a y)}{a\pi^2 y^2}.
\end{aligned}$$

Then

$$\int_{-a}^{a}\left(1-\frac{|y|}{a}\right)\mathcal{F}f(y)\,dy = \frac{1}{a\pi^2}\int_{-\infty}^{\infty} f(x)\frac{\sin^2(\pi a x)}{x^2}dx$$

is immediate from Parseval's formula. If $\mathcal{F}f$ is integrable, the left side tends to $\int_{-\infty}^{\infty}\mathcal{F}f(y)\,dy$ as $a \to \infty$ while the right side tends to $f(0)$, as in the proof of Theorem 7.22, and the result for $x \neq 0$ follows in the same way.

7. The function is integrable and (Lipschitz) continuous on $\mathbb{R}$ and

$$\mathcal{F}f(y) = \frac{2a}{(2\pi ya)^3}(\sin(2\pi ya) - 2\pi ya\cos(2\pi ya))$$

after two integrations by parts.

8.* (i) (a) Notice that $\int_{-\infty}^{\infty}\psi = 0$ and that ψ is a test function by considering how fast $\int_t^{\infty}\psi$ goes to zero as $t \to \infty$. (b) Given T and U as specified, then for any ϕ, $U'\phi = -U(\phi') = -\left[k\int_{-\infty}^{\infty}\phi' - T\chi\right]$ where χ is formed from ϕ' in the same way as ψ is from ϕ, i.e. $\chi(t) = \int_{-\infty}^{t}(\phi'(u) + d\phi_0(u))\,du$ and $d = -\int_{-\infty}^{\infty}\phi'/\int_{-\infty}^{\infty}\phi_0 = 0$ so that in fact $\chi = \phi$. Then $U'\phi = -\left[k\int_{-\infty}^{\infty}\phi' - T\chi\right] = T\chi = T\phi$ as required.

(ii) If $U_1' = U_2' = T$ then for any ϕ, $(U_1 - U_2)\phi' = -(U_1' - U_2')\phi = 0$. Put $\psi(t) = \int_{-\infty}^{t}(\phi + c\phi_0)$ as in (i), so that $\psi' = \phi + c\phi_0$. It follows that for any ϕ, $(U_1 - U_2)\phi = (U_1 - U_2)\psi' - (U_1 - U_2)(c\phi_0) = 0 - \left\{(U_1 - U_2)\phi_0/\int_{-\infty}^{\infty}\phi_0\right\}\int_{-\infty}^{\infty}\phi$ is a constant multiple of $\int_{-\infty}^{\infty}\phi = E_0\phi$ as required.

9. Using Definition 7.43 and the result of Example 7.41 we have for instance

$$\begin{aligned}
E^{-1}\phi &= P^{-1}\phi + RP^{-1}\phi = P^{-1}\phi + P^{-1}R\phi \\
&= \int_0^1 \frac{\phi(t) + \phi(-t) - 2\phi(0)}{t}dt + \int_1^{\infty}\frac{\phi(t) + \phi(-t)}{t}dt, \\
O^{-1}\phi &= P^{-1}\phi - RP^{-1}\phi = P^{-1}\phi - P^{-1}R\phi \\
&= \int_0^{\infty}\frac{\phi(t) - \phi(-t)}{t}dt
\end{aligned}$$

and the other results are similar.

10. From Definition 7.40(iii) and Example 7.41(iii) we have

$$\begin{aligned}
P^{-3}(\phi) &= \frac{-1}{2}DP^{-2}(\phi) = \frac{1}{2}P^{-2}(\phi') \\
&= \frac{1}{2}\left(\int_0^1 \frac{\phi'(x) - \phi'(0) - x\phi''(0)}{x^2}dx\right. \\
&\quad \left. + \int_1^{\infty}\frac{\phi'(x)}{x^2}\,dx - \phi'(0) - \phi''(0)\right) \\
&= \int_0^1 \frac{\phi(x) - \phi(0) - x\phi'(0) - x^2\phi''(0)/2}{x^3}dx + \int_1^{\infty}\frac{\phi(x)}{x^3}\,dx \\
&\quad - \frac{1}{2}\left(\phi(0) + 2\phi'(0) + \frac{3}{2}\phi''(0)\right)
\end{aligned}$$

after another integration by parts. The general formula is of the form

$$P^{-k}(\phi) = \int_0^1 \frac{\phi(x) - p_k(x)}{x^k} dx + \int_1^\infty \frac{\phi(x)}{x^k} dx + \sum_{j=1}^{k-1} c_j(k) \phi^{(j)}(0)$$

where p_k is the kth Taylor polynomial of ϕ about 0, and the coefficients $c_j(k)$ are determined from the recurrence $P^{-k-1} = -DP^{-k}/k$.

11. If $a > 0$, $DP^a(\phi) = -P^a(\phi') = -\int_0^\infty x^a \phi'(x)\, dx = -[x^a \phi(x)]_0^\infty + a\int_0^\infty x^{a-1}\phi(x)\, dx = aP^{a-1}(\phi)$.

    Hence if $a < -1$, $a + n > 0$, $a + m > 0$ with $n > m$ then

$$\begin{aligned}
\frac{1}{(a+1)\cdots(a+n)} D^n P^{a+n} &= \frac{a+n}{(a+1)\cdots(a+n)} D^{n-1} P^{a+n-1} \\
&= \frac{(a+n)\cdots(a+m+1)}{(a+1)\cdots(a+n)} D^m P^{a+m} \\
&= \frac{1}{(a+1)\cdots(a+m)} D^m P^{a+m}.
\end{aligned}$$

    Once the consistency is proved, the relation $DP^a = aP^{a-1}$ for other values of a follows from Definition 7.40(ii) or (iii) as appropriate.

12. We have

$$\frac{1}{h}(\mathcal{F}f(y+h) - \mathcal{F}f(y)) = \int_{-\infty}^{\infty} f(x) e^{-2\pi ixy} \frac{e^{-2\pi ixh} - 1}{h} dx$$

    and since for real t, $\left|e^{it} - 1\right| = 2|\sin(t/2)| \le |t|$ the integrand is dominated by $|xf(x)|$ which is given to be integrable, we can let $h \to 0$ to get

$$D\mathcal{F}f(y) = -2\pi i \int_{-\infty}^{\infty} f(x) e^{-2\pi ixy} x\, dx = -2\pi i \mathcal{F}g(y).$$

13. With $\varepsilon > 0$,

$$\begin{aligned}
P^{-1-\varepsilon}(\phi) &= \int_0^\infty \frac{\phi(x) - \phi(0)}{x^{\varepsilon+1}} dx \qquad \text{as in Example 7.41(i)} \\
&= \int_0^1 \frac{\phi(x) - \phi(0)}{x^{\varepsilon+1}} dx + \int_1^\infty \frac{\phi(x)}{x^{\varepsilon+1}} dx - \frac{1}{\varepsilon}\phi(0)
\end{aligned}$$

    so $\lim_{\varepsilon \to 0+} P^{-1-\varepsilon} - \delta/\varepsilon \to P^{-1}$ also.

14. With the given conditions, Example 7.41 (iii) gives $P^{-2}(\phi) = \int_0^1 (\phi(x) - \phi(0))x^{-2} dx - \phi(0)$ which is clearly < 0.

15. (i) The results for $a \neq 0$ follow at once from Exercise 11. Also $DE^0(\phi) = -E^0(\phi') = -\int_{-\infty}^{\infty} \phi'(x)\, dx = 0$ and DO^0 is similar.

(ii) From (i), $D(pE^a) = D(O^{a+1})$, i.e. $E^a + paO^{a-1} = (a+1)E^a$, or $pO^{a-1} = E^a$ for $a \neq 0$ and $pE^{a-1} = O^a$ similarly. When $a = 0$ we have for instance

$$\begin{aligned} pE^{-1}(\phi) &= E^{-1}(p\phi) \\ &= \int_0^1 \frac{p(x)\phi(x) + p(-x)\phi(-x) - 2p(0)\phi(0)}{x} dx \\ &\quad + \int_1^\infty \frac{p(x)\phi(x) + p(-x)\phi(-x)}{x} dx \\ &= \int_0^\infty \frac{x(\phi(x) - \phi(-x))}{x} dx = O^0(\phi). \end{aligned}$$

(iii) If $pT = 0$ then from Theorem 7.50(iii) it follows that $D\mathcal{F}T = -2\pi i\mathcal{F}(pT) = 0$ and so $\mathcal{F}T = kE^0$. Hence $T = k\mathcal{F}E^0 = k\delta$.

(iv) $L_1 = \Lambda \ln - R\Lambda \ln$.

(v) If $\varepsilon > 0$ then

$$O^{-1+\varepsilon}(\phi) = \int_0^\infty \frac{\phi(x) - \phi(-x)}{x^{1-\varepsilon}} dx \to \int_0^\infty \frac{\phi(x) - \phi(-x)}{x} dx = O^{-1}(\phi)$$

and the other parts are similar.

16. Here $e^{\cos x}$ has equal maxima at $0, 2\pi$ and since n is an integer, the same is true for $\cos nx$. This contributes a factor of 2 in the result since we have two equal maxima. Expanding $e^{\cos x} = e^{1-x^2/2+\cdots} = e(1 - x^2/2 + \cdots)$ shows that in the formula (7.29) we have $l = e$, $a = e/2$, $p = 2, b = 1$, and so (after checking that the other necessary properties of f, g are satisfied) we have

$$\int_0^{2\pi} e^{t\cos x} \cos nx\, dx \sim 2.\frac{1}{2}.e^t \left(\frac{e}{te/2}\right)^{1/2} \Gamma(1/2) = e^t \sqrt{\frac{2\pi}{t}}.$$

17.* (i) $a_{n+1} - b_{n+1} = (a_n + b_n)/2 - \sqrt{a_n b_n} < (a_n + b_n)/2 - b_n = (a_n - b_n)/2$ so the difference is more than halved at each step, while (b_n) increases and (a_n) decreases. Hence if we take N so that $a_N - b_N < 8b_0$ then $d = b_N$ has the required property. Then for $n \geq N$,

$$\begin{aligned} a_{n+1} - b_{n+1} &= \frac{1}{2}(a_n + b_n) - \sqrt{a_n b_n} = \frac{1}{2}\left(\sqrt{a_n} - \sqrt{b_n}\right)^2 \\ &= \frac{1}{2}\left(\frac{a_n - b_n}{\sqrt{a_n} + \sqrt{b_n}}\right)^2 \leq \frac{(a_n - b_n)^2}{8d}, \\ \frac{a_{n+1} - b_{n+1}}{8d} &\leq \left(\frac{a_n - b_n}{8d}\right)^2. \end{aligned}$$

(ii) Given

$$G(a_0, b_0) = \int_0^{\pi/2} \frac{dt}{\sqrt{a_0^2 \cos^2 t + b_0^2 \sin^2 t}}$$

put $\tan t = x$ to get

$$G(a_0, b_0) = \int_0^{\infty} \frac{dx}{\sqrt{(1+x^2)(a_0^2 + b_0^2 x^2)}}.$$

Write a, b for a_0, b_0 and make the substitution $bx^2 = ay^2$ to show that

$$\begin{aligned} G(a,b) &= \int_0^{\infty} \frac{dy}{\sqrt{(b+ay^2)(a+by^2)}} \\ &= \int_0^{\infty} \frac{dy}{\sqrt{ab(1+y^4) + (a^2+b^2)y^2}} \\ &= 2\int_0^{1} \frac{dy}{\sqrt{ab(1+y^4) + (a^2+b^2)y^2}} \\ &= 2\int_0^{1} \frac{dy}{(1-y^2)\sqrt{ab + (a+b)^2 y^2 (1-y^2)^{-2}}}. \end{aligned}$$

Now put $v = 2y/(1-y^2)$ to get

$$G(a,b) = \int_0^{\infty} \frac{dv}{\sqrt{(1+x^2)(a_1^2 v^2 + b_1^2)}} = G(a_1, b_1).$$

18. If $x_n \sim cn^d$ then $x_{n+1} = \ln(1+x_n) = x_n - x_n^2/2 + \cdots$ and so $c = 2, d = -1$ by comparing terms. To make this precise, follow the method of Example 7.66 using

$$\begin{aligned} x - \frac{x^2}{2} &< \ln(1+x) < x - \left(\frac{1}{2} - \varepsilon\right) x^2, \\ 1 + x &< \frac{1}{1-x} < 1 + (1+\varepsilon)x \end{aligned}$$

for small enough x.

19. The question outlines all the necessary steps. The only point which might be overlooked is that the equations

$$\begin{aligned} 0 &= 2C^2 - 2S^2, \\ 0 &= \frac{\pi}{2} - 4CS \end{aligned}$$

determine C, S only up to sign, so it is necessary to say why the positive values should be chosen. But this is easy for S since

$$\begin{aligned} S &= \int_0^\infty \sin\left(x^2\right) dx = \int_0^\infty \frac{\sin t}{2\sqrt{t}} dt = \sum_0^\infty \int_{n\pi}^{(n+1)\pi} \frac{\sin t}{2\sqrt{t}} dt \\ &= \int_0^\pi \sin t \sum_0^\infty \frac{(-1)^n}{\sqrt{t+n\pi}} dt \end{aligned}$$

which is positive since the terms of the series decrease strictly.

Bibliography

[1] T.M. Apostol. *Mathematical Analysis.* Addison-Wesley, Reading, MA, USA, 1957.

[2] R.P. Boas (Jr.). *A Primer of Real Functions.* Mathematical Association of America, Washington, DC, USA, 1981.

[3] H.-D. Ebbinghaus, H. Hermes, F. Hirzebruch, M. Koecher, K. Mainzer, J. Neukirch, A. Prestel, and R. Remmert. *Numbers.* Springer, New York, NY, USA, 2nd edition, 1990.

[4] G.H. Hardy. *Divergent Series.* Oxford University Press, Oxford, UK, 1949.

[5] E. Hewitt and K. Stromberg. *Real and Abstract Analysis.* Springer, Berlin, Germany, 1965.

[6] M.J. Lighthill. *Fourier Analysis and Generalised Functions.* Cambridge University Press, Cambridge, UK, 1962.

[7] G. Pólya. *Mathematics and Plausible Reasoning.* Princeton University Press, Princeton, NJ, USA, 1954.

[8] G. Pólya and G. Szegö. *Problems and Theorems in Analysis.* Springer, New York, NY, USA, 1978.

[9] J.I. Richards and H.K. Youn. *Theory of Distributions.* Cambridge University Press, Cambridge, UK, 1990.

[10] W. Rudin. *Principles of Mathematical Analysis.* McGraw-Hill, New York, NY, USA, 2nd edition, 1964.

[11] W. Rudin. *Real and Complex Analysis.* McGraw-Hill, New York, NY, USA, 1966.

[12] W. Rudin. *Functional Analysis.* McGraw-Hill, New York, NY, USA, 1973.

[13] L. Schwartz. *Theorie des Distributions.* Hermann, Paris, France, 1957.

[14] G. Smith. *Introductory Mathematics: Algebra and Analysis.* Springer Undergraduate Mathematics Series. Springer, London, UK, 1998.

[15] P.L. Walker. A bijection from **Z** to **Q**. *Math Gazette*, page 119, 1995.

[16] P.L. Walker. *Elliptic Functions, a Constructive Approach.* J. Wiley and Sons, Chichester, UK, 1996.

[17] E.T. Whittaker and G.N. Watson. *A Course of Modern Analysis.* Cambridge University Press, Cambridge, UK, 4th edition, 1927.

[18] R. Wong. *Asymptotic Approximation of Integrals.* Academic Press, London, UK, 1989.

[19] A. Zygmund. *Trigonometric Series.* Cambridge University Press, Cambridge, UK, 2nd edition, 1968.

Index